KU-354-768

Mechanical Engineering Science

An introduction

Mechanical Engineering Science

An introduction

Val Ivanoff

B.Sc.(Tech.), M.Eng.Sc.
Head Teacher of Mechanical Engineering,
North Sydney Technical College

McGRAW-HILL BOOK COMPANY Sydney

New York St Louis San Francisco Auckland Bogotá
Caracas Hamburg Lisbon London Madrid Mexico Milan
Montreal New Delhi Oklahoma City Paris San Juan
São Paulo Singapore Tokyo Toronto

Cover illustration

A set of gears to control a counterweight. A design and sketch by Leonardo da Vinci.

Reprinted 1985, 1988, 1991, 1992

A51005

National Library of Australia
Cataloguing-in-Publication data:

Ivanoff, Val.
Mechanical engineering science.

ISBN 0 07 451005 3.

1. Mechanical engineering. I. Title.

621

Produced in Australia by McGraw-Hill Book Company Australia Pty Limited
4 Barcoo Street, Roseville, NSW 2069
Typeset in Australia by Monoset Typesetters
Printed and bound in Singapore by Kyodo Printing Co (S'pore) Pte Ltd

Sponsoring Editor: Stuart Laurence
Copy Editor: Daphne Rawling
Designer: George Sirett
Cover Designer: Eric Prior
Technical Illustrator: Colin Bardill

My dear young friends:

It is not enough that you should understand about applied science in order that your work may increase man's blessings. Concern for man himself and his fate must always form the chief interest of all technical endeavours, concern for the great unsolved problems of the organisation of labour and the distribution of goods—in order that the creations of our mind shall be a blessing and not a curse to mankind.

Never forget this in the midst of your diagrams and equations.

Albert Einstein

From an address before students of California Institute of Technology 1938

Contents

Part Five Strength of materials

Preface

This book has been written as an introductory text in mechanical engineering science for middle-level or technician students in technical colleges. Principally the book covers elementary topics in statics, dynamics, mechanics of machines, strength of materials, fluid mechanics and thermodynamics.

Many excellent books have already been written for technician level courses. Unfortunately, they often aim at the senior stages of the courses and do not serve the beginner very well. There are other books on the market which, while suitable in level, do not give a balanced introduction to the various branches of engineering science. I believe there is room for a book that could provide an introduction to mechanical engineering science, covering all its major branches in a single volume, complementary to the more advanced and specialised technical college textbooks.

The main objective which I had in mind during the preparation of this text was to present the fundamental principles and current terminology of mechanical engineering science in a clear and unpretentious manner, acceptable to the student, without sacrificing the technical correctness or usefulness of its subject matter. The book makes no claim to excellence or originality and if usefulness and simplicity prove to be its most redeeming features, my aims will be quite satisfied.

Contrary to my initial intention, the material of the book has been arranged along traditional lines dictated by the logical and historical divisions between the various branches of engineering science and by the common categories of technical college subjects. This should make the book readily adaptable within usual technician course syllabuses.

It is intended that the book may serve as a text for two college years at about two and a half hours per week. The material can be worked through progressively, with statics, dynamics and mechanics of machines covered in the first year, followed by the strength of materials, fluid mechanics and thermodynamics in the second year. Alternatively, a concentric approach can be adopted with the earlier, more elementary parts of some chapters covered in the first year, followed by the more advanced topics in the second year.

There are numerous worked examples in the text. In addition, a liberal quantity of practice problems, with answers, are furnished for private working. Each section of every chapter, consisting of explanatory text with worked examples, followed by about ten graded practice problems, is intended as a teachable lesson unit. There are also some general review questions at the end of each chapter.

It would be surprising if there were no errors in the book. I hope that they are few and relatively insignificant in nature. Any errors or omissions found in the book and brought to my attention, as well as suggestions for future improvements to the book, will be gratefully received.

I am indebted to all my students, past and present, who provided the inspiration without which the book might never have been written, and to my colleagues who helped in many ways. My special thanks are due to my wife for her patience and for aid in typing the manuscript.

V. Ivanoff

PART ONE

Introduction and units

If I have seen further, it is by standing on the shoulders of giants.

Sir Isaac Newton

1
Introduction

It is difficult to imagine our world without all the fascinating machines, which over the centuries helped to shape and transform society into what it is today, through the development of modern industry and technology. Machines are mechanical devices that augment or replace human effort for the accomplishment of physical tasks. The number and complexity of machines are increasing all the time, and each new machine is a combination of the knowledge of the scientist, the ingenuity of its designer and the skill of its builder.

This is a book about mechanical engineering science written for the technical college student. Its primary objective is to provide a study guide that will help the student learn the fundamental concepts, units and laws of engineering science, in a context of applied examples, with only a basic minimum mathematical skill. In addition, it is the author's wish that the book will also reveal to the student some of the history of people whose ideas created the technology that permeates and affects so many aspects of our lives.

1.1 Engineering

Engineering is the art and science of applying the knowledge of the physical world to the conversion of the resources of nature, in order to improve or control our environment.

It is said that in the ancient world there were only two kinds of engineering—military and civilian. The former was concerned with the building of such engines of war as assault towers, catapults and floating bridges, while the latter with roadworks, water supply and sewerage systems.

Imhotep, the builder of the famous Egyptian pyramid near Memphis about 2500 BC, was the first engineer known to history by name. The construction methods used by him must have combined the best of the practical engineering skills of his time. His successors in Greece, Rome, medieval Europe and other regions of the world developed many sophisticated techniques of metallurgy, construction and hydraulics, which helped to establish and maintain advanced civilisations, responsible for a variety of creative forms of art, literature and government.

The methods used in those days were based principally on trial and error, supported by the engineer's knowledge of arithmetic, geometry, draughtsmanship and of materials used. A reflection of this knowledge survived not only in large ancient structures, such as the Roman aqueducts or Persian road systems, some of which endure to this day, but also in many ancient manuscripts, intended by their writers to preserve the accumulated knowledge for the education of future generations of engineers. One such work was written by a Roman engineer and architect, Marcus Vitruvius, in the first century after Christ, and contained ten volumes covering city planning, building materials, construction methods, mensuration, clocks, military machines and hydraulics.

The gradual growth of specialised engineering knowledge and the formulation of scientific theories relevant to engineering practice necessitated the establishment of systematic engineering education. The first of the engineering schools, the National School of Bridges and Highways, was founded in France in 1747, at about the same time as the term *civil engineer,* separate from *military engineer*, came into general use. The British Institution of Civil Engineers, incorporated by royal charter in 1828, became the world's first engineering society.

The birth of the second branch of modern engineering, mechanical engineering, was largely the result of the Industrial Revolution in England and Scotland in the eighteenth century, with the invention of the steam engine and the development of machines for the textile industry, coalmining and later for transportation. The formal recognition of mechanical engineering, dealing with machinery of all types, came with the founding in 1847 of the Institution of Mechanical Engineers in England.

Modern developments in the knowledge of electricity, chemistry, electronics and nuclear physics gave an impetus for further growth and branching of engineering into separate specialised fields.

1.2 Mechanical engineering

Mechanical engineering is a branch of engineering concerned with the design, construction and use of machines for the production of goods and power, and for environmental control. The areas of mechanical engineering comprise the development of machine tools, materials handling equipment, furnace and boiler technology, automotive and power generating plant, pneumatic and hydraulic systems, refrigeration and air conditioning.

The mechanical engineer is particularly concerned with harnessing natural sources of **energy**, such as the chemical energy of fuels, and its conversion into heat and work. In doing so he makes use of **forces** and **motion**. He employs two kinds of materials—solids and fluids. Solid materials form stationary and moving parts of mechanisms used to transmit or change forces and motion. Fluids, i.e. liquids and gases, are used as working agents to transform force into pressure and to facilitate conversion of energy from one form to another.

The functions of the mechanical engineer include research, development, design, construction, production, operation and management of mechanical plant and associated services, all of which require some degree of knowledge and understanding of the physical world and the ability to adapt this knowledge to practical problem solving. The types of occupational requirements in industry vary greatly in the degree of emphasis placed on the applications of science. Between the research and development engineer, who is largely involved in the application of scientific methods in search of new ideas, and the tradesman who is using manual skills in the construction, operation and maintenance of machines, there are engineers, draughtsmen and technicians, whose work brings the knowledge of some aspects of engineering science to bear on practical problems.

1.3 Mechanical engineering science

Mechanical engineering science is based principally on mathematics and physics and their extension into mechanics of solids, fluid mechanics and thermodynamics. Its purpose is to explain and predict physical phenomena and to serve as a basis for engineering analysis and design.

Solid mechanics, which is the oldest of the physical sciences, is usually subdivided into statics and dynamics. Statics deals with bodies at rest and enables the engineer to predict forces in the solid members of a machine or structure. Dynamics, on the other hand, deals with bodies in motion and considers relations between the forces involved and the motion. Other topics in mechanics include friction and elasticity.

Fluid mechanics is concerned with liquids and gases at rest, or in motion, and is important for the design of tanks, pumps, turbines and pipework systems. Fluid mechanics deals with the relationships between pressure and the mechanical properties of fluids, such as density and viscosity. It can also be subdivided into fluid statics, which is the study of pressure measurement, fluid forces on submerged surfaces and buoyancy, and fluid dynamics, dealing with fluid flow and fluid machinery.

Thermodynamics is the science of temperature and heat, which is particularly concerned with the theory of heat engines, i.e. mechanical devices for the conversion of thermal energy into mechanical work.

It should be noted that the division of mechanical engineering science into its component parts is purely a matter of logical convenience and, to some extent, historical tradition. Its applications, particularly in the area of energy systems design, usually involve a combination of factors and principles derived from different parts of engineering science. The automotive engine and the steam power plant are two good examples.

1.4 The foundations

The foundations of mechanical engineering science rest firmly on two principles—mathematical reasoning and experimental observation. The former provided the tools and the latter the material from which the great body of specialised, systematised and verified knowledge was developed.

The practical use of arithmetic for the measurement of land areas was, as far as is known, first made by the Egyptians, without any mathematical proof that the ideas they used were correct. The true development of mathematical deduction, which is the cornerstone of all science and technology, started with

Pythagoras in the sixth century BC. Two centuries later, Euclid's "Elements" were written, in which he introduced such basic concepts as point, line, plane and angle, and related these to physical space. Mathematics had reached its climax in ancient Greece with Archimedes, third century BC, who did a great amount of work with the sphere and cylinder, and was able to calculate the value of π with remarkable accuracy, sufficient even today for most practical purposes.

In the seventeenth century AD, the century of genius, wonderful discoveries followed one another in rapid succession. John Napier of Scotland published his discovery of logarithms. René Descartes, the French philosopher, invented analytic geometry, in which a point can be represented by its distances from two perpendicular axes. Sir Isaac Newton and the German mathematician Gottfried Leibniz, working independently of each other, developed the methods of differential and integral calculus, which were destined to become an invaluable tool for engineering science, as well as the mainspring of further developments in higher mathematics, which go beyond the scope of this book.

The names of Archimedes and Newton are also associated with the second fundamental principle of science—that of objective experimental observation. The well-known story—perhaps it is legend—of how Archimedes studied the effects of buoyancy, and arrived at what is still called Archimedes' Principle, while in the bathtub, illustrates his ability to observe, to understand what is observed, and to use the observation to discover new ideas. In the words of Newton "the best and safest way of doing scientific work seems to be, first to inquire diligently into the properties of things, and of establishing these properties by experiment, and then to proceed slowly to theories for the explanation of them". Another early exponent of experimental science was Galileo, who lived in Italy in the sixteenth century, and is best known for his defence of the Copernican concept of the solar system. He is rightly regarded as the founder of modern experimental scientific method, who had set up the basis for Newton's developments. In an age when equipment had not yet been developed to make really accurate measurements, Galileo conducted experiments to confirm his reasoning on motion, acceleration and gravity, particularly in relation to the problems associated with falling bodies. He also anticipated the idea of inertia, which was later used and refined by Newton in his laws of motion.

Each area of engineering science is based on the work of many great scientists throughout the centuries. Archimedes, Galileo, Hooke and Newton in solid mechanics; da Vinci, Torricelli, Pascal and Bernoulli in fluid mechanics; Boyle, Rumford, Carnot and Joule in thermodynamics, are only some of them.

1.5 From science to applications

The progress of the machine from its early primitive forms to the modern marvels of the technological age is the result of brilliant inventions followed by patient work of development, adaptation and improvement. The relationship between scientific knowledge and the ability to apply it to practical purposes

Fig. 1.1 *Leonardo da Vinci, 1452–1519*

has varied from person to person, and from one historical period to another. Some excelled in the depth of their theoretical perception, while others succeeded in building and perfecting useful machines. In general, most scientists have been aware of the practical possibilities of their scientific knowledge—some were also engineers. On the other hand, most great inventors have had a good understanding of engineering science, on which their inventions were based.

Leonardo da Vinci, born in Italy in 1452, is considered by many historians to have been the greatest genius of all time—a great artist, scientist and the greatest inventor of his age. Throughout his life, Leonardo was an inventive builder, for whom an interest in pure science merged increasingly with an interest in applied mechanics. His model book on the elementary theory of mechanics contains thousands of beautifully illustrated pages describing his observations and outlining inventions of all kinds. He was particularly interested in problems of frictional resistance, and described various combinations of machine elements, such as screw threads, gears, hydraulic jacks, designed to overcome friction and transmit or modify forces. His mechanical inventions are interesting and varied—a machine gun, a military tank, a submarine, an anemometer, a pump, a flying machine. Many mechanisms invented by Leonardo are in use in similar forms today. He anticipated variable speed drives, roller bearings and screw-cutting machines. As a scientist-inventor he was so far ahead of his time that many of his inventions, all feasible, had to wait for centuries before they could be realised. The importance of Leonardo's contribution to engineering science lies in his ability to apply the principles of mechanics to invention of practical machines.

Fig. 1.2 *James Watt, 1736–1819*

Another name which marks a turning point in successful combination of science and engineering is that of James Watt, born in Scotland in 1736. Watt was not the original inventor of a steam engine. His predecessors included Hero of Alexandria in the first century AD, Thomas Savery (1698) and Thomas Newcomen (1705). However, Hero's novelty was not much more than a scientific toy, while Savery's and Newcomen's engines were very inefficient. Once while repairing a model of Newcomen's engine, Watt was impressed with its waste of steam and became interested in steam engines. In 1765 he invented the separate condenser for steam engines, and later developed the more efficient double-acting engine, in which the piston both pushed and pulled. He also adapted the steam engine for rotary motion. In 1769, James Watt took out his famous patent for "A New Invented Method of Lessening the Consumption of Steam and Fuel in Fire Engines". This event ushered in the age of practical efficient technology and introduced the social change known as the Industrial Revolution.

From the time of James Watt, mechanical engineering has developed in response to the demand for increased efficiency, accuracy and complexity. Production machinery has been developed for highly automated manufacturing processes; thermal efficiency and output of power generating plants throughout the world has steadily increased, making possible other advances in industrial development; steam and internal combustion engines completely revolutionised transport; and mechanical refrigeration has been applied to food preservation and to comfort air conditioning.

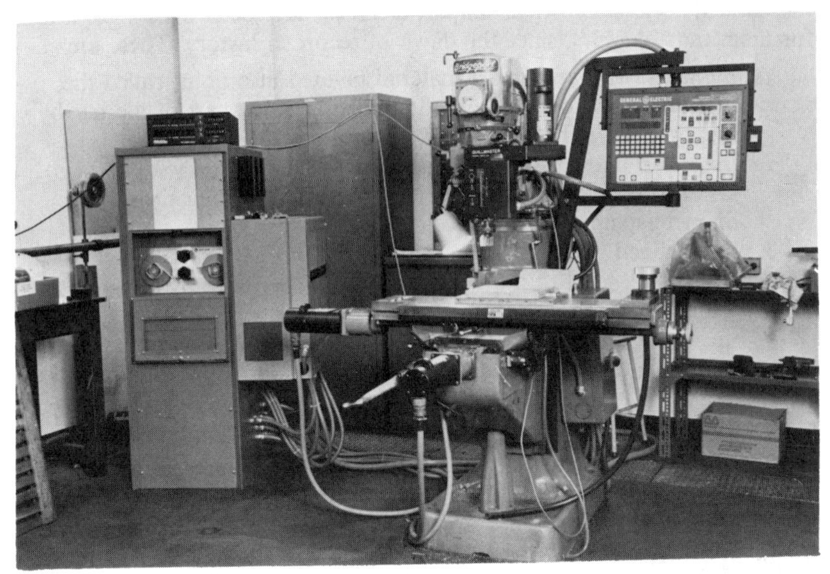

Fig. 1.3 *Numerically controlled milling machine*

There have also been some undesirable side effects of the products of mechanical engineering, such as noise, pollution of air and water, and the rapid depletion of the natural sources of energy, particularly oil.

In the future, the demand for mechanical engineering skills will continue, with an emphasis on conservation and efficient utilisation of scarce material resources and on maintaining a satisfactory environment.

1.6 Machines and machine components

This is not a book about machines. It is a book about mechanical forces at work in the world of machines and machine components. It is about the transformation of heat energy into mechanical energy, or vice versa. In general, it will be assumed that the student has an elementary knowledge of how mechanical things work, or that he or she is learning about machines and machine components concurrently in another subject of the engineering course. However, it would be useful to briefly define and summarise some of the more common mechanical devices before proceeding to the study of engineering science as such.

Machines

A **machine** can be defined as a mechanical device for overcoming a resistance at one point by the application of a force at some other point. It is usual to call the output resistance force the **load**, and the input force the **effort**. Some basic mechanical devices, known as **simple machines**, have been in use in some form for thousands of years, since the dawn of recorded history. These are:

1. the **lever**—a bar of rigid material pivoted at a point called the **fulcrum**, used to move a load applied at some part of the bar by means of an effort applied at another part.
2. the **wheel** and **axle**—a cylindrical axle on which a wheel is fastened concentrically, the difference between their respective diameters supplying the leverage.
3. the **inclined plane**—a plane surface inclined at a small angle to the horizontal, used to lift a heavy mass up the plane by applying an effort along the plane.
4. the **screw**—a cylinder with a spiral groove round its outer surface, called the **thread**, used for conveying motion or bringing pressure to bear.
5. the **pulley**—a wheel with a grooved rim, or a combination of such wheels mounted in a block, for a cord or chain to run over, for changing the direction or magnitude of a force.

Mechanisms

One of the distinctive characteristics of a machine is that its parts are interconnected and constrained to move only in a particular predetermined way relative to each other. The way in which the parts are interconnected and guided is called the **mechanism** of the machine. The most common **mechanisms** include:

1. **linkages**—assemblies of solid members, or links, connected to each other by hinges or by sliding joints, for the transmission of motion in a machine. The piston, connecting rod and crankshaft, for example, constitute the mechanism of a reciprocating pump or engine.
2. **cams**—eccentric projections on revolving shafts, shaped so as to give some desired linear motion to a follower, which is usually returned by a spring.
3. **gears**—operating in pairs, to transmit rotational motion by means of successively engaging projections called **teeth**. Gear boxes containing multiple pairs, or trains, are often used to obtain speed ratios that cannot be obtained with a single pair of gears.
4. **friction drives**—in which one wheel causes rotation of a second wheel with which it is pressed into contact, by means of friction force at the point of contact.
5. **wrapping connectors**—such as belt, rope or chain drives used for transmitting rotational motion over some distances.

Machine components

In addition to what is usually referred to as mechanisms and simple machines described above, there are various machine components that a student of

Fig. 1.4 *Section through a centrifugal pump*

engineering will learn about in the course of studies. The following is a brief summary of the more common machine components:

1. **shafts**—cylindrical bars, rotating and transferring rotational motion.
2. **flywheels**—heavy wheels attached to rotating shafts for the purpose of storing up energy or moderating fluctuations in the speed of the machine.
3. **bearings**—connectors that support a rotating shaft relative to stationary parts of the machine.
4. **couplings**—devices for connecting the ends of two adjacent rotating shafts, conveying a drive from one to the other.
5. **springs**—elastic members, usually of bent or coiled metal, used for a variety of purposes in many mechanical devices.

1.7 About this book

Mechanical engineering science is essentially a mathematical science, based on a few fundamental principles and laws. The primary objective of this book is to enable the student to understand these fundamental principles and laws. In order to assist the student in achieving this goal, a number of worked examples have been provided throughout the text. In addition, each chapter contains problems which should be used by the student to reinforce and test understanding of the material covered. Most of the examples and problems are of a practical nature and should be relevant to the student's interests. However, care has been exercised to ensure that using engineering jargon would not obscure the basic concepts presented. As far as possible the problems are graded.

Since this book is intended as an introductory text in mechanical engineering science, new concepts have been presented in elementary terms and, as far as practicable, one at a time. The emphasis throughout is on learning fundamental principles. For this reason, applications and methods which do not have an immediate and direct relationship to a concept to be learned have been avoided. It has also been assumed that in the majority of cases the student will pursue the study of engineering science beyond the scope of this text. This book can, therefore, be regarded as a basic introduction to the subject-matter covered in greater depth in books such as *Applied Mechanics* by D. Watkins *et al., Applied Strength of Materials* by B. Parker *et al., Applied Heat* and *Applied Fluid Mechanics* by R. Kinsky.

The material presented in this book requires from students a reasonable degree of competence in elementary mathematics. This includes competence in the four basic arithmetical operations: addition, subtraction, multiplication and division, with the aid of an electronic calculator. Students must be competent in manipulating numbers presented in either scientific or engineering, exponential notation. The geometry and trigonometry of the right-angled triangles are also required. Mensuration of areas and volumes of basic geometrical shapes must be understood. Above all, transposition of terms in an equation and the solution of linear equations must be handled confidently and correctly. An elementary understanding of vectors would be helpful, but not essential. Knowledge of differential and integral calculus is not required.

Graphical methods are used in the book where appropriate for vector addition of forces. They are one of the analytical tools of the engineer, and should not be regarded as inferior to mathematical methods. One should recognise that some engineering problems are better solved mathematically, while others lend themselves better to graphical solutions. This means that a small drawing board, a pair of set-squares, a rule, a protractor and a compass are required for solving some of the problems.

In engineering, the possible accuracy of answers is limited by the accuracy of the original data contained in the statement of a problem. Unless stated otherwise, this must be assumed to be known with a degree of accuracy comparable with that of ordinary engineering measurements, which is seldom greater than three or four significant figures, especially in heat and fluid flow measurements. Students must resist the temptation of false accuracy when using electronic calculators with eight, or more, digit display. Usually three or four significant figures are quite sufficient for an answer to an engineering problem. As a general rule, answers given to three significant figures have been preferred in this book. In the case of graphical solutions, two significant figures are adequate.

On the other hand, crude approximations in a number of successive intermediate steps can often lead through accumulation of errors to a very inaccurate answer. In the majority of worked examples in this text the answers have been calculated with the aid of a pocket calculator and not approximated until the final answer has been obtained. However, the explanations often show intermediate steps with only approximate intermediate answers. Therefore, students may observe a slight discrepancy if they use these approximate intermediate values to check the final results.

Where standard computed or natural constants such as π (3.14159...) or g (9.80665) are required, it is usually quite sufficient to simplify their values to three or four significant figures. However, if the appropriate function key is available on the calculator it may be used without affecting the essential accuracy of the results from the engineering point of view. In this book the value of the gravitational constant is taken to be $g = 9.81$ and calculator function keys used for all mathematical constants and trigonometric functions without approximation.

Unlike problems in pure mathematics, engineering problems are about structures and machines which always have some practical purposes and limitations. In problem solving, students must strive to appreciate the engineering significance of each problem by drawing on their own experience and by relating engineering science to other subjects in their engineering course. Considerable importance should be attached to a proper understanding of the problem and to the logical, clear and precise recording of its solution. A neat diagram showing only the essential information, such as equipment layout or forces acting on a body, is always very helpful in analysing the given information.

Finally, a word of advice for the student, from an old Chinese proverb:

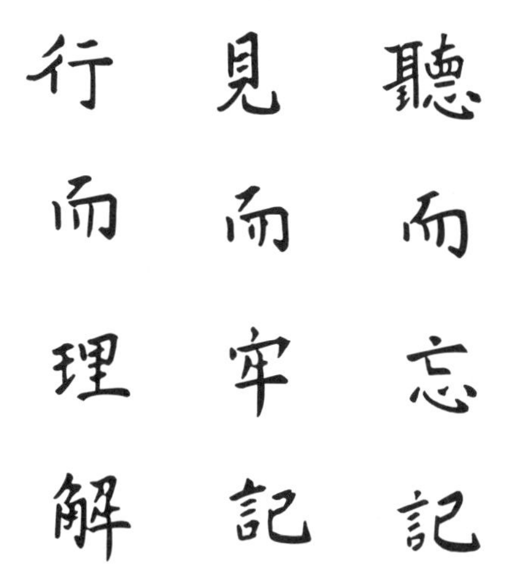

which translated means:

I hear and I forget,
I see and I remember,
I do and I understand.

which translated again means:

Do not just read this book,
Work through the problems and
you will learn.

2

Fundamental concepts and units

Engineering is concerned with the world of physical objects and their properties. In engineering analysis and design we are continually confronted with questions such as: how long? how much? how fast? how hot? how strong? and many others. Answers to questions of this kind require an understanding of the physical nature of the objects or phenomena the questions are about. They also require a common basis of measurement, involving a standardised system of units. This chapter introduces the science of measurement and the few fundamental concepts of physical science on which the modern international system of units is based.

2.1 Physical quantities and dimensions

Our experience of the world around us reveals qualities of different kinds, some simple and obvious, others complex and difficult to describe or define. Some of the qualities are thought to belong to the objective reality of the physical world, while others are subjective judgments found "in the eye of the beholder". Size, speed, colour, beauty and justice represent in some ways certain aspects of a particular object or event. However, only some of these categories can be regarded as physical quantities.

An important characteristic of a physical quantity is its ability to be described quantitatively in terms of precise physical measurements. In other words, a **physical quantity** can be defined as a measurable attribute of a physical object, substance or process. When you drive a car, the speed of the car is a physical quantity, because it has to do with a physical object, the car, and it can be measured. On the other hand, if you are a passenger in a car driven at 150 kilometres per hour, the feeling of exhilaration or of fear is not a physical quantity, because it is subjective and does not allow description in terms of physical measurements.

Through experience, the scientist and the engineer have learned that certain physical quantities are **fundamental**, in a sense that they can be used to define or describe all other physical quantities and relationships. A limited number of such fundamental physical quantities have been selected arbitrarily to form the fundamental **dimensions** on which our system of measurement is based. The fundamental dimensions with which we are concerned in this book are **length, mass, time** and **temperature**. The first three of these are discussed in some detail in this chapter, while temperature is introduced in Chapter 25.

Somewhat related to the concept of length, and supplementary to it, is the geometrical concept of **plane angle**. For our purposes, it is possible to consider angular measure as an independent fundamental quantity. However, unlike length, plane angle is a dimensionless quantity.

In addition to the abovementioned fundamental quantities, it is also convenient to include electric current, amount of chemical substance, luminous intensity of light and solid angle as independent fundamental quantities. However, these are outside the scope of this book.

Fundamental dimensions can be combined in numerous ways to form **derived dimensions**. For example, two linear measurements of a rectangular shape can be combined to express an essentially different physical quantity called **area**. Similarly, three measurements combine to express **volume**. Throughout the book, we will introduce derived dimensions of various physical quantities, such as force, energy, pressure, density, acceleration, enthalpy, etc.

All equations involving physical quantities must be dimensionally *homogeneous*, i.e. they must be balanced dimensionally as well as numerically. For example, volume of a rectangular prism is equal to the product of its three sides, for which a simple formula is: $V = a \times b \times c$.

When the dimensions of a particular prism are given in a problem such as, "What is the volume of a room 4 m long, 2.5 m wide and 2.8 m high?", the solution is: $V = 4\text{ m} \times 2.5\text{ m} \times 2.8\text{ m} = 28\text{ m}^3$. Note that regardless of the magnitudes or units used, dimensional homogeneity of this equation must always be true and can be shown as:

$$l \times l \times l = l^3$$

where l represents a single linear dimension, while l^3 stands for the derived dimension of volume, or "cubic measure".

2.2 Measurement and systems of units

All physical quantities, no matter how complex, are measurable, using instruments and techniques appropriate to the quantity to be measured. Some physical quantities can be measured with relative ease and a high degree of precision. Others can only be measured indirectly, by measuring some other physical quantities associated with the object or process under investigation and then computing the magnitude of the required quantity as a function of these measurements. Some measurements are made with precise instruments, while others are only crude approximations.

The science of engineering measurement is called **metrology** and deals with the essential requirements of uniformity of units and standards, as well as the instruments and techniques used for practical engineering measurements.

Uniformity of measurement requires reliable units and standards, as well as accuracy in collecting and recording of results. A **unit** is an agreed-on part of a physical quantity, defined by reference to some arbitrary material standard or to a natural phenomenon. In essence, the process of measurement involves making a comparison between the quantity being measured and a standard unit appropriate for the particular physical quantity. The result of

Fig. 2.1 *A lesson in metrology*

measurement is, therefore, a statement of magnitude as a number of standard units, or a fraction of a standard unit. Thus of a surface measuring 2 metres long by 0.7 metre wide, we can say that in comparison with the standard unit, the metre, its length is equal to two units and its width is equal to seven tenths of the unit.

The early units devised and used by man were closely associated with the practical trades, such as carpentry and surveying, and with commercial transactions. They were based on common, but ill-defined and unreliable standards such as the digit, or a finger's breadth, the foot, which was later redefined as "one third of the Iron Yard of our Lord the King", the wine-jar and the barley corn. These and similar units were inaccurate and unrelated to each other and could hardly provide a sound basis for methodical and precise measurement of physical quantities.

Throughout history, a number of systems of measurement emerged, competing with each other and gradually replacing each other. Two types of systems can be distinguished: an evolutionary system, which grows haphazardly out of usage, and a planned system. The British system, which survived with some local variations in the English-speaking world, is an example of the former. The metric system, which originated in the Netherlands and France, is an example of a planned system.

The latest worldwide move is towards a universal, simple and internationally accepted system of units known as the *International System of Units*, with the abbreviation "SI" (from *S*ystème *I*nternational d'Unités). In Great Britain, by an Act of Parliament in 1963, all units of the British system were redefined in terms of the metric system, with a national changeover to SI beginning two years later. The progress in the United States has been slower, but quite discernible. In the meantime the traditionally metric countries of Europe have also been adjusting to the new International System, which is essentially an expansion of the metric system, incorporating scientific and technological developments of the twentieth century.

In Australia, conversion to SI units began with the Metric Conversion Bill introduced in the Commonwealth Parliament in 1970 with the aim of progressively introducing the use of metric measurement as the sole system of measurement of physical quantities in this country. The inherent advantages of the metric system, particularly its decimal nature and coherent relationships between units, as well as the fact that all major countries were already using metric or were in the process of changing, suggested that it was desirable and practical for Australia to make the change.

The Metric Conversion Act 1970 provided for a Metric Conversion Board to be established to coordinate the conversion program, and to compile and disseminate appropriate information and advice. The Act also defined the metric system to be used in Australia as measurement in terms of basic and derived SI units, with certain additions of special units declared as part of Australia's metric system (see Appendix B).

2.3 The International System of Units (SI)

A decimal system of units was originally proposed by the Flemish mathematician, engineer and public servant Simon Stevin in a small pamphlet *La Thiende* (*The Tenth*), published in 1585, in which he declared that the universal introduction of decimal coinage and measures would be only a matter of time. Other European scientists discussed the desirability of a new rational and uniform system of units to replace a multitude of local inconsistent weights and measures, which made scientific communication difficult. The establishment of the metric system is often associated with the social and political upheavals of the French Revolution, being one of its most significant and lasting results.

In 1790, the French National Assembly requested the French Academy of Sciences to establish a set of units suitable for international use. The French formally adopted the metric system of measurement in 1840. By that time the Netherlands, Belgium, Luxembourg and Greece had already done so. Internationally, the Metre Convention was signed in 1875, leading to the establishment of the International Bureau of Weights and Measures. The United States, along with fourteen other nations, was a party to the original treaty, Great Britain became a signatory in 1884, and Australia joined formally in 1947.

The original metric system was based on the metre, gram and second as base units, but in 1873 the centimetre replaced the metre to define the Centimetre–Gram–Second (CGS) system. Later the Metre–Kilogram–Second (MKS) system was introduced. The International System was developed from the MKS system and adopted by the General Conference on Weights and

Table 2.1 SI base and supplementary units

Physical quantity	*Name of unit*	*Symbol*
Length	metre	m
Mass	kilogram	kg
Time	second	s
Temperature	kelvin	K
Plane angle	radian	rad

Measures in 1960 as "a practical system of units of measurement suitable for adoption by all signatories of the Metre Convention".

The International System of Units comprises a set of **base units** and **derived units**, corresponding to the fundamental and derived dimensions as discussed previously. One of the main advantages of the system is that for each physical quantity there is only one SI unit, with its decimal multiples and sub-multiples, without odd multiplying factors to be remembered. Furthermore, the system is coherent, i.e. one in which the product or quotient of any two unit quantities in the system is the unit of the resultant quantity.*

Altogether, there are seven SI base units. Of these, we are only concerned with the units of length, mass, time and temperature. In addition, there are two SI supplementary units for angular measurement, of which we shall only use the radian as a measure of plane angle.

It should be noted, that as a general rule names of units (e.g. metre, kelvin) and prefixes (e.g. kilo) when written in full are written in lower case letters.** Unit symbols are also written in lower case letters (m, kg), except the symbols for units named after people (K). Unit symbols do not change in the plural and a full stop should not be used after a symbol, except at the end of a sentence. However, when spelled out, unit names take a plural "s", e.g. 3 kilograms *but* 3 kg.

In order to express decimal multiples and sub-multiples of SI units, the system provides a number of decimal prefixes, which when attached to a particular unit indicate the relationship with the parent unit. Preference is given to the use of prefixes related to the parent unit by a factor of 1000.†

When stating the value of any measurement, the appropriate unit should be chosen so that the numerical value of the statement is between 0.1 and 1000. For example, the distance between Sydney and Canberra is stated as 309 km, not 309 000 m, and shaft diameter as 55 mm, not 0.055 m. However, some engineering disciplines have their own special rules, e.g. all dimensions on mechanical engineering drawings are shown in millimetres, regardless of the size of the component shown.

---

* The full significance of this characteristic will become apparent when derived units of force and pressure are discussed:
$1\ \text{N} = 1\ \text{kg} \times 1\ \text{m/s}^2$ (see Chap. 8); $1\ \text{Pa} = 1\ \text{N}/1\ \text{m}^2$ (see Chap. 21).

** The sole exception is "degree Celsius" (see Chap. 25).

† This is not a strict rule and does not preclude the use of such prefixes as centi $= 10^{-2}$ or deci $= 10^{-1}$, as in centimetre and decimetre, if necessary.

Table 2.2 SI prefixes

Prefix	*Symbol*	*Value*
giga	G	10^9
mega	M	10^6
kilo	k	10^3
milli	m	10^{-3}
micro	μ	10^{-6}

2.4 Fundamental dimensions and units

It has already been stated that a limited number of physical quantities can be selected as fundamental dimensions on which a coherent and comprehensive system of units can be built. Let us now consider length, angle, mass and time as such independent basic concepts and their associated units.

The concept and units of length

Definition and measurement of length is based on our intuitive appreciation of the notion of space and of our position in it in relation to other material things with which we share the space. Questions relating to spatial measurements are usually of two kinds: "Where?" (in relation to some known reference point or a system of coordinates) and "How long?" (referring to a distance between two points). Answers to both types of question require a suitable measuring instrument and an agreed standard unit of length. The concept of **length** can be defined as **a measure of distance between two points**, one of which may be the origin of a coordinate system.

It should be noted that length, or linear dimension, is always measured along a single line. This line, however, does not have to be a straight line. Thus, while the diameter of a circle is a convenient description of its size, the length of its circumference is an equally valid linear measure.

The base unit of length is the **metre**, originally conceived as one ten-millionth part of the distance from the equator to the pole along the meridian line through Paris, introduced in 1801 by law of the French National Assembly. Historically, the metre was the first unit in the metric system of units, both words having a common origin in the Greek *metron* for measure. As a result of the International Metric Convention of 1875, the metre was redefined as the distance between two defining marks on the International Prototype Metre, which became the physical embodiment of the unit, and is a standard bar of platinum-iridium alloy held at specified temperature of 0°C on two roller supports in a horizontal position, at the International Bureau of Weights and Measures at Sèvres near Paris, France. Because of its inaccessibility, secondary standards, or copies of the primary standard, were made and distributed to the standardising agencies of various countries. The accuracy of measuring tapes and other length measuring devices is usually fixed by manufacturing specifications derived from the secondary standards.

In 1960 the definition of the metre was further refined in terms of new techniques of measuring light waves, which made the standard more accurate and more readily reproducible. The metre is now defined as equal to exactly 1 650 763.73 wavelengths of the orange line in the spectrum of the krypton-86 atom in an electrical discharge. This incredibly precise definition was the result of very careful comparison being made between the light emitted from a krypton-86 lamp under specified laboratory conditions and the distance between the defining lines on the prototype metre. All precise measurements of length in scientific laboratories are now made with light waves.

In engineering practice, linear dimensions are the most common measurements made. The instruments used vary with the size and nature of the item and the degree of accuracy required. They range from a simple rule or tape to vernier calipers, precision micrometers, dial indicators and non-contacting electronic probes. The measurements can be made directly from a scale, or indirectly by transferring a calipered dimension to separate standards or gauge blocks. In transfer measurements with calipers used by a skilled mechanic, accuracies in the order of 0.01 mm can be achieved. However, direct micrometer and vernier-caliper measurements are more reliable and are capable of accuracies from 0.01 to 0.001 mm.

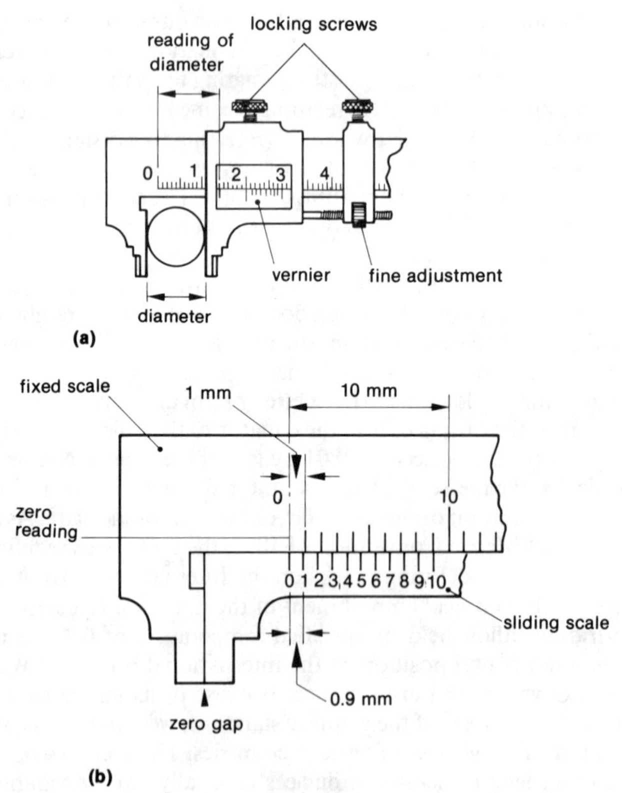

Fig. 2.2 (a) *Vernier calipers* (b) *Principle of 0.1 mm vernier*

For day-to-day work in engineering measurement which does not require high degree of accuracy, a steel rule or tape, usually graduated in millimetres, is used. Dimensions of solid objects, to which an ordinary rule cannot be applied, are measured with the aid of engineer's calipers consisting of a pair of hinged steel jaws which are closed until they touch the object in the desired position. The distance between the jaws is afterwards compared with an ordinary scale. For many purposes, slide or vernier calipers (Fig. 2.2), capable of more accurate direct reading, are preferred. The vernier scale, named after Pierre Vernier, a seventeenth century French technician who patented this ingenious device, enables an accurate reading of the last significant decimal place in the measurement without the necessity to estimate fractions of a division by eye. Besides calipers, many other measuring instruments are fitted with vernier scales.

For measuring the diameter of a small shaft, a piece of wire or similar small dimensions which nevertheless require a relatively high order of accuracy, a micrometer screw gauge as shown in Figure 2.3 may be used. The main feature of the instrument is a screwed spindle fitted with a graduated thimble. When the object is gripped gently between the spindle and the anvil, an accurate reading is obtained from the graduated sleeve and the thimble.

Many attachments are used to adapt these instruments to round, square, inside, outside, screw thread and other special measurements. Non-contacting electronic micrometers can be used for dynamic measurements, such as those on a rotating shaft. A digital display eliminates the need for skilful contact measurements.

Dimensional control of mass-produced items in many branches of manufacturing technology makes use of measurement by comparison of the item with appropriate standard gauges, e.g. slip gauges, used in association with dial test indicators or other forms of comparators.

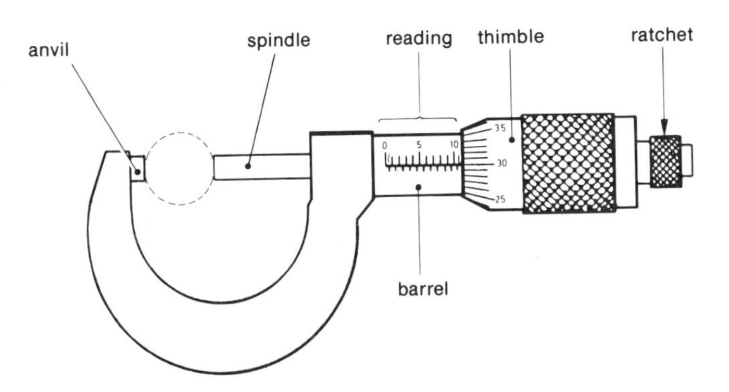

Fig 2.3 (a) *Micrometer*

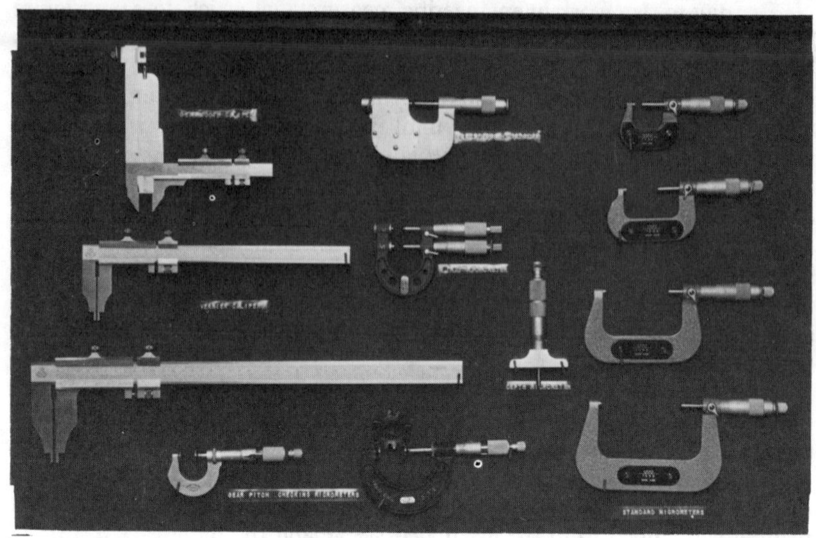

Fig 2.3 (b) *Micrometers*

The student should keep in mind that the most convenient unit of measurement for mechanical engineering components is the millimetre (= 0.001 m) and that it is common practice to show all dimensions on mechanical engineering drawings in millimetres. On the other hand, the requirement of dimensional homogeneity of physical equations often demands that base units be used for calculations, especially if quantities other than purely linear dimensions are also involved. Therefore, care should be exercised in converting the measured dimensions in millimetres to base unit, the metre, when necessary for calculations.

The concept and units of plane angle

In plane geometry the **angle** is defined as the **inclination of one straight line to another,** i.e. the angle between two straight lines is determined by the amount of turning about the point of intersection required to bring one line into coincidence with the other.

The SI unit of angular measure is the **radian**, defined as the angle subtended at the centre of a circle by an arc of length equal to the radius. By virtue of this definition, the radian is a ratio of two equal lengths, and therefore dimensionless. On the other hand, angular measure is a useful geometrical concept. The radian is therefore incorporated into the set of fundamental units with the status of a supplementary unit. There are 2π (= 6.283 approx) radians in a circle.

In addition to the radian, the use of degrees, minutes and seconds is allowed for measurement of plane angle. The circle is divided into 360 degrees

(°), the degree into 60 minutes (′) and the minute into 60 seconds (″). The degree is sometimes decimally divided. The degree–minute–second system was codified by Ptolemy of Alexandria (about AD 130) and has remained unchanged for almost 2000 years. Within Australia's metric system, degrees, minutes and seconds have been declared non-SI units which may be used without restriction. It should be remembered, however, that the radian is mathematically more fundamental than the degree, and must be used where dimensional homogeneity demands.

The relationship between degrees and radians follows from the fact that in a full circle there are 2π radians or 360 degrees. Therefore, one degree is equal to $2\pi/360$ (= 0.01745 approx) radians. Conversely, one radian is equal to $360/2\pi$ (= 57.3 approx) degrees.

Example 2.1

In a centre lathe, the cutting tool movement required for accurate control of the amount of metal removed is achieved by the use of a graduated indexing dial attached to the lead screw, so that for one complete revolution of the lead screw the tool has a linear movement equal to the pitch of the lead screw thread.

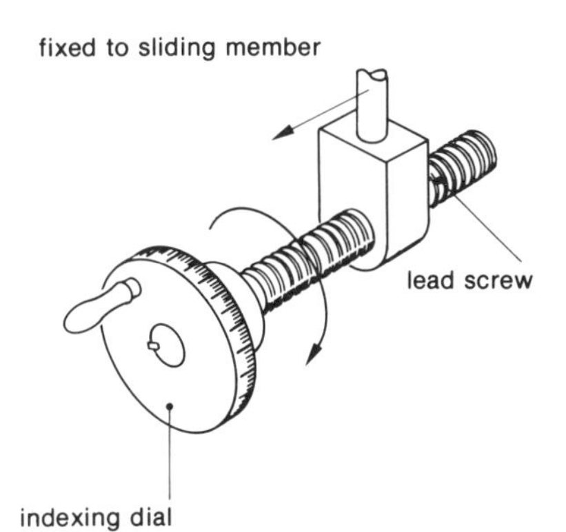

Fig. 2.4

Assuming that the lead screw shown in Figure 2.4 has a pitch of 2.5 mm and an indexing dial with 100 divisions, determine the angle in degrees through which the dial must turn to move the cutting tool a distance of 1 mm. What will this angle be if expressed in radians? What is the angle turned per division on the dial?

Solution

Distance moved by the tool for one revolution $= 2.5$ mm

Distance moved by the tool for one division on the dial $= 2.5\ \text{mm}/100 = 0.025$ mm

Number of divisions required $= 1\ \text{mm}/0.025\ \text{mm} = 40$

Angle of revolution required $= 360° \times \dfrac{40}{100} = 144°$

$$\text{Angle expressed in radians} = 2\pi \times \frac{40}{100} = 2.51 \text{ rad}$$

$$\text{Angle per division on the dial} = 360^\circ / 100 = 3.6^\circ = 3^\circ\, 36'$$

Plane angles are usually measured with a protractor. A plain protractor, as used by a draughtsman or an engineer, is graduated in degrees, with an accuracy for each reading of plus or minus half a degree. When a protractor cannot be used directly to check the taper of a component, a bevel gauge is used to transfer the angle from the component to the protractor. For more accurate readings, a vernier scale can be incorporated into an instrument such as the universal bevel vernier protractor shown in Figure 2.5 or a variety of optical instruments such as the surveyor's level.

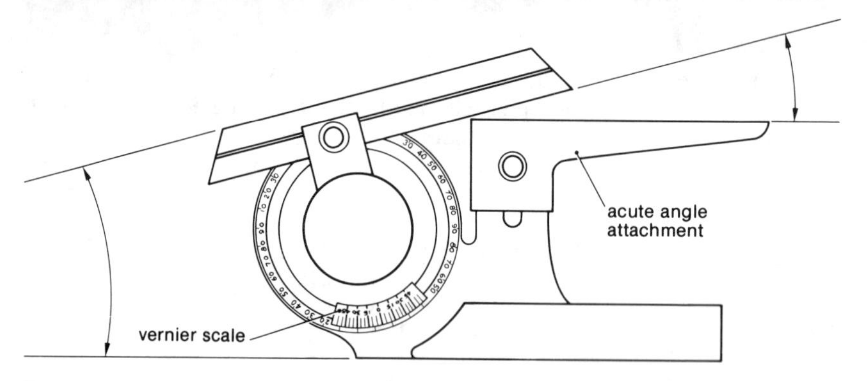

Fig. 2.5

The concept and units of mass

In engineering science we have to deal with material things, or matter. Steel, aluminium, concrete, water, oil, steam and air are typical examples of physical substances or matter. Some are used to manufacture engineering items, such as steel girders, concrete beams and aluminium components for aircraft. Others, like steam and air, are used as working agents in turbines and compressors. Although all kinds of material substances occupy space, the volume they occupy is not a constant property, being subject to changes with variations in temperature and pressure, as will be discussed in Chapters 20 and 27. Likewise the three physical states of matter, i.e. solid, liquid and gaseous phases, are not permanent and can change from one to the other when heat is added or removed (see Chap. 26).

The concept of mass was developed gradually over many centuries, with the greatest contributions coming from Leonardo da Vinci, Galileo and Newton. Our understanding of mass comes from the experience of observing how different objects respond to our attempts to set them in motion. The property of the object which determines its resistance to change of motion is called the **inertia** of the object. It is reasonable to assume that inertia of a body

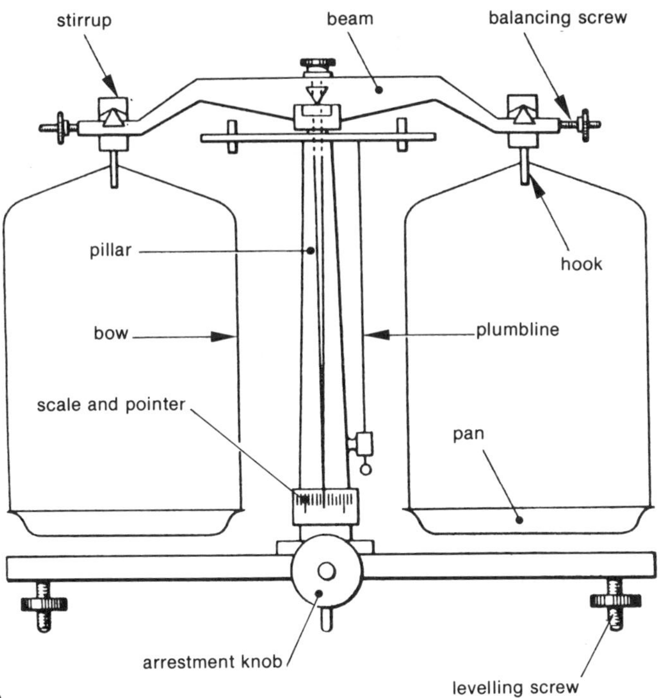

Fig. 2.6

depends on the quantity of matter contained in that body, i.e. the greater the quantity of matter, the greater the tendency to resist a change of motion. Unlike volume, the quantity of matter in a given body is constant, not subject to variation with location, temperature or pressure, for so long as the body retains its physical identity.

This leads to the definition of **mass** as **the measure of the quantity of matter in a body** as evidenced by its inertia. In a scientific laboratory a device called the **ballistic balance** is sometimes used to compare an unknown mass to a set of graded reference masses, by means of observing their inertial effects during accelerated motion.

Another, more convenient, method by which one can determine the mass of an object is to compare its mass with that of known masses, using an equal-arm platform balance as shown in Figure 2.6. This method is based on the fact that equal masses experience equal attraction towards the Earth, a phenomenon discussed in more detail in Chapter 3. Unfortunately, in everyday conversation this procedure is called "weighing", even though it is mass, i.e. the quantity of matter, and not weight which is being measured.*

* In science the word "weight" means the force of gravity, which is quite different from the concept of mass, and the student must carefully distinguish between them (see Chap. 3).

Table 2.3 SI units of mass

Name	*Symbol*	*Relationship to base unit*
milligram	mg	1 mg = 10^{-6} kg
gram	g	1 g = 10^{-3} kg
kilogram	kg	SI base unit
tonne	t	1 t = 10^{3} kg
kilotonne	kt	1 kt = 10^{6} kg

(*Note:* 1 tonne = 1 Mg and 1 kilotonne = 1 Gg)

The base SI unit of mass is the **kilogram**. The kilogram is defined as the mass of the International Prototype Kilogram, kept at the International Bureau of Weights and Measures in Sèvres. The Prototype Kilogram is a cylinder 39 mm in diameter and 39 mm high, made from an alloy containing 90 per cent platinum and 10 per cent iridium.

Special attention should be drawn to the anomaly that, although it contains the prefix "kilo" for historical reasons, the kilogram is nevertheless a base unit and not a multiple. In order to avoid the use of two adjacent prefixes, the SI prefixes having their standard values are attached to the stem word "gram" when forming units of mass larger and smaller than the kilogram. Furthermore, a unit equal to 1000 kg is often referred to as the tonne. The correct names, symbols and relationships between units of mass are summarised in Table 2.3.

The concept and units of time

Of all fundmental concepts of science, time is probably the most difficult to define satisfactorily. Time can be described as a measure of the sequence of events taking place in the physical world. It has to do with such notions as beginning and end, before and after, past and future, but thoroughly eludes any attempt at simple explanations. From ancient times of Egypt, Greece and Rome, time has been associated with periodic natural phenomena such as the duration of an average day and the duration of an average year, i.e. the movements of the Earth as observed by the astronomers. The problem with this approach is that such phenomena are not in fact constant in duration and have very inexact and awkward relationships between themselves. Besides, they do not provide the answer to the fundamental question, "What is time?"

Our standard of time originated from the astronomical observation of the spinning of the Earth, as reflected in the apparent motion of the sun. The duration of the mean solar day was divided into twenty-four hours. Each hour was itself divided into sixty minutes, each containing sixty seconds.* In essense, this system of time measurement has not changed in the Internationl System of Units and is still the legal standard. The SI base unit of time,

* The system of sexagesimal fractions (1/60, 1/3600, etc.), which survived also in the angular measure, was derived from the Chaldeans, became standard in Greek arithmetic and continued in general use until the introduction of decimal notation about AD1600.

Table 2.4 Units of time

Name	*Symbol*	*Definition in terms of SI unit*
second	s	SI base unit
minute	min	1 min = 60 s
hour	h	1 h = 3.6×10^3 s
day	d	1 d = 86.4×10^3 s

however, is now the **second**, redefined in 1967 as the time interval occupied by 9 192 631 770 cycles of a specified energy change in the caesium atom, as measured by the caesium atomic clock. The atomic clock is similar to a radio transmitter giving out short waves, the frequency of which is controlled by energy changes of gaseous caesium atoms. The accuracy of such a clock is considerably better than results of astronomical observations, even with the most modern instruments possible. The added advantage is that, unlike the movements of the Earth, the frequency of the atomic radiation is constant and can be accurately reproduced in any suitably equipped laboratory anywhere in the world.

Accordingly, the day, hour and minute have been redefined in terms of the standard second, as shown in Table 2.4. It should be remembered, however, that these units do not differ significantly from the mean solar day, hour, minute and second.

Practical measurement of time, depending on the accuracy required, is made with pendulum clocks, chronometers and stop watches, electric clocks and quartz clocks. All these instruments depend for their operation on some form of uniformly repeated motion: pendulums of suitable length swinging in the gravitational field of the Earth, vibrations of coiled springs, electric motors which operate synchronously on alternating current of standard frequency, or oscillating quartz crystals.

2.5 Derived dimensions and units

Dimensions and units of many physical quantities can be derived from those of the fundamental quantities: length, angle, mass and time. In this section, we examine area, volume and some time-related quantities.

The concept and units of area

Area is defined as a **measure of the extent of a surface**. The SI unit of area is the **square metre**, defined as the area enclosed by a square each side of which is one metre in length. This can clearly be seen as a derived unit having a dimension of a linear measure squared. The square metre does not have a special symbol and is an example of an SI unit with compound name. The symbol used is m^2, which reflects its derivation.

The preferred multiples and submultiples of the square metre are formed from the preferred units of length, the millimetre and the kilometre, with

Table 2.5 SI units of area

Name	*Symbol*	*Relationship to the parent unit*
square millimetre	mm^2	$1\ mm^2 = 10^{-6}\ m^2$
square metre	m^2	SI unit
hectare	ha	$1\ ha = 10^4\ m^2$
square kilometre	km^2	$1\ km^2 = 10^6\ m^2$

compound symbols mm^2 and km^2. Due to the second order relationship between the dimensions of length and area, the multiplier between one preferred unit of area and the next is 1000^2, or one million, as seen in Table 2.5.

In addition to preferred units described above, there is the hectare, equal to 10 000 square metres (100 m × 100 m), which has been declared an acceptable unit. The use of the hectare, however, is generally limited to computation of land areas smaller than one square kilometre, and is not of particular interest to the mechanical engineer.

In general, areas are not measured directly, but are calculated from measured linear dimensions of surfaces, e.g. sides, diameters, etc. The formulae for calculating the areas of most common geometrical shapes are given in Appendix C. One interesting exception is the planimeter, an instrument for measuring mechanically the area of a plane surface of irregular shape. A tracing point on an arm is moved round the closed curve, whose area is then given to scale by the revolutions of a small wheel supporting the arm.

Example 2.2

How many parquet flooring blocks, 125 mm × 25 mm each, are required to cover an area 10 m × 8 m?

Solution

$$\text{Area to be covered} = 10\ m \times 8\ m = 80\ m^2$$
$$\text{Area of each block} = 0.125\ m \times 0.025\ m = 0.003125\ m^2$$
$$\text{Number of blocks required} = 80\ m^2 \div 0.003125\ m^2 = 25600$$

Note that where different units are involved, e.g. metres and millimetres, it is convenient to convert all given information to base units and then solve. This is not an absolute rule, so long as consistent units are used.

The concept and units of volume

Volume is defined as a **measure of the amount of space occupied by an object or matter**.

The SI unit of volume is the **cubic metre**, defined as the volume of a cube each side of which is one metre in length. This is a derived unit having a dimension of linear measure cubed. The cubic metre does not have a special symbol and is another example of an SI unit with compound name. The symbol used is m^3.

Table 2.6 SI units of volume

Name	Symbol	Alternative Name	Alternative Symbol	Relationship to the parent unit
cubic millimetre	mm^3	microlitre	μL	$= 10^{-9}\ m^3$
		millilitre	mL	$= 10^{-6}\ m^3$
		litre	L	$= 10^{-3}\ m^3$
cubic metre	m^3	kilolitre	kL	SI unit $= m^3$
		megalitre	ML	$= 10^3\ m^3$
		gigalitre	GL	$= 10^6\ m^3$
cubic kilometre	km^3			$= 10^9\ m^3$

The preferred multiples and submultiples of the cubic metre are the cubic millimetre (mm^3) and the cubic kilometre (km^3). In this case the third order relationship between the dimensions of length and volume make the multiplier equal to 1000^3, or one thousand million (10^9), which is a very large gap between one preferred unit and the next.

It is for this reason that an additional decimally related unit called the **litre**,* equal to one thousandth part of a cubic metre, has been declared for general use. The SI prefixes with their standard values can be attached to the word "litre" to form its multiples and submultiples, resulting in some units having two possible names, as can be seen from Table 2.6.

The volumes of regular geometrical shapes, such as prisms, spheres and cylinders, are usually calculated from their linear dimensions. The appropriate formulae can be found in Appendix C. For determining the volumes of liquids and gases we use special calibrated containers. Volumes of solids of irregular shape can be found indirectly by measuring the volume of a liquid displaced when the solid is immersed in the liquid, a method suggested by Archimedes.

Example 2.3

The pressure vessel shown in Figure 2.7 has hemispherical ends. Determine the volume of gas contained in the vessel, in cubic metres and in litres.

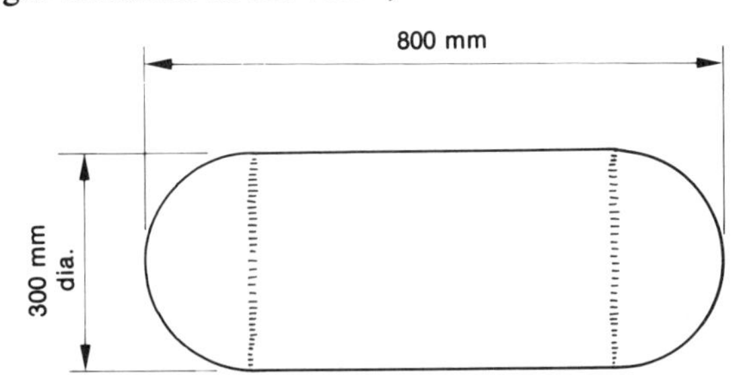

Fig. 2.7

* The litre has a long history, having been related to the volume of one kilogram of pure water at 4° C. This relationship has now been abandoned as being inaccurate and fundamentally inconvenient, and the litre redefined in terms of the cubic metre, as above.

Contrary to the earlier practice, the accepted symbol for the litre is now the upper-case "L", in order to avoid confusion between lower-case "l" and the numeral "1", particularly in typewritten material.

Solution

The length of the cylindrical part (l) $= 0.8 \text{ m} - 0.3 \text{ m} = 0.5 \text{ m}$

The volume of the cylindrical part $= \dfrac{\pi D^2}{4} . l = \dfrac{\pi \times 0.3^2}{4} \times 0.5 = 0.03534 \text{ m}^3$

The volume of the hemispherical ends $= \dfrac{\pi D^3}{6} = \dfrac{\pi \times 0.3^3}{6} = 0.01414 \text{ m}^3$

Total volume of the vessel $= 0.03534 \text{ m}^3 + 0.01414 \text{ m}^3 = 0.04948 \text{ m}^3$

Converting to litres, volume $= 0.04948 \text{ m}^3 \times \left[\dfrac{1000 \text{ L}}{\text{m}^3}\right] = 49.48 \text{ L}$

Say *49.5 litres*

Proportional measure: Rate and ratio

Many engineering concepts represent a comparative measure of one physical quantity or magnitude with another of the same or different kind. The units used to express such proportional measures are formed by the division of two different units, resulting in a new compound unit. There are too many derived concepts and units for all to be introduced here. However, a few examples are given below to illustrate the idea.

The term **rate** usually refers to a quantity or process in which time is a factor. Thus, we speak of water flow rate in a pipe, which can be measured in litres per second, i.e. volume per unit time. Note that the quotient of two units is indicated by the word "per" immediately in front of the unit forming the denominator. The symbol of the compound unit is formed using a horizontal line, an oblique stroke or the use of a negative index,

e.g. litre per second $\dfrac{\text{L}}{\text{s}}$, L/s or L.s^{-1}

Sometimes, the dimension of time does not appear in the definition of a concept, but a continuous process, such as travelling by car, is involved. Under these circumstances it is appropriate to use the term "rate" to describe relationships between physical quantities, e.g. petrol consumption rate in litres per kilometre, or volume of fuel used per unit distance travelled.

Quite often, a comparative measure does not involve or imply time or process, but describes some other physical property of a substance or object, in which case it is usually called **ratio**. For example, volume occupied by a unit mass of substance can be described as the ratio of the volume occupied by a given quantity of matter to its mass. The appropriate derived unit could be litre per kilogram, or cubic metre per tonne, or some other quotient of volume and mass units.

If the quantities being compared are of the same kind and measured in the same units, the ratio is dimensionless and represents the number of times one quantity is contained by the other. For example, if the area of room x is 18 m^2 and the area of room y is 12 m^2, the ratio of the areas is

$$\frac{A_x}{A_y} = \frac{18\ \not{m}^2}{12\ \not{m}^2} = 1.5,$$

meaning that room x is one-and-a-half times larger in area than room y. Notice how the units cancel out, making the ratio dimensionless.

Example 2.4

Determine the cost of heating oil used during the three winter months, if the price of fuel is 35 cents/litre, the average fuel consumption rate is 2.3 litres/hour and the average duration of heater use is 5 hours/day.

Solution

The number of days in the three winter months

$$= 30\ \text{(June)} + 31\ \text{(July)} + 31\ \text{(August)}$$
$$= 92\ \text{days}$$

The total number of hours the heater was in use

$$= 92\ \not{d} \times \frac{5\ \text{h}}{\not{d}} = 460\ \text{h}$$

$$\text{Total fuel consumption} = 460\ \not{h} \times 2.3\frac{\text{L}}{\not{h}} = 1058\ \text{L}$$

$$\text{The total cost of fuel used} = 1058\ \not{L} \times 0.35\frac{\$}{\not{L}} = \$370.30$$

Note that if consistent units are used and the relationship between all variables is well understood, it is possible to combine the entire solution into one line of calculations, as follows:

$$\text{Cost of fuel used} = (30 + 31 + 31)\not{d} \times 5\frac{\not{h}}{\not{d}} \times 2.3\frac{\not{L}}{\not{h}} \times 0.35\frac{\$}{\not{L}}$$
$$= \$370.3$$

Observe how the units cancel out, leaving dollar as the unit of the final answer.

Problems

2.1 Express the following quantities in SI base units:

(a) 0.4 km (b) 53 g
(c) 75.3 mm (d) 45 ms
(e) 357 000 mg (f) 0.08 Mg
(g) 734 μm (h) 0.54 Gs

2.2 Express the following quantities using the most appropriate multiples or submultiples of SI units:

(a) 12 300 m (b) 7500 kg
(c) 0.079 s (d) 0.0047 m

(e) 0.03 kg (f) 85×10^3 kg
(g) 3×10^{-6} m (h) 4.7×10^{-6} kg

2.3 Convert:
(a) 5 m^2 to mm^2 (b) 750 mm^2 to m^2
(c) 663 mm^3 to L (d) 500 000 mm^3 to m^3
(e) 1.35 m^3 to L (f) 0.75 m^3 to mm^3
(g) 632 L to m^3 (h) 47 L to mm^3

2.4 Convert:
(a) 35° 15′ to degrees (b) 150° to radians
(c) 1.57 rad to degrees (d) 2.75° to radians
(e) 2.5 rad to degrees and minutes
(f) 25° 50′ to radians

2.5 Convert:
(a) 3 days 5 hours to minutes
(b) 0.75 hour to seconds
(c) 5000 seconds to hours, minutes and seconds
(d) 6.7 minutes to seconds

2.6 A heating panel is 750 mm long by 350 mm wide. What is its area in square metres?

2.7 It is proposed to replace a round duct of 400 mm diameter by a square duct of equivalent cross-sectional area. Determine the required dimensions of the new duct.

2.8 A cylindrical vessel is 340 mm diameter and 900 mm long. Determine its total surface area in square metres and its volume in litres, neglecting the thickness of the walls.

2.9 A rotating shaft makes 2100 revolutions in one minute. How many times does it rotate per second?

2.10 During a test of a diesel engine, the average time to consume 50 mL of oil was 35.4 s. What was the fuel consumption in litres per hour?

2.11 A rectangular swimming pool has a length of 50 metres and is 20 metres wide. The average depth of water is 3.6 metres. How long would it take to fill it with a pump which delivers 12 cubic metres of water per minute?

2.12 A rectangular room has a wall to wall measurement of 5 m by 3.5 m. Which is cheaper? To cover it with broadloom carpet 3.5 m wide at $75 a metre length, or to cover it with 0.5 m × 0.5 m carpet tiles at $5 per tile? What is the difference in total cost?

2.13 A piece of lead 200 mm × 100 mm × 100 mm, having a mass of 22.6 kg, is used to manufacture 2 mm diameter spherical shot pellets. Determine the mass of each pellet.

2.14 A boiler generates 720 kg of steam per hour, while consuming 0.3 t of coal. What is the mass of steam generated per kilogram of coal?

2.15 A car consumes 11 litres of petrol every 100 km. The price of petrol is 40 cents per litre. What is the cost of petrol for a 750 km trip?

2.16 A reciprocating pump has a 75 mm diameter piston, with a 90 mm stroke, and produces 5 pumping strokes per second. Determine the number of litres of water pumped per hour.

Review questions

1. What is a physical quantity? Give examples.
2. Length, mass, time and temperature are regarded as fundamental quantities.
   (a) Explain the difference between plane angle and the four fundamental quantities.
   (b) Explain the difference between volume and the four fundamental quantities.
3. What is a unit of measurement?
4. What does SI stand for?
5. List names and symbols of SI units of length, mass, time, temperature and plane angle.
6. Which of the following are SI base units? metre, hour, tonne, square metre, kilogram, kilometre, litre, cubic metre, second, gram
7. Which of the following are non-preferred prefixes? giga, mega, kilo, deci, centi, milli, micro.
8. Is the litre a unit of mass or volume?
9. What is your mass in kilograms?
10. What is your height in metres?

PART TWO

Statics

Mechanics is the only science in which we know exactly what the word force signifies.

Friedrich Engels in
Dialectics of Nature

3

Force and gravity

The concept of force is one of the most important for the engineer. Design of structures and machine components requires a thorough understanding of the action and the effects of forces in the connected elements of bridges, engines, machine tools and other structures and mechanical devices. In this chapter the concept of force is introduced, together with some important principles associated with the mathematical and graphical analysis of forces.

3.1 The concept of force

In everyday experience we usually associate force with muscular effort. When a weight-lifter tries to lift a heavy barbell, he uses the strength of his arms to first move the barbell and then to hold it above his head by applying a force. Another athlete uses force to throw a heavy metal sphere in a shot put event, while in a circus a strongman amuses by bending a metal rod with the force of his hands. Note that in all of these instances a force is applied, but the results are quite different: lifting, holding, throwing and bending. In all cases, however, the force can be described as push or pull, i.e. effort. Furthermore, force is seen to be an interaction between two bodies.

In mechanics **force** is defined as **any action that tends to maintain** (e.g. support or hold) **the position of a body, to alter** (e.g. lift, throw, etc.) **the position of a body, or to distort** (e.g. stretch or bend) **it.** It is very useful to continue thinking about forces in terms of push or pull, or effort.

There are many kinds of forces—such as muscular effort, tractive effort at the wheels of a vehicle, cutting force at the tip of a cutting tool, forces due to gas pressure in an engine cylinder—which are a type of push or pull, produced as a result of interaction between two or more bodies.

Another very important type of force is gravity, or gravitational attraction, i.e. the pull exerted by the Earth on every physical object located on or near its surface. The law of gravity is discussed in some detail in Sections 3.8–3.10.

There is also a group of forces which are the result of friction (see Chap. 7), often described as resistance to motion, e.g. friction between two surfaces sliding relative to each other, air resistance, etc. These forces usually oppose some applied force such as tractive effort, which tends to cause or maintain motion.

Forces are often discussed and classified with respect to their effects. The science of **statics** deals with the equilibrium of bodies at rest, under the combined action of several balanced forces. When all forces acting on a body are balanced, the body is said to be in static equilibrium. The equilibrium of forces is discussed in Chapters 4 and 6. A bridge is a structure subjected to a large number of forces. The forces acting on the bridge at any one time must be in equilibrium for the bridge to remain in its position.

If a force, or forces, acting on a body are not balanced the condition of rest or motion will be altered, the body will accelerate or slow down. For example, an object dropped from a height falls with an increasing speed under the influence of the unbalanced force of gravity. The part of mechanics dealing with the analysis of bodies in motion under the influence of various forces acting on the bodies is called **kinetics**. The relationships used to predict the motion caused by given forces or to determine the forces required to produce a given motion will be explained in Chapter 8.

Another possible effect of a force is to deform the material of the body to which the force is applied. The branch of engineering science which studies the relations between forces and the amount of deformation produced in the material is called **strength of materials**. An introduction to the concept of elasticity, fundamental to strength of materials, is given in Chapter 15. In many practical problems, the ability of a material to withstand the applied forces without failure or excessive deformation is of paramount importance. Selection of suitably sized components by the designer is usually based on the analysis of these forces and of the elastic behaviour of the material.

However, as the first step towards understanding the concept of force, let us start with the proposition that **force is a push or a pull exerted on a body.**

3.2 Measurement and unit of force

The SI unit of force is the **newton**, with symbol N. The exact definition of the newton is based on the ability of a unit force, when applied to a unit mass, to impart to that mass a unit of acceleration. In order to fully appreciate the significance of this definition, one must have a good understanding of the laws of linear motion. We shall therefore return to this definition in Chapter 8, when the exact relationship between the forces acting on a body, the mass of the body and the motion of the body will be discussed.

It is more important for the student at this time to get some idea of the actual magnitude of the unit force, rather than to grapple with its definition.

Force is a push or pull. The unit of force—the newton—is therefore a push or pull of a particular strength, which is taken to be the standard for comparing forces. It is not possible to describe the newton as we can describe some material object such as an apple. However, an average size apple* is a convenient aid with which to provide a recognition point for learning purposes. Put the apple on the palm of your hand and you can feel a small downward force acting on your hand (see Fig. 3.1(a)). The force is approximately one newton. The newton, as can be seen from this simple

* A mass of just over 100 grams.

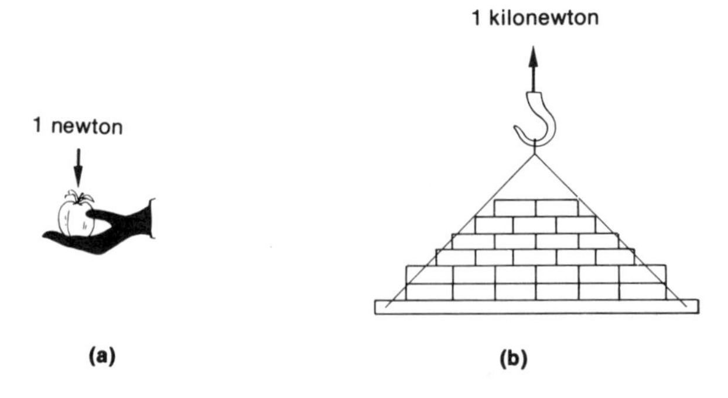

Fig. 3.1

demonstration, is not very large. In engineering practice, a force one thousand times stronger than the newton is a more useful unit. With the aid of SI decimal prefixes such unit is called **kilonewton**, with the symbol kN. The force required to support 26 common house bricks* is approximately equal to one kilonewton (Fig. 3.1(b)).

Forces are usually measured by means of a spring balance or dynamometer. A simple dynamometer (Fig. 3.2) consists of a spring with one end restrained and the other movable, at which the force is applied. The extension of the spring is proportional to the applied force, within the normal operating range of a given instrument. As the force stretches the spring, a marker attached to the movable end of the spring moves along a graduated scale, indicating the magnitude of the applied force. As a force-measuring device, a spring balance is calibrated in the units of force, i.e. newtons or kilonewtons.** Simple spring balances are not very accurate. More sophisticated ones are capable of accuracies in the order of ±0.2 per cent.

Other force-measuring instruments, such as hydraulic and pneumatic load cells and strain gauges, are used extensively in many varied engineering applications, particularly where the force cannot be transmitted through a

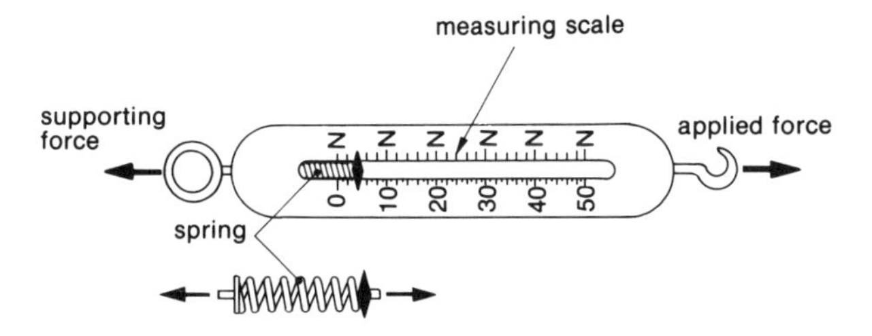

Fig. 3.2

* The mass of one brick is about 3.9 kg.

** It should be noted that many mass-measuring devices, e.g. butcher's scales, are in fact spring balances calibrated to read in kilograms and/or grams. However, their use is strictly limited to measuring a quantity, e.g. meat, and should not be extended to measures of force. Always remember, kilogram is **not** a unit of force.

dynamometer,* and must therefore be measured indirectly. Strain gauge load cells are probably the most common devices for measuring forces of large magnitudes. Accuracies of ±0.1 per cent are common. Strain gauges are devices built around a strip of elastic material, attached to a member of structure or machine that is compressed or stretched by a force. Connected to this strip is an electric device that measures its compression or elongation but provides an output calibrated in units of force.

3.3 Characteristics of a force

A force is characterised by its **magnitude**, its **direction** and its **point of application**.

The magnitude of a force is a measure of the strength of the pushing or pulling effort, expressed in standard units of force, usually newtons or kilonewtons, e.g. 560 N, 3.75 kN.

The direction of a force is defined by the **line of action** and the **sense** of the force. The line of action is a straight line along which the force acts. It can be described by the angle it forms with some reference axis, e.g. 30° to the horizontal. The sense can be indicated descriptively as to the right or to the left, up or down, etc.

Finally, when applied to an actual object or component, a force must have a point of application, e.g., the point at which a cable is attached to a mast.

Graphically a force can always be represented by a straight line drawn to scale, in the direction corresponding to that of the force. For example, a force of 50 N acting down and to the right along a line inclined to the horizontal at 30° and applied at point A can be represented by a line 50 mm long, i.e. a scale of 1 mm = 1 N, as shown on Figure 3.3. Notice how the graphical representation is clear and precise in comparison with the somewhat unwieldy verbal description.

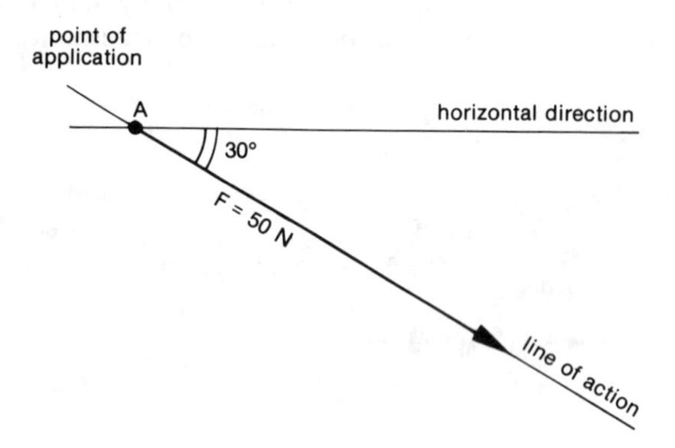

Fig. 3.3

* The term "dynamometer" is also used for a device for testing power output of engines or electric motors.

Those familiar with the mathematical concept of vectors will easily recognise that, as a quantity which has both magnitude and direction, force is a vector. As a vector, a force can not only be represented graphically as shown, but can also be manipulated according to the rules of vector algebra.

3.4 Basic principles

There are three main principles concerning forces which are necessary for the solution of problems involving forces in mechanics. These are action and reaction, transmissibility of a force and the parallelogram of forces.

Action and reaction forces

The principle of action and reaction was formulated by Sir Isaac Newton, who pointed out that, whenever a force acts on a body, there must be an equal and opposite force or reaction acting on some other body. Expressed in a simple form this principle states that **action and reaction are equal and opposite**.

To take an example, a car is pulling a trailer with a force of 0.2 kN (Fig. 3.4). The trailer experiences a pull of 0.2 kN exerted by the car. At the same time, the trailer exerts an equal, but opposite, reaction equal to 0.2 kN on the car. Note that if a spring dynanometer was fitted between the car and the trailer, it would indicate a force of 0.2 kN

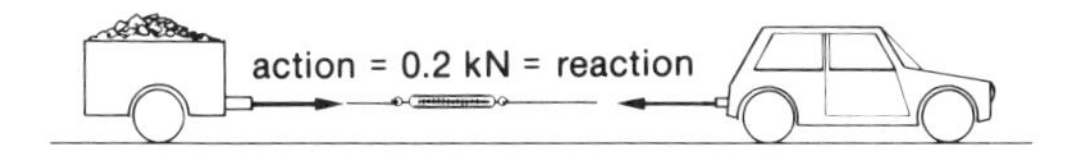

Fig. 3.4

Action and reaction forces are always collinear, i.e. they act along the same line, always equal in magnitude but opposite in sense. However, they are applied to two different bodies and should always be considered with respect to the body to which they are applied, e.g. the action applied to the trailer and the reaction applied to the car in the example above.

Transmissibility of a force

The principle of transmissibility states that the effect of a force on a body to which it is applied is not altered when the point of application of the force is moved to some other position on the line along which the force acts. In short, **a force can be moved along its line of action without changing its effect.** For

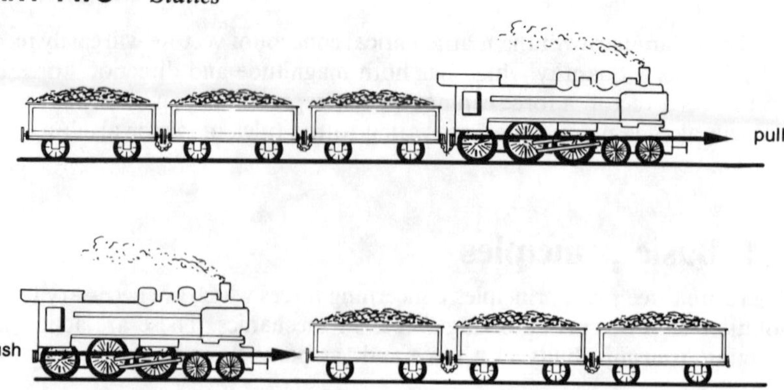

Fig. 3.5

example, a shunting locomotive can pull a train, or it can push a train with equal force and equal effect (Fig. 3.5).

It must be pointed out that the principle of transmissibility applies only to what are called rigid bodies, i.e. bodies which do not deform. One will readily appreciate that for elastic bodies such as a coiled spring a pull and a push produce different effects, namely stretching and compression. However, since most of the bodies considered in elementary mechanics are practically rigid, transmissibility of a force remains a very useful principle.

Parallelogram of forces

The principle of the parallelogram of forces, formulated by Stevin in 1586, states that if two forces intersecting at a point are represented in magnitude and direction by the adjacent sides of a parallelogram, their combined action is equivalent to the action of a single force, represented both in magnitude and direction by the diagonal of the parallelogram.

The single force, which has exactly the same effect as the two given forces, is called the **resultant** force.

Example 3.1

Two cables are attached to a bracket as shown in Figure 3.6(a).

Determine the resultant force acting on the bracket, if the force in the horizontal cable is 4.2 kN and that in the inclined cable is 2.4 kN.

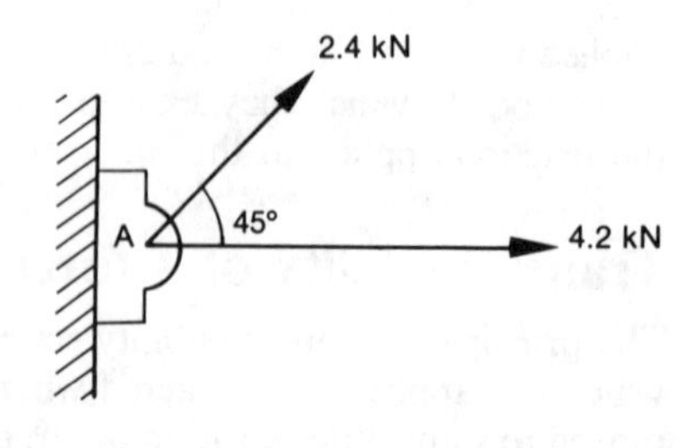

(a)

Solution

The resultant can be found by constructing the parallelogram of forces as shown in Figure 3.6(b), using a suitable scale, such as 10 mm = 1 kN.

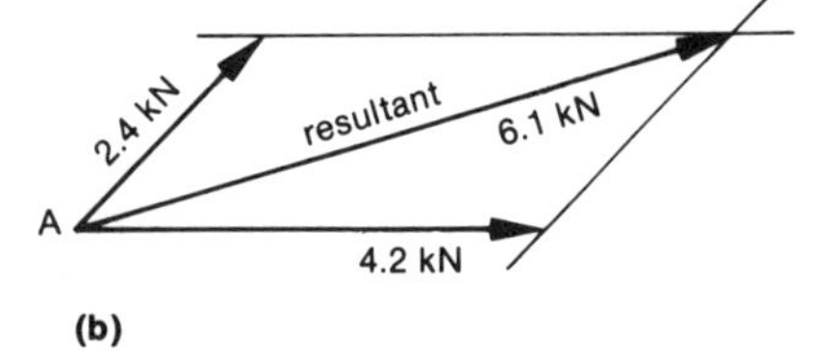

(b)

The answer is a force of 6.1 kN acting at 16° to the horizontal, as shown in Figure 3.6(c).

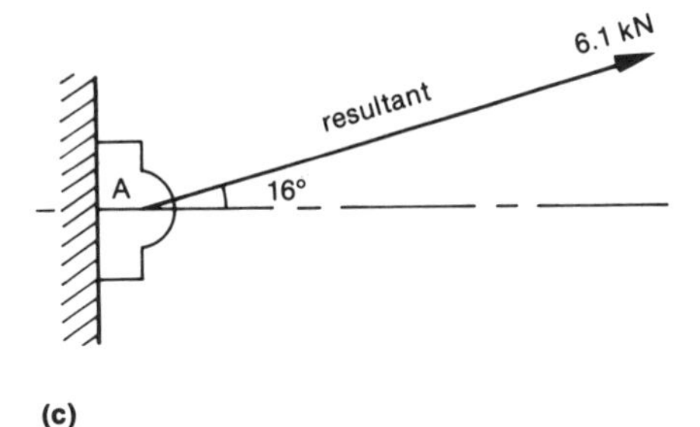

(c)

Note that in a simple example like this the three diagrams can be superimposed to form a combined diagram as in Figure 3.6(d). However, as the complexity of force systems increases in other problems, it is more convenient to draw separate force diagrams as in Figure 3.6(b).

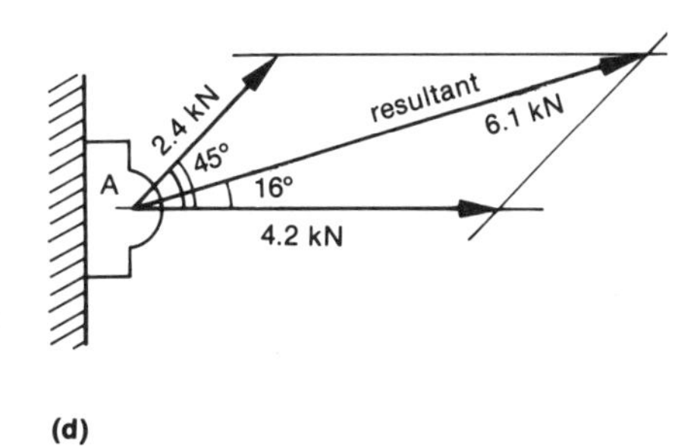

(d)

Fig. 3.6

3.5 Rectangular components of a force

In the previous section it was shown how two forces could be combined into a single resultant force. The reverse problem, called **resolution** of a force into components, is the separation of a single force into two component forces acting in different directions on the same point. The previous example could be reversed as follows.

Example 3.2

It is necessary to exert a force of 6.1 kN on the bracket at 16° to the horizontal by means of two cables, one horizontal, the other at 45° to it. What should the force in each cable be?

Solution

The information given in the problem is sufficient for us to construct a parallelogram of forces, which in fact is identical to that in Figure 3.6(b). The answers now are 4.2 kN in the horizontal cable and 2.4 kN in the other cable. Note that the forces in the two cables are the components of the required force of 6.1 kN.

In many engineering problems, the two components into which a single force must be resolved are at right angles to each other. Such components are called **rectangular components** of a force, as the parallelogram of forces becomes a rectangle. Problems involving the resolution of a force into rectangular components can be solved graphically by constructing a rectangle of forces, or mathematically by using trigonometric relationships.

Example 3.3

A force of 5 kN is acting up and to the right at 30° to the horizontal. Determine its horizontal and vertical components.

Solution

(a) Graphical method: Draw the parallelogram, i.e. rectangle of forces to scale, as in Figure 3.7. The answers are: the horizontal component is 4.3 kN to the right and the vertical component is 2.5 kN upwards.

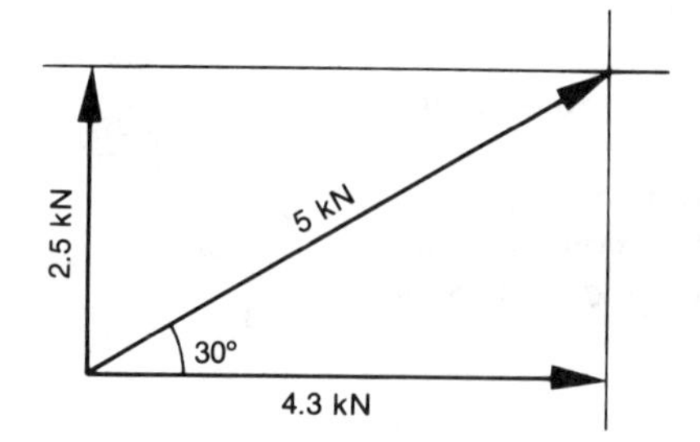

Fig. 3.7

(b) Mathematical method: Sketch a diagram similar to that in Figure 3.7, but not necessarily to scale. From the geometry of the triangles involved, the horizontal component is

$$F_H = 5 \times \cos 30° = 5 \times 0.866 = 4.33 \text{ kN} \rightarrow$$

and the vertical component is

$$F_V = 5 \times \sin 30° = 5 \times 0.5 = 2.5 \text{ kN} \uparrow$$

The relationships between a force F and its rectangular components in the mutually perpendicular x and y directions, F_x and F_y, are

$$F_x = F \cos \theta \quad \text{and} \quad F_y = F \sin \theta$$

where θ is the angle between the force and the x direction.*

* θ (theta) is a letter of the Greek alphabet often used to represent angular measurement.

Conversely, if the two components F_x and F_y are known, the force itself and the angle it makes with the x direction can be calculated from

$$F = \sqrt{F_x^2 + F_y^2} \quad \text{and} \quad \tan\theta = \frac{F_y}{F_x}$$

The x and y directions are usually horizontal and vertical, but they may also be chosen in any two mutually perpendicular directions, as in Figure 3.8.

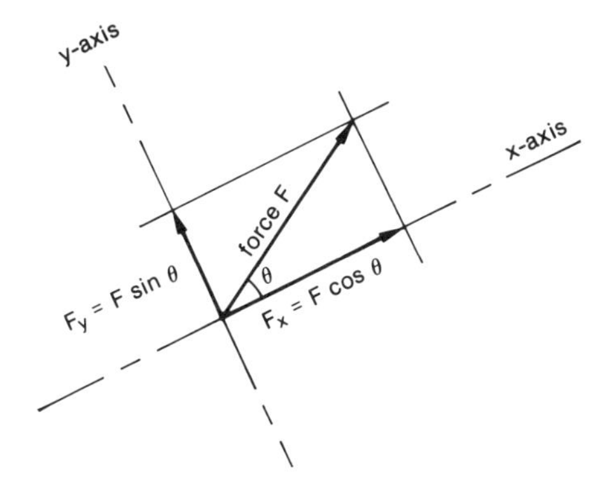

Fig. 3.8

Example 3.4

If the components of the force acting on a gear tooth are 54.6 N in the radial direction and 150 N in the tangential direction, determine the total force and the angle between the force and tangential direction (see Fig. 3.9).

Solution

Total force $F = \sqrt{F_x^2 + F_y^2} = \sqrt{150^2 + 54.6^2} = 159.6$ N

Angle $\tan\theta = F_y/F_x = 54.6 \div 150 = 0.364$

$\therefore \theta = 20°$

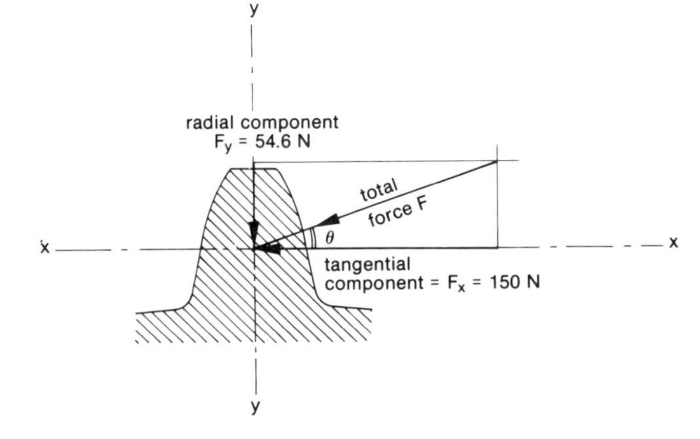

Fig. 3.9

Problems

3.1 What will the reading be on each of the spring dynamometers shown in Figure 3.10, if a force, or forces, are applied as indicated?

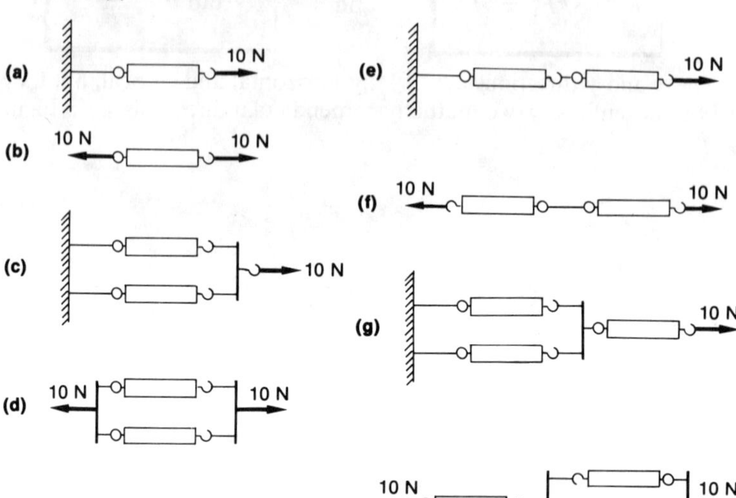

Fig. 3.10

3.2 Determine the resultant of the two forces acting on point A in each of the examples in Figure 3.11.

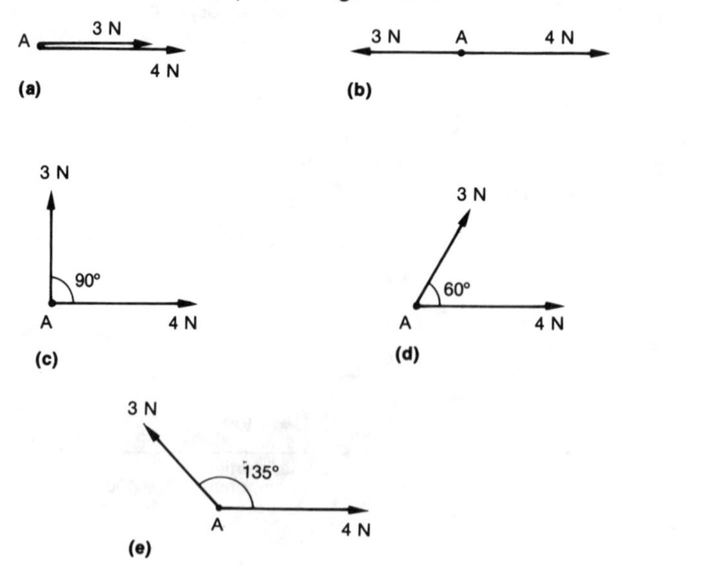

Fig. 3.11

3.3 Two cranes are attempting to right an overturned truck by each applying a pull of 50 kN. The two cables form an angle of 30° between each other. What is the resultant force on the truck?

3.4 A man pulls a loaded wagon with a force of 300 N acting along the handle of the wagon which makes an angle of 30° with the ground. What are the magnitudes of the horizontal and vertical components?

3.5 Four forces are acting at a point as shown in Figure 3.12. For each of the forces determine the horizontal and vertical components:
(a) graphically.
(b) mathematically.

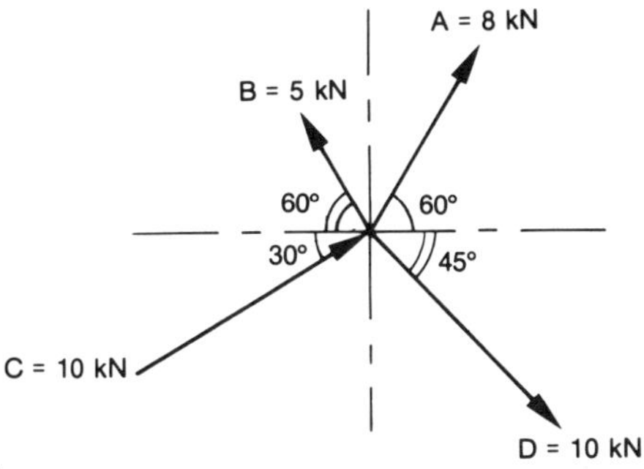

Fig. 3.12

3.6 A wagon is being pushed up an incline by a horizontal force of 500 N as shown in Figure 3.13.

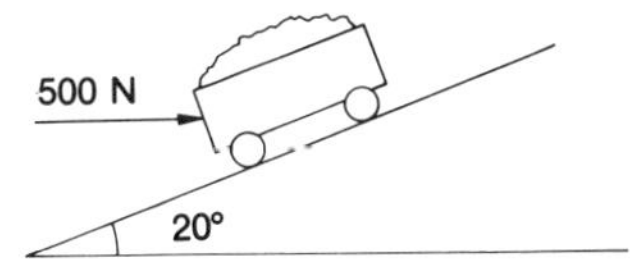

Fig. 3.13

Determine the components of the force acting along the plane and perpendicular to the plane.

3.7 A structural member is subjected to a load of 10 kN as shown in Figure 3.14. Determine the components of the load:
(a) horizontal and vertical, and
(b) along and perpendicular to the axis of the member.

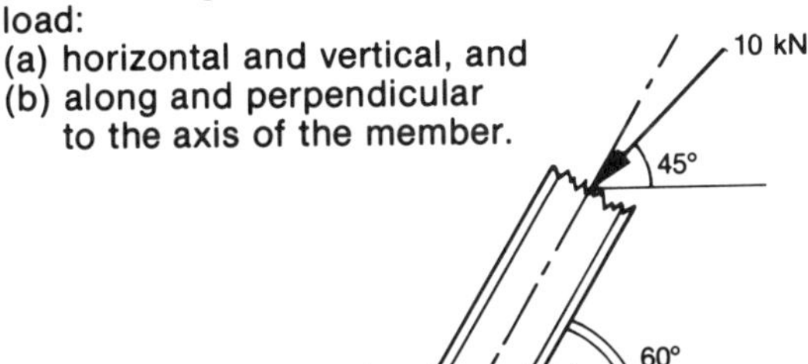

Fig. 3.14

3.8 Resolve the 800 N force acting on a cutting tool into components in directions perpendicular and parallel to the axis of the tool (Fig. 3.15).

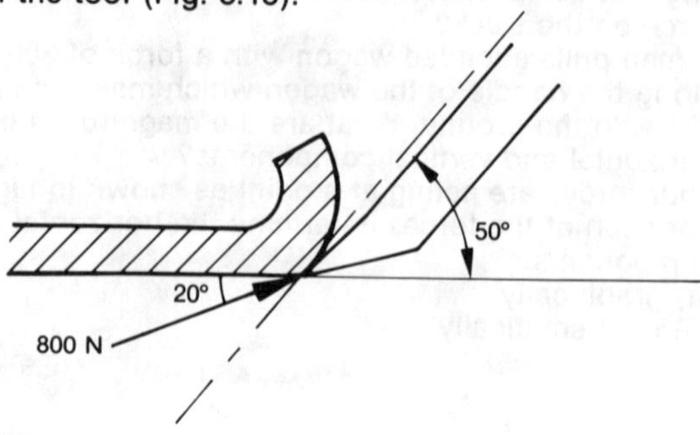

Fig. 3.15

3.6 Graphical addition of forces

In many engineering problems there are systems of forces consisting of more than two forces. If a resultant of such a system is required, it is possible to apply the parallelogram of forces principle to find the resultant of any two forces in the system, then combine that resultant with another force, and then repeat the procedure until all forces have been included. As one can imagine, the construction required is rather complicated, involving too many construction lines.

The solution can be simplified by introducing the triangle of forces rule, and then extending this rule to the polygon of forces method.

The force triangle rule

The triangle of forces rule is derived simply from the parallelogram principle. When two forces are added by the parallelogram method, the opposite sides of the parallelogram are always equal and any one of the two opposite sides can represent a force in magnitude, resulting in two possible triangles of forces, as in Figure 3.16.

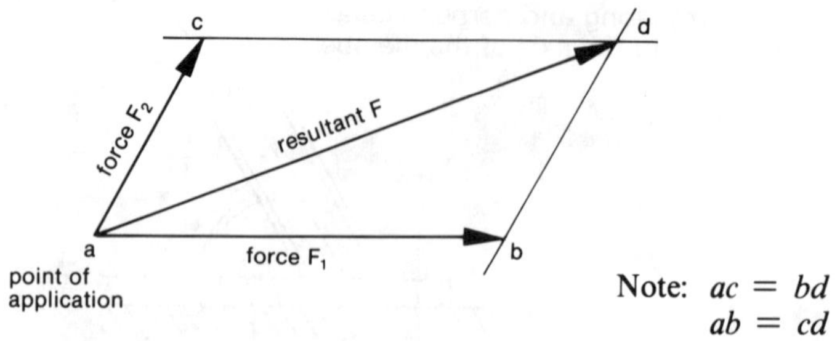

Note: $ac = bd$
$ab = cd$

Fig. 3.16(a) *Parallelogram of forces*

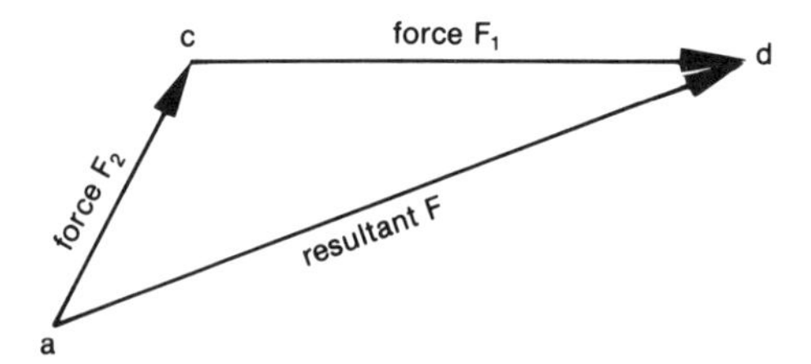

Fig. 3.16(b) *Triangle of forces (first alternative)*

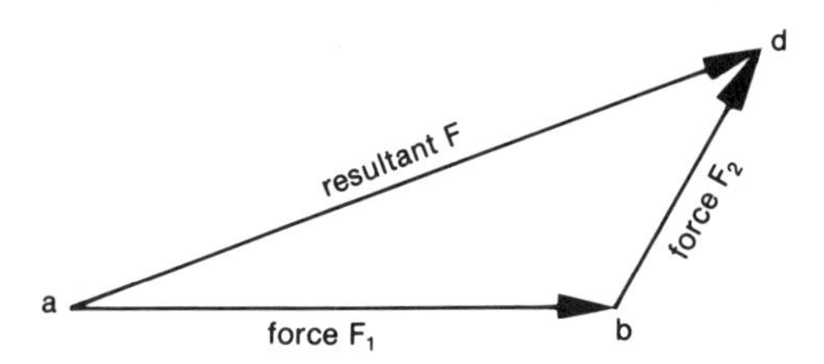

Fig. 3.16(c) *Triangle of forces (second alternative)*

The force triangle rule states that by arranging given forces F_1 and F_2 in a tip-to-tail fashion, taking into account the scaled magnitude and direction of the forces, the resultant of the two forces is found by connecting the tail of one, with the tip of the other, as shown in Figure 3.16.

Although closely related, the parallelogram and the triangle of forces differ in one very important respect. The parallelogram of forces shows all given and resultant forces as passing through a common point, the actual point of application of the forces. In this respect, the parallelogram is a geometrical representation of a fundamental principle showing true relationships between all forces. In a triangle of forces, one of the forces does not pass through the point of application. The triangle construction can only be regarded as a graphical rule or method for determining the magnitude and direction of the resultant force. As such, the triangle of forces should always be drawn as a separate "force diagram", and not superimposed on the diagram showing the actual layout of forces in relation to the point of application.

Example 3.5

A damaged vehicle is being pulled by means of two ropes as shown in Figure 3.17(a). Determine the resultant force in magnitude and direction, by the triangle of forces rule.

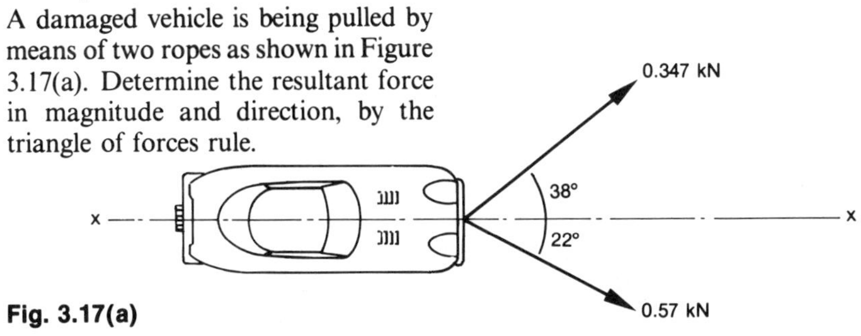

Fig. 3.17(a)

Solution

To solve the problem, construct the triangle of forces according to the rule, and scale-off the magnitude and direction (angle).

The answer is 0.8 kN in the x direction, acting to the right.

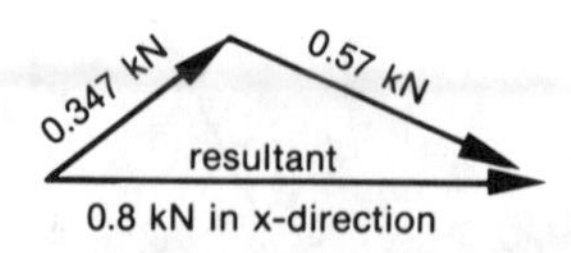

Fig. 3.17(b)

The force polygon method

The rule or method for finding resultants of force systems involving more than two forces consists of repeated applications of the force triangle rule to successive pairs of forces, until all the given forces are reduced to a single resultant force.

Example 3.6

Determine the resultant force on the eye-bolt used to anchor four guy wires as shown in Figure 3.18(a).

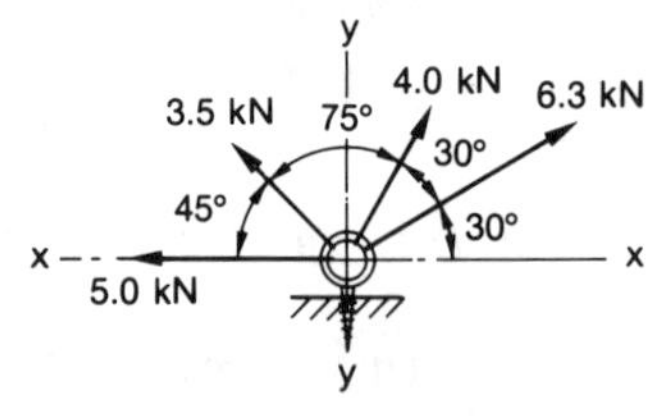

(a)

Solution

Construct the polygon of forces by successive addition of forces, as shown in Figure 3.18(b).

Answer is a vertical resultant force of 9.1 kN.

Note that the results of intermediate addition steps shown by dotted lines are usually omitted, only the outline of the force polygon is needed.

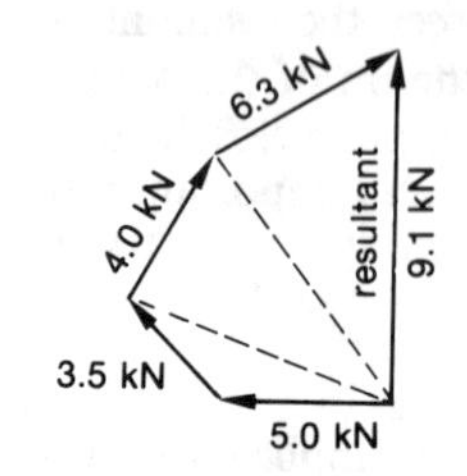

(b)

Fig. 3.18

Before leaving the subject, the student must be warned against losing sight of the real physical significance of the answers obtained. This should not be obscured by whatever method or construction is used to obtain the answers. Thus in Example 3.5, the combined effect of the applied forces is that the vehicle is being pulled forward with a force equal to 0.8 kN. Likewise, in Example 3.6, the combined pull of the guy wires produces an upward force of 9.1 kN acting on the bolt. One must always try to visualise the answers in tangible physical terms, as illustrated in Figure 3.19, and not think of the method or construction used to obtain the answers as the end in itself.

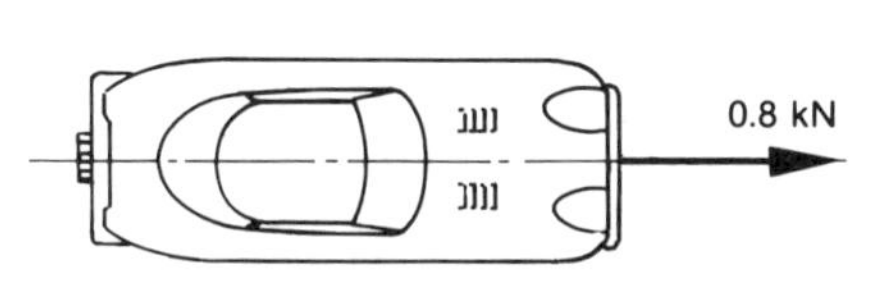

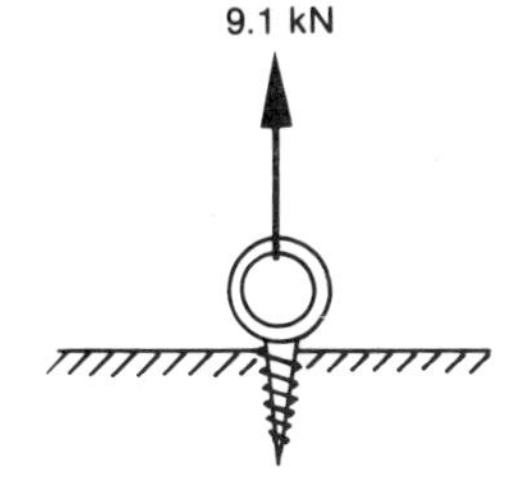

Fig. 3.19

3.7 Mathematical addition of forces

Addition of forces, i.e. solving for the resultant of a system of forces, can also be achieved mathematically by summing their x and y components. The method of solution consists of several steps, as follows:

Step 1 Resolve given forces into x and y components (usually horizontal and vertical), using $F_x = F \cos \theta$ and $F_y = F \sin \theta$, where θ is the acute angle between each force and the x axis (horizontal axis).

Step 2 Assign positive and negative signs to each component, according to the usual mathematical sign convention, i.e. to the right—positive, to the left—negative, upwards—positive, downwards—negative.

Step 3 Add all x components $= \Sigma F_x$,* taking into account the positive and negative signs, then add all y components $= \Sigma F_y$, taking signs into account.

Step 4 The two sums can now be used to determine the resultant force using $F = \sqrt{(\Sigma F_x)^2 + (\Sigma F_y)^2}$.

Step 5 Determine the angle the resultant makes with the x direction using $\tan \theta = \dfrac{\Sigma F_y}{\Sigma F_x}$.

In summary, this method involves resolving forces into rectangular components (steps 1 and 2), reducing all components to a single force in each of the two directions x and y (step 3), and adding these two remaining forces into the resultant (steps 4 and 5), according to the rules explained in Section 3.5.

The procedure is greatly assisted by the use of a table for recording all intermediate results, as illustrated in the following example.

* Σ is a mathematical sign meaning "the sum of", pronounced "sigma". It is a letter of the Greek alphabet.

Example 3.7

Determine the resultant force acting on the barge due to the combined effort of four tug-boats as shown in Figure 3.20(a).

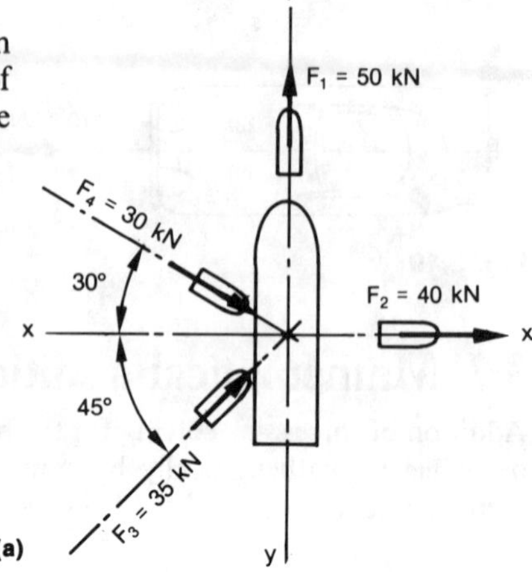

Solution

The solution is carried out according to the procedure outlined above, with the aid of a diagram as illustrated in Figure 3.20(b), if necessary.

The results are tabulated as shown in Table 3.1.

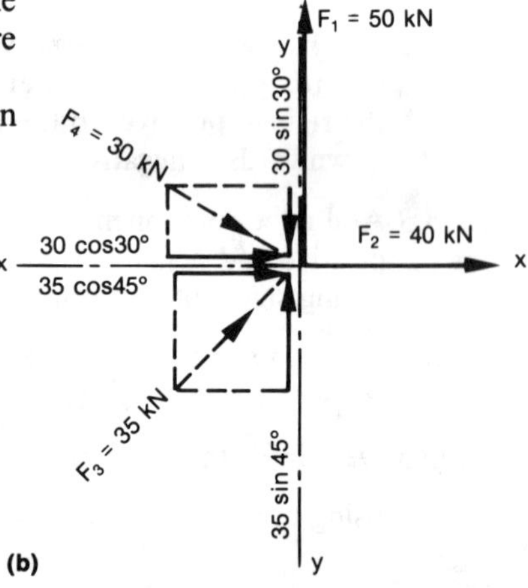

Resultant $F = \sqrt{90.7^2 + 59.7^2} = 108.6$ kN

Angle $\tan\theta = \dfrac{59.7}{90.7} = 0.658$, $\theta = 33.4°$

The resultant force is as shown in Figure 3.20(c).

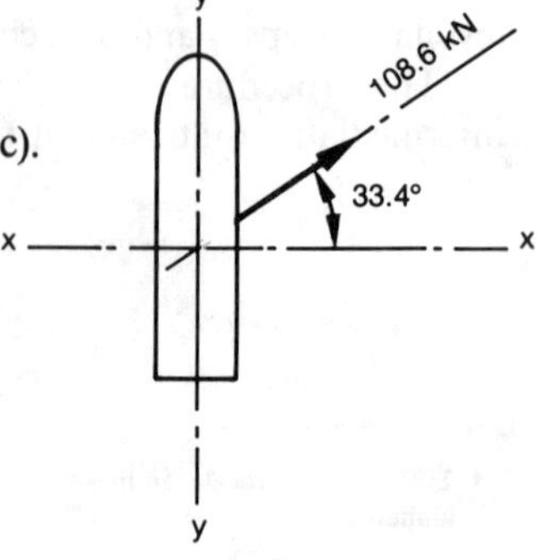

Fig. 3.20 **(c)**

Table 3.1

Force	*Magnitude*	*X component*	*Y component*
F_1	50	0	50.0
F_2	40	40.0	0
F_3	35	24.7	24.7
F_4	30	26.0	-15.0
		$\Sigma F_x = 90.7$	$\Sigma F_y = 59.7$

Finally, it is always a good idea to cross-check the solution by an independent alternative method. A mathematical solution can be verified graphically, and vice versa. Can you solve this example graphically? What about Example 3.6, mathematically?

Problems

3.9 For the systems of forces acting at a point in Figure 3.21, determine the magnitude and direction of the resultant by constructing polygons of forces.

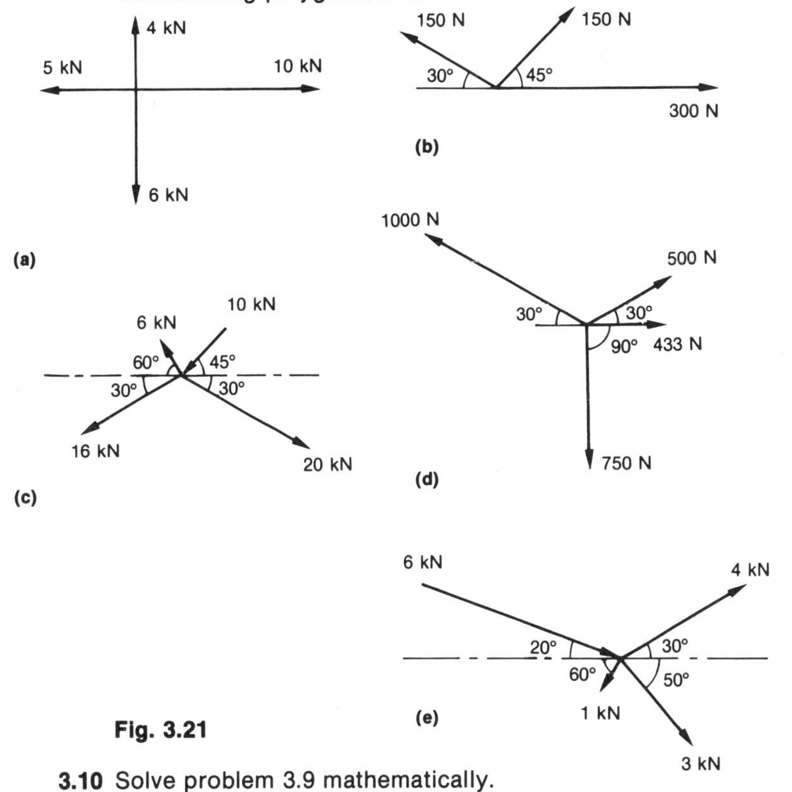

Fig. 3.21

3.10 Solve problem 3.9 mathematically.

3.8 Universal gravitation

Gravity is one of the most common physical phenomena seen in nature, which manifests itself as a **force** of mutual attraction between masses.

Historically, our understanding of the Law of Universal Gravitation was a result of many centuries of astronomical observations culminating in the work of Johannes Kepler (1571–1630) who discovered important regularities in the motion of the planets, lending support to the Copernican heliocentric theory of the solar system.

However, it was the mathematical insight of Isaac Newton that enabled the formulation of the law relating the force of mutual attraction between two bodies of known masses to the square of the distance between them.

Newton's Law of Universal Gravitation states that the force of gravitational attraction (F_g) between two bodies having masses m_1 and m_2 separated by a distance d is given by

$$F_g = G\frac{m_1 . m_2}{d^2}$$

It is important to understand that the constant G, known as the universal gravitational constant, has a constant value which is the same for any pair of bodies, regardless of their masses or the distance separating them.

The actual value of G was measured experimentally by Henry Cavendish in 1798. Significant improvements in the accuracy of the measurement were achieved in the nineteenth and twentieth centuries. The present accepted value, determined at the United States National Bureau of Standards is

$$G = 66.7 \times 10^{-12}\ \frac{\text{N.m}^2}{\text{kg}^2}$$

Example 3.8

Determine the force of mutual attraction between the following pairs of bodies:

(a) The Earth and the Moon, given mass of Earth 5.97×10^{24} kg, mass of Moon 73.7×10^{21} kg and distance between them 0.38×10^6 km.
(b) Two ships 30 000 tonnes each at a centre to centre distance of 50 m.
(c) Two 1 kg masses at a distance of one metre.

Solution

(a) $F_g = 66.7 \times 10^{-12}\ \frac{\text{N.m}^2}{\text{kg}^2} \times \frac{5.97 \times 10^{24}\ \text{kg} \times 73.7 \times 10^{21}\ \text{kg}}{(0.38 \times 10^9\ \text{m})^2}$

$= 0.203 \times 10^{21}$ N

This is the huge pull between the Earth and the Moon which keeps the Moon in its orbit.

(b) $F_g = 66.7 \times 10^{-12} \dfrac{\text{N.m}^2}{\text{kg}^2} \times \dfrac{30 \times 10^6 \text{ kg} \times 30 \times 10^6 \text{ kg}}{(50 \text{ m})^2} = 24 \text{ N}$

In comparison with other forces acting on the ships, a force of 24 N is totally insignificant.

(c) $F_g = 66.7 \times 10^{-12} \dfrac{\text{N.m}^2}{\text{kg}^2} \times \dfrac{1 \text{ kg} \times 1 \text{ kg}}{(1 \text{ m})^2} = 66.7 \times 10^{-12} \text{ N}$

This is a tiny force which is of no particular interest to the engineer.

It may appear at this point that the Law of Universal Gravitation is of no particular practical use to the engineer, since forces of attraction between objects, even as large as ships, are very small indeed. However, as we shall see in the next section, the understanding of the law is necessary for developing the concept of weight.

3.9 Weight of a body

The engineer is concerned with structures and machines which are located on, or very near, the surface of the Earth, i.e. at nearly constant distance from the centre of the Earth, equal to its mean radius of 6370 km.

If this distance is taken as constant for all such objects, the universal law of gravity can be reduced to a special case applicable at or near the Earth's surface, as follows:

$$F_g = G \frac{m_e \times m_o}{r_e^2}$$

where m_e is mass of Earth $= 5.97 \times 10^{24}$ kg
r is mean radius of Earth $= 6.37 \times 10^6$ m
m_o is mass of a given object

Substitution and combining of constants yields:

$$F_g = 66.7 \times 10^{-12} \frac{\text{N.m}^2}{\text{kg}^2} \times \frac{5.97 \times 10^{24} \text{ kg} \times m_o \text{ kg}}{(6.37 \times 10^6 \text{m})^2}$$

$$F_g = 9.81 \frac{\text{N}}{\text{kg}} . m_o \text{ kg}$$

The new constant we have obtained is called the **local gravitational constant** with the symbol g.

In science and engineering the force of gravity exerted by the Earth on an object is often referred to as the **weight of the object,** F_w, given by

$$\boxed{F_w = m . g}$$

where m is the mass of the object and

$$\boxed{g = 9.81 \frac{\text{N}}{\text{kg}}}$$

Being a force, weight is measured in units of force, i.e. newtons. In this regard, one must not be misled by the common usage of the word "weight" as equivalent to mass, or quantity. To the engineer, mass is a measure of quantity expressed in kilograms, and **weight is a measure of gravitational force expressed in newtons.**

The force of gravity on an object, or weight, is always acting towards the centre of the Earth, i.e. vertically downwards, and is applied to the object at the centre of its mass distribution known as its centre of gravity. When the mass of an object is distributed uniformly throughout its volume, the centre of gravity coincides with the geometrical centre of the shape, e.g. the centre of gravity of a uniform solid sphere is at its geometrical centre.

Example 3.9

A man has a mass of 79 kg. What is his weight?

Solution

$$F_w = m.g = 79\ \text{kg} \times 9.81\ \frac{\text{N}}{\text{kg}} = 775\ \text{N}$$

Example 3.10

What is the force in a cable supporting a load of 1.5 tonne?

Solution

$$F = F_w = m.g = 1500\ \text{kg} \times 9.81\ \frac{\text{N}}{\text{kg}} = 14\,715\ \text{N} = 14.7\ \text{kN}$$

3.10 Local variations in gravity

In the previous section the value of the local gravitational constant g was assumed to be the same for all places on Earth's surface. However, this is true only if the distance to the centre of the Earth is the same for all locations, which is not strictly correct.

Because the Earth is slightly ellipsoidal in shape and not a perfect sphere, there is a small gradual variation in g with latitude, as can be seen from Table 3.2.

These values are average values only because there are other local influences owing to the nature of the underlying rocks, etc., at different locations of the same latitude.

Gravity also varies with altitude above sea level, as shown in Table 3.3 for a latitude of 45°.

Table 3.2 Variation of g at sea level

Latitude	*Equator*	*10°*	*20°*	*30°*	*40°*	*50°*	*60°*	*70°*	*80°*	*90°*
g (N/kg)	9.780	9.782	9.786	9.793	9.802	9.811	9.819	9.826	9.831	9.832

Table 3.3 Variation of g with altitude

Altitude (km)	*Sea level*	*1*	*2*	*5*	*10*	*30*	*100*
g (N/kg)	9.806	9.803	9.800	9.791	9.776	9.714	9.598

The overall average, known as the International Standard value has been defined as $g = 9.80665$ N/kg at sea level. The local value for Sydney (Australia) is 9.79683 N/kg.

It is common practice to use the value of 9.81 N/kg for most engineering calculations. It can easily be seen that the error involved in using this value instead of the more accurate local value is quite insignificant, except for very high altitudes.

Problems

3.11 Determine the magnitude of the gravitational attraction between the Sun and the Earth, given that the mass of the Sun is 1.99×10^{30} kg, that of the Earth is 5.97×10^{24} kg and the distance between them is 1.5×10^{8} km.

3.12 Determine the weight of the following:
(a) one-gram mass
(b) one-kilogram mass
(c) one-tonne mass
(d) a brick of mass 4 kg
(e) a 50 g egg

3.13 Determine the weight of a truck which has a mass of 2.3 tonnes.

3.14 Determine the reading in newtons on each of the spring balances shown in Figure 3.22.

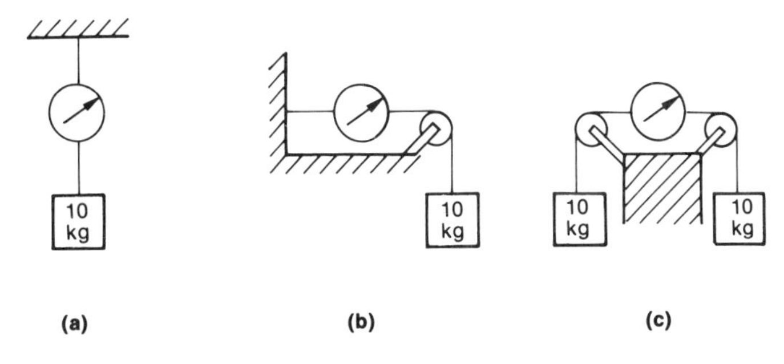

Fig. 3.22

3.15 A vehicle having a mass of 1.35 tonne is standing on an incline at 15° to the horizontal. Determine the components of its weight parallel and perpendicular to the road surface.

3.16 If the tractive effort applied at the wheels of the vehicle in problem 3.15 is equivalent to a force of 4 kN acting up along the incline, determine the resultant of the weight and the tractive effort.

3.17 In a laboratory experiment forces are applied to a bar by means of attaching different masses as shown in Figure 3.23. The mass of the bar itself is 600 g.

Draw a diagram of the bar showing all forces acting on the bar, including the weight of the bar, indicating correct magnitude and direction of each force.

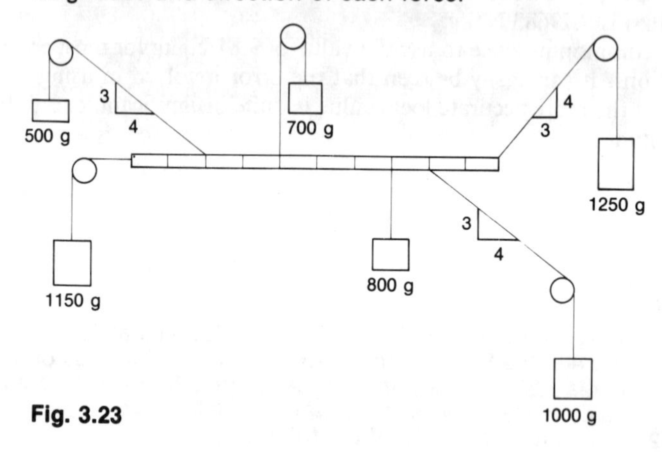

Fig. 3.23

Review questions

1. What is meant by force?
2. What is the SI unit of force?
3. Which one of the following cannot be measured in newtons? muscular effort, gravitational attraction, quantity of matter, frictional resistance. Why?
4. Describe a simple spring dynamometer.
5. Name the three main characteristics of a force.
6. What is the principle of action and reaction?
7. What is meant by transmissibility of a force?
8. Explain the principle of the parallelogram of forces.
9. What is meant by the resultant of two or more forces?
10. State the formulae for resolving a force into rectangular components.
11. State the formulae for combining two forces at 90° to each other into a resultant force.
12. Explain how the force triangle and the force polygon can be used for graphical addition of forces.
13. Outline the steps for the mathematical addition of forces.
14. If the mathematical solution of a problem is compared with the graphical solution, would you expect the answers to be the same?
15. State the Law of Universal Gravitation.
16. What is the value of the universal gravitational constant?
17. What is greater, the attraction of the Earth for the Moon, or the attraction of the Moon for the Earth?
18. What is meant by weight?
19. What is the SI unit of weight?
20. If a body is taken from the equator to the south pole, what will be the effect on its mass and on its weight?
21. What is the value of the local gravitational constant generally used in practical engineering calculations? Are there any limitations on its accuracy?
22. Is it possible to use a spring balance to measure mass? Explain.

4

Equilibrium of concurrent forces

When a structure or a machine is at rest, it is said to be in static equilibrium. In general, the state of **equilibrium** can be defined as a state of rest or balance under the action of forces which counteract each other.

This chapter is concerned with conditions of equilibrium of **concurrent forces**. A system of forces is called concurrent when the lines of action of all forces intersect at a common point, the point of concurrency.

4.1 Conditions of equilibrium

For a system of concurrent forces to be in equilibrium, the resultant force, that is the result of summation of all forces acting through that point, must be equal to zero. In other words, all forces balance each other in such a way that there is no resultant push or pull acting on the body at the point of application of the forces.

To prove that a given system of forces is in fact in equilibrium, we must demonstrate graphically or mathematically that the forces add up to zero. Naturally, since forces are vector quantities, the addition must be vectorial addition as explained in Chapter 3. Consider the following example.

Example 4.1

A joint of a pin-jointed truss is as shown in Figure 4.1(a) and has internal and external forces as shown. Prove that the joint is in equilibrium.

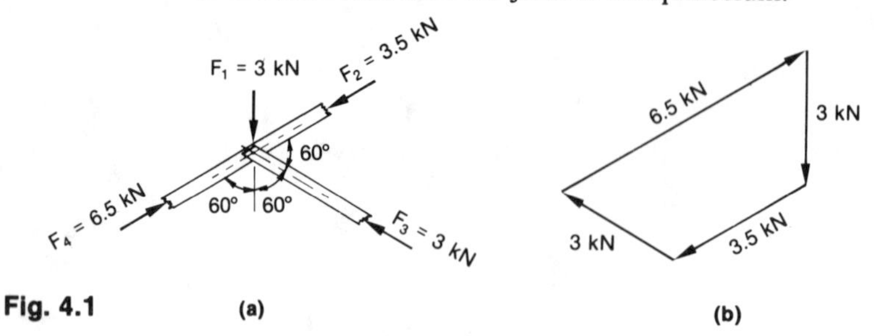

Fig. 4.1 (a) (b)

Solution

In order to prove the equilibrium of this system of concurrent forces graphically, it is necessary to construct the polygon of forces, using all of the applied forces, external (F_1) as well as internal forces (F_2, F_3 and F_4), and to show that the starting point of the construction coincides with the end point, i.e. the polygon must close, leaving no room for a gap representing a resultant force (remember: the resultant must be zero). See Figure 4.1(b).

Mathematically the summation of forces is signified by ΣF, meaning that the resultant force is the sum of all the applied forces. For a system in equilibrium, the resultant is equal to zero, or

$$\Sigma F = 0$$

This is the equation of forces in equilibrium, which is a very useful equation. However, this equation is often interpreted in terms of the perpendicular components of the resultant force. If the resultant is equal to zero, its components must also be equal to zero, i.e. no force—no components. It follows that, for a given system of forces, the sum of all components in any direction must be zero. This is usually expressed in terms of mutually perpendicular directions x and y, often horizontal and vertical.

$$\Sigma F_x = 0 \quad \text{and} \quad \Sigma F_y = 0$$

Now let us consider a mathematical solution to the previous example.

Alternative solution

To prove the equilibrium of forces mathematically, it is necessary to demonstrate that both vertical and horizontal components add up to zero. We can use a table similar to Table 3.1 to record all the force components, including their positive and negative signs. The solution will appear as in Table 4.1.

The sum of the horizontal, or x direction, components and the sum of the vertical, or y direction, components have been shown to be zero, i.e. the two equations $\Sigma F_x = 0$ and $\Sigma F_y = 0$ have each been satisfied. The conclusion is that the system of forces is in equilibrium, as we would expect in a structure such as a roof truss.

Table 4.1

Force	*Magnitude*	*X component*	*Y component*
F_1	3.0	0	-3.0
F_2	3.5	-3.03	-1.75
F_3	3.0	-2.60	1.50
F_4	6.5	5.63	3.25
		$\Sigma F_x = 0$	$\Sigma F_y = 0$

4.2 Free-body diagrams

When problems involving the equilibrium of bodies under the action of force systems are being solved, a method of setting out the essential details in the form of a diagram is necessary.

We often start with a semi-pictorial sketch, or space diagram showing the physical conditions of the problem, i.e. the layout of its mechanical or structural components such as pulleys, supports, cables, rollers, etc.

The next step is to isolate the essential facts about the forces involved and to draw a free-body diagram. Such a diagram usually shows the point of concurrency acted upon by all the forces, indicating the magnitudes and directions of the forces.

A force polygon is then constructed, based on the information contained in the free-body diagram.

Example 4.2

Consider the equilibrium of forces and hence determine the force in each cable, when two cranes are supporting a 5 tonne mass as shown in Figure 4.2(a).

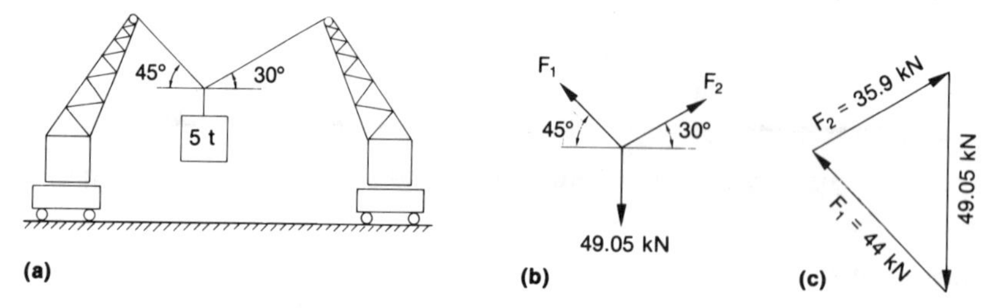

Fig. 4.2 *(a) Space diagram* *(b) Free-body diagram* *(c) Force triangle*

Solution

The weight of the load is:

$$F_w = m.g = 5000 \text{ kg} \times 9.81\frac{\text{N}}{\text{kg}} = 49\,050 \text{ N} = 49.05 \text{ kN}$$

The free-body diagram, Figure 4.2(b), is drawn to represent all forces acting on the point of concurrency. Note that each force is represented as a vector, showing its magnitude and direction in relation to the point.

Knowing one force, i.e. the weight, and the direction of the other two forces enables us to construct the force triangle. (Note that a triangle is just a special case of force polygon where the number of forces involved is three.)

If the triangle of forces is drawn to scale, the answers can easily be scaled off. Alternatively, the sine rule can be used to calculate the unknown forces as side-lengths of the force triangle.

$$F_1 = 49.05 \times \frac{\sin 60°}{\sin 75°} = 43.98 \text{ kN}$$

$$F_2 = 49.05 \times \frac{\sin 45°}{\sin 75°} = 35.91 \text{ kN}$$

Example 4.3

An elastic member *ABC* is stretched as shown in Figure 4.3(a) by three forces F_1, F_2 and F_3.

Determine the forces in *AB* and *BC*.

Solution

In this example more than three forces are involved. Therefore, a polygon of forces, rather than a triangle would have to be constructed, as in Figure 4.3(b).

The procedure is to construct as much of the polygon as possible using all known forces and then to complete it by drawing two lines, parallel to the unknown forces, through the first and last points.

The answers are scaled off the force polygon and are equal to $F_{AB} =$ 148 N and $F_{BC} = 167$ N.*

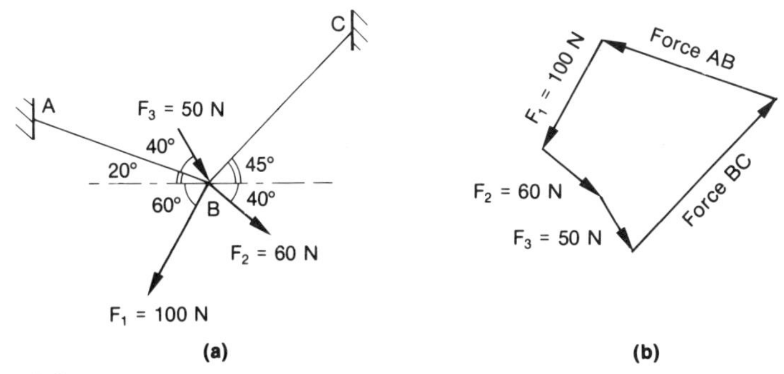

Fig. 4.3

4.3 The equilibrant force

If a system of concurrent forces is not balanced, i.e. not in equilibrium, it is possible to determine the additional force required to produce equilibrium. Such a force is called the **equilibrant**. In other words, the equilibrant of a system of concurrent forces is that force which when added to the system produces equilibrium.

In the event of a balanced system of forces, the equilibrant force would be equal to zero.

Example 4.4

Determine the equilibrant force for the system of forces shown in Figure 4.4(a).

* An alternative mathematical solution is also possible, but is rather complex, involving two simultaneous equations in terms of unknown forces. It is not recommended at this level.

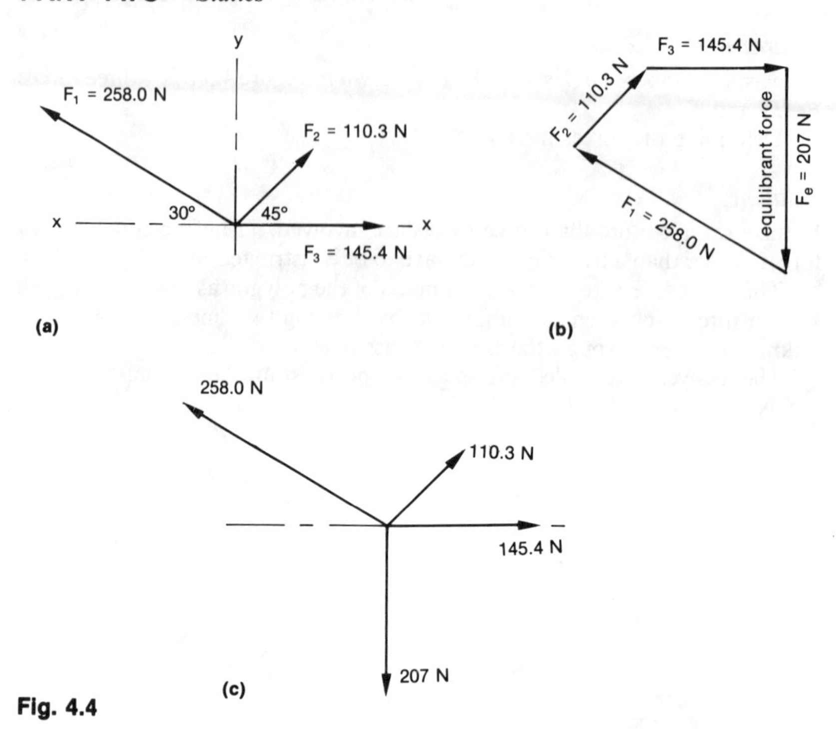

Fig. 4.4

Solution

To solve this problem graphically, we must recall that for a system of forces in equilibrium the force polygon must close. If we attempt to construct the force polygon, using only the given forces F_1, F_2 and F_3, arranging the forces in head-to-tail order, we find that the first and last points do not coincide. In order to close the gap an additional line is required, as shown in Figure 4.4(b). This line will close the polygon and represent the required equilibrant force, both in magnitude and direction.

The system of forces in equilibrium is shown in Figure 4.4(c).

It is important to remember that the closed polygon of forces has all its forces, including the equilibrant force, follow the head-to-tail order. This enables us to determine the correct direction of the equilibrant force, e.g. in this example the equilibrant is a vertical downward force.

It should also be understood that the equilibrant and the resultant forces are always equal in magnitude and opposite in direction. (Compare Examples 3.6 and 4.3.)

Alternative solution

The equilibrant force can also be found mathematically by means of addition of rectangular components, including those for the unknown equilibrant force.

The solution is best set out in tabular form as in Table 4.2.

Table 4.2

Force	Magnitude	X component	Y component
F_1	258.0	-223.4	129.0
F_2	110.3	78.0	78.0
F_3	145.4	145.4	0
Equilibrant	F_e	F_x	F_y

For equilibrium: $\Sigma F_x = 0 - 223.4 + 78.0 + 145.4 + F_x = 0$
Hence $F_x = 0$
Also $\Sigma F_y = 0 + 129.0 + 78.0 + 0 + F_y = 0$
Hence $F_y = -207$ N

The magnitude of the equilibrant force is therefore

$$F_e = \sqrt{F_x^2 + F_y^2} = \sqrt{0 + 207^2} = 207 \text{ N}$$

Consideration of the directions of components F_x and F_y indicates that the equilibrant is acting vertically downwards.

Problems

4.1 A light fitting having a mass of 1.5 kg is hanging from the ceiling on wire *AB* and tied to the wall by string *BC* (Fig. 4.5). Determine the forces in *AB* and *BC*.

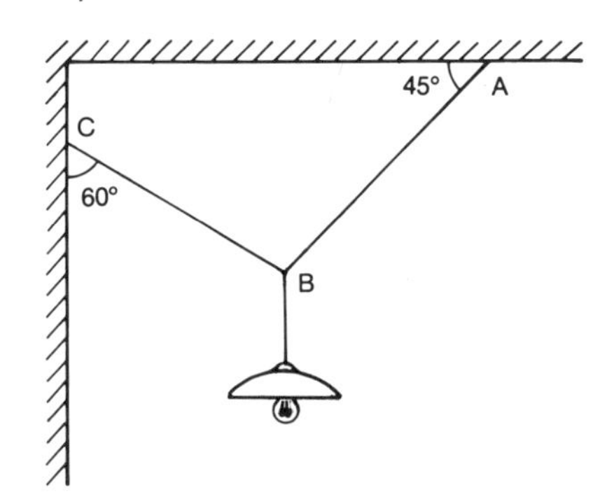

Fig. 4.5

4.2 A street light of mass 15 kg is supported at mid-point between two poles by a cable *ABC* (Fig. 4.6). If the length of the cable *ABC* is 20 m and deflection *BD* at mid-point is 0.2 m, determine the force in the cable.

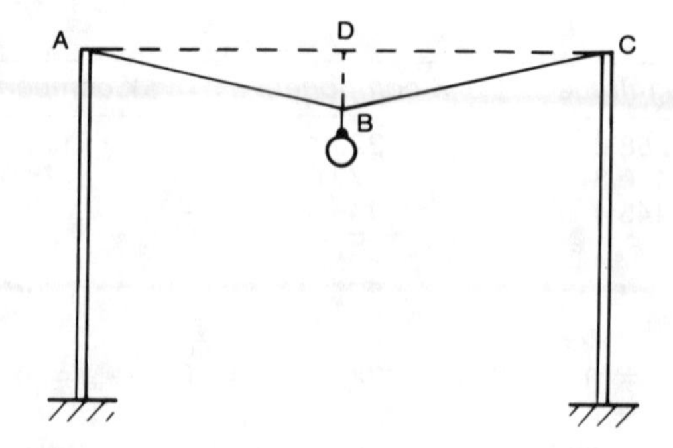

Fig. 4.6

4.3 The jib-crane in Figure 4.7 carries a load of 1.4 tonne. Determine the forces in the jib and the tie.

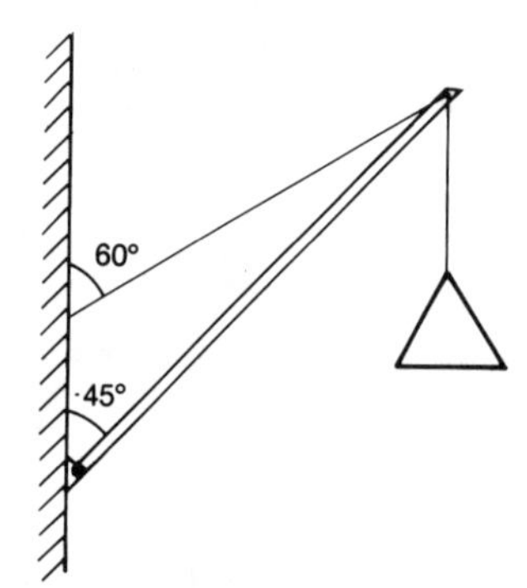

Fig. 4.7

4.4 Determine the forces in *AB, BC, CD* and *BE* (Fig. 4.8) if the mass m = 500 kg.

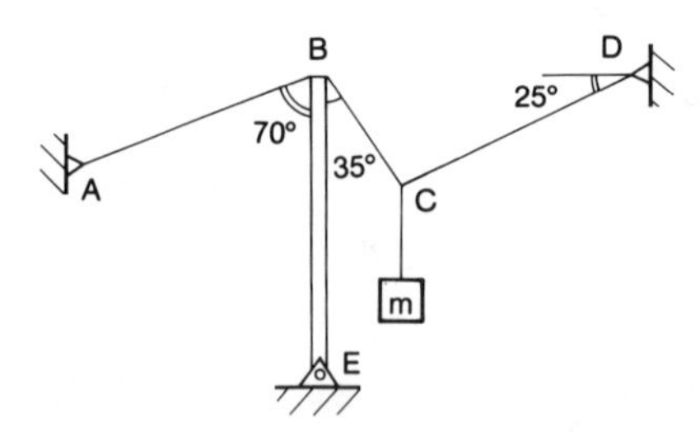

Fig. 4.8

4.5 Determine the force acting through the axis of the pulley shown (Fig. 4.9). Does the magnitude or direction of the force depend on pulley diameter?

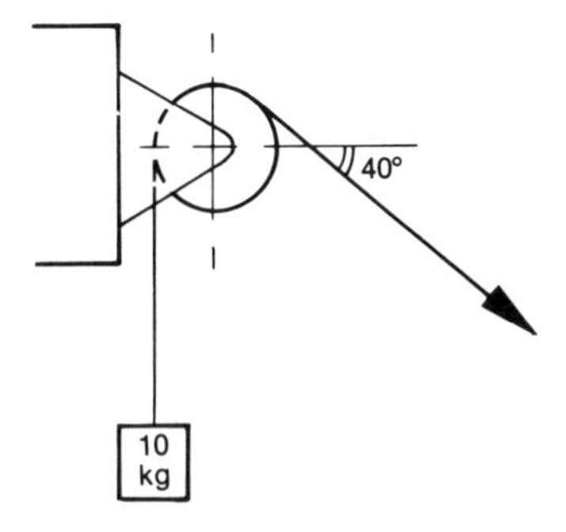

Fig. 4.9

4.6 Determine the mass lifted and the forces in members *AB* and *BC* (Fig. 4.10) when the tension in the cable is 20 kN.

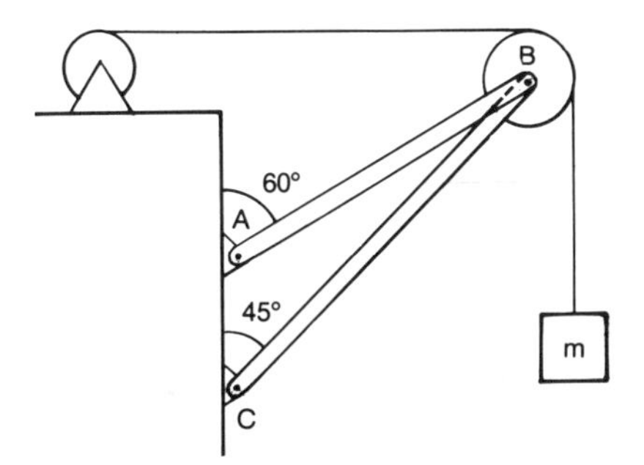

Fig. 4.10

4.7 Figure 4.11 represents joints in a simple pin-jointed roof truss. Using the information available for each joint, determine the unknown forces by constructing a separate polygon of forces for each joint.

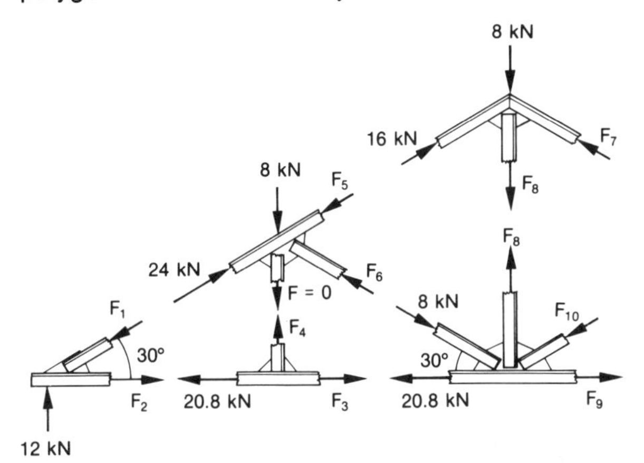

Fig. 4.11

4.8 Determine the equilibrants of the systems of forces in Figure 4.12, graphically and mathematically.

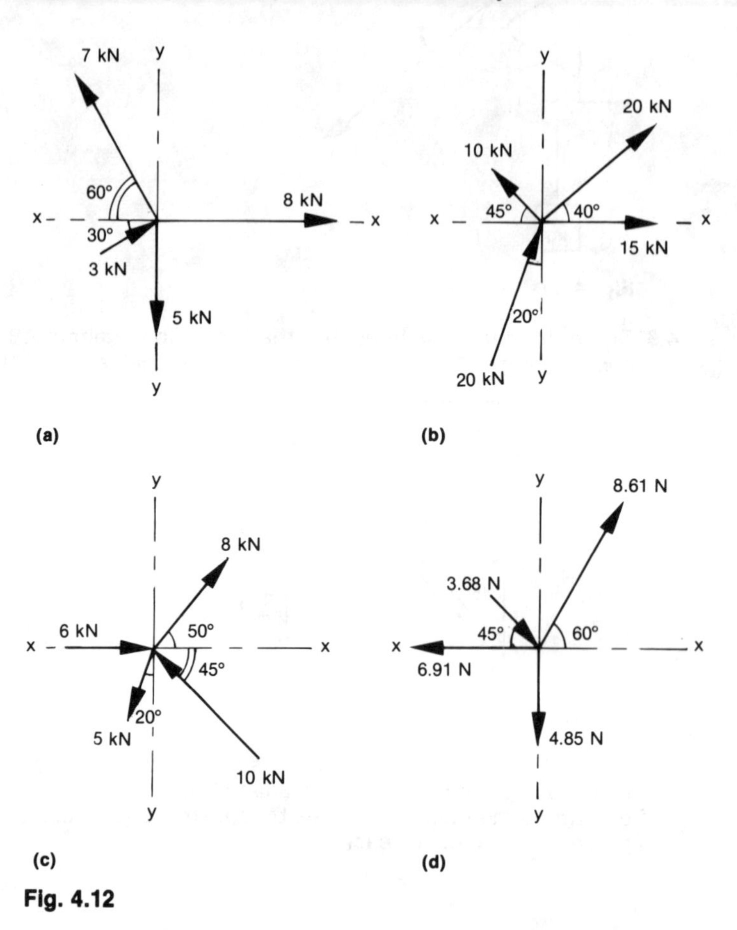

Fig. 4.12

4.4 Support reactions

Before proceeding to the next section, it is advisable to discuss the types of contact surfaces or supports in relation to the kind of reaction force that may exist between a body, such as a beam or a truss, and its supporting surface.

In this chapter we need to consider two categories of supports: those at which the reaction force is always normal to the supporting surface, and those which may support a force at any angle to the supporting surface.

The first category includes a *smooth*, frictionless* surface contact between a body and a supporting surface, such that the body can slide along the surface at the point of contact without any resistance. A support on rollers, such as one can see under some long bridge sections, has a similar force action, i.e. it provides normal** reaction only.

The second category includes a rough surface contact between a body and a supporting surface in which friction prevents relative sliding motion, producing a reaction force in any direction. A fixed hinge or bearing has a similar force action capable of supporting a force at any angle to the supporting surface.

The rule to remember is that, at a smooth surface or at a roller support, the reacting force is always perpendicular to the supporting surface.

In addition to the above-mentioned reaction forces, it is helpful to recognise that forces in links pivoted at each end and in cables connecting two points of a structure always act along the axes of such members (see Table 4.3).

Table 4.3 Classification of supports, connections and contact surfaces

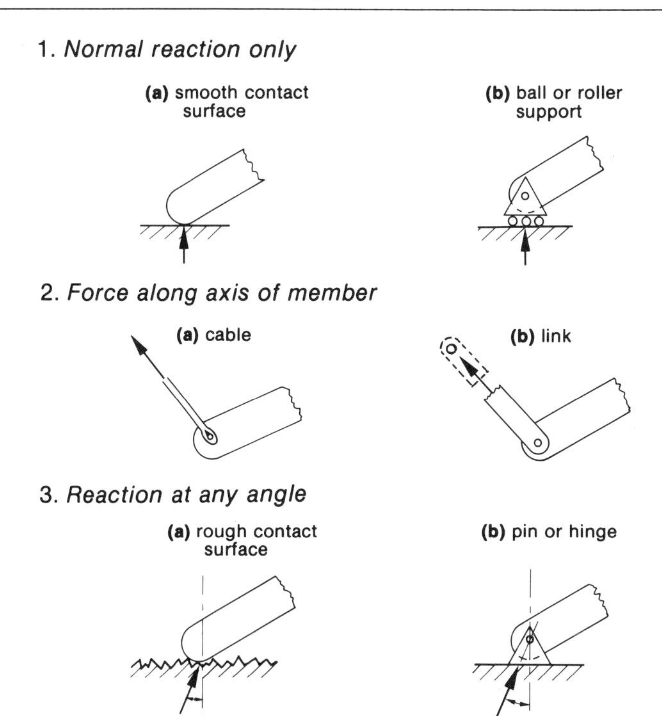

* Friction is discussed in Chapter 7.

** The *normal* to a line or surface is a direction perpendicular to it, e.g. normal force is perpendicular to the supporting surface.

4.5 The three-force principle

A particular case of equilibrium which is of considerable interest is that of a body in equilibrium under the action of three non-parallel forces. Such a body is usually called a **three-force body**.

It can be shown that if a three-force body is in equilibrium, the lines of action of the three forces must intersect at a common point.* This principle, known as the **three-force principle**, is very useful in the solution of many engineering problems, as it helps to determine easily the direction of an unknown force without recourse to more complex mathematical methods.

Example 4.5

A 2.5 m long ladder of mass 10 kg rests on the floor and against a smooth wall as shown in Figure 4.13(a). Determine the reactions.

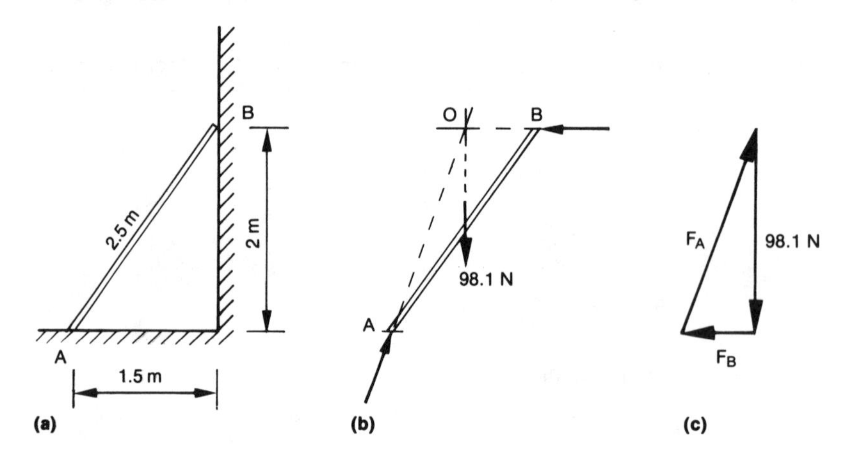

Fig. 4.13

Solution

Draw a free-body diagram (Fig. 4.13(b)) showing weight, $F_w = m.g = 10 \times 9.81 = 98.1$ N, acting through the mid-point, and a horizontal reaction on the smooth wall at *B*.

Extend their lines of action to intersect at point *O* and join point *O* with point *A*. This determines the direction of the reaction at *A*.

A triangle of forces can now be constructed (Fig. 4.13(c)) and the reaction forces found.

$F_B = 36.8$ N and $F_A = 105$ N at 21° to vertical.

Example 4.6

The truss shown in Figure 4.14 is subjected to wind loads equivalent to a force of 10 kN applied at 90° to the member at the joint as shown. Determine the reactions at *A* and *B*.

* The only exception to this principle is when the forces are parallel.

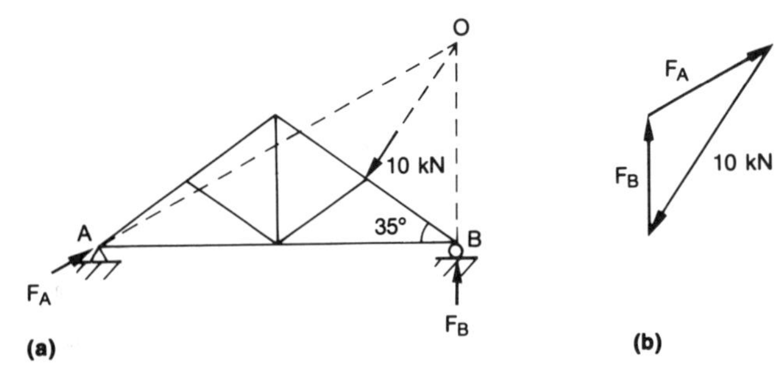

Fig. 4.14

Solution

The normal reaction will be at the roller support at B. Therefore, draw a vertical line through B to intersect with the line of action of the load at point O.

Join O and A and construct the force triangle.

From the force triangle $F_A = 6.5$ N and $F_B = 5.14$ N.

Problems

4.9 A 120 N force is required to operate the foot pedal shown in Figure 4.15. Determine the force on the connecting link and the reaction at the bearing.

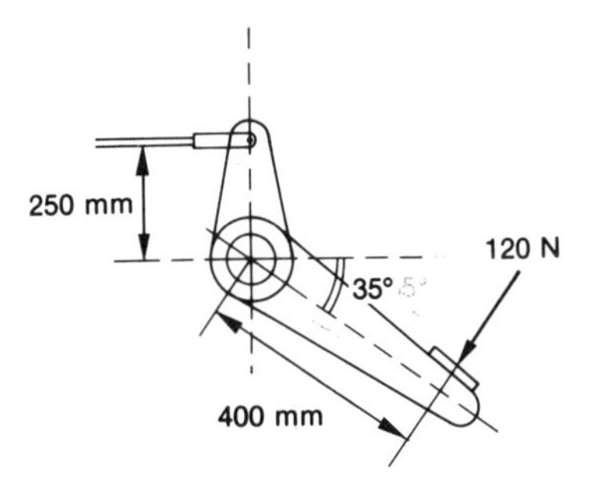

Fig. 4.15

4.10 From Figure 4.16, determine the force in the supporting cable AB and the reaction at C for the following conditions of loading on the jib-crane shown.

(a) m = 1 tonne, x = 1 m (b) m = 1 tonne, x = 4 m
(c) m = 2 tonne, x = 2 m (d) m = 2 tonne, x = 3 m

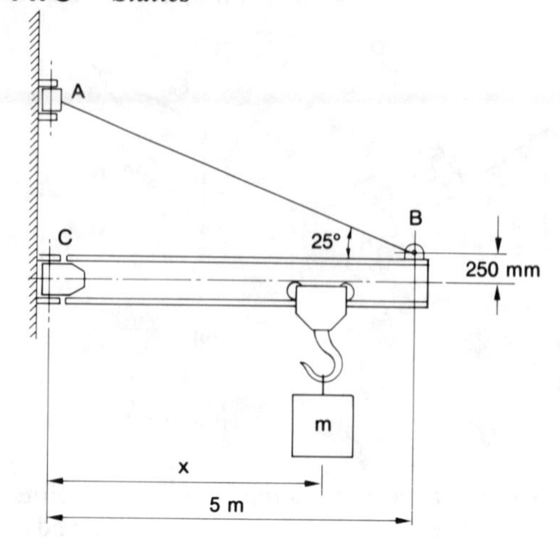

Fig. 4.16

4.11 A beam is subjected to a load of 10 kN as shown in Figure 4.17. Determine the support reactions at each end for $\theta = 20°, 40°$ and $60°$.

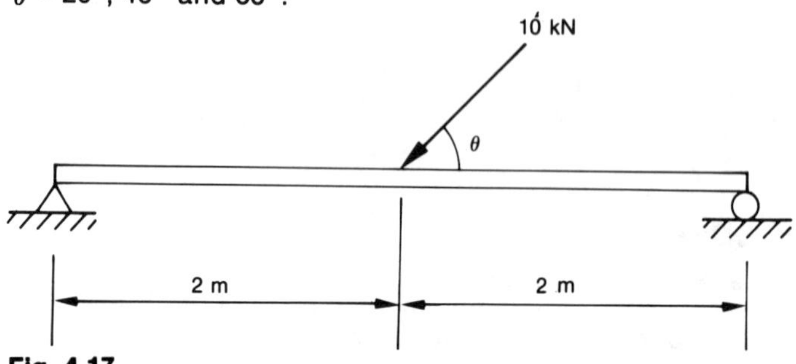

Fig. 4.17

4.12 A street light has a mass of 20 kg and is supported as shown in Figure 4.18. Determine the reactions at *A* and *B*.

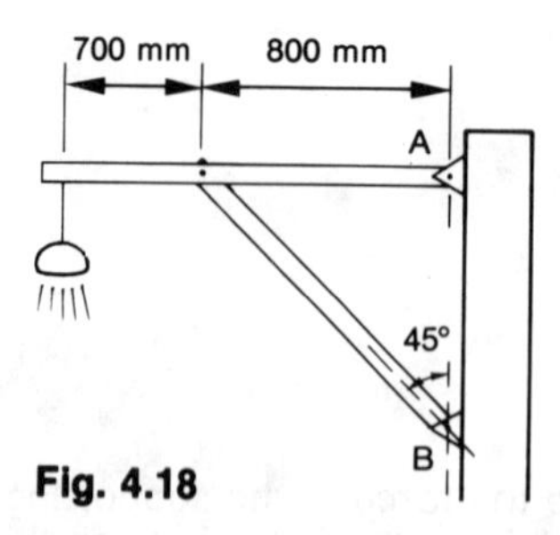

Fig. 4.18

4.13 An advertising sign is supported as shown in Figure 4.19. Determine the force in *AB* and the reaction at *C* if the mass of the sign is 30 kg.

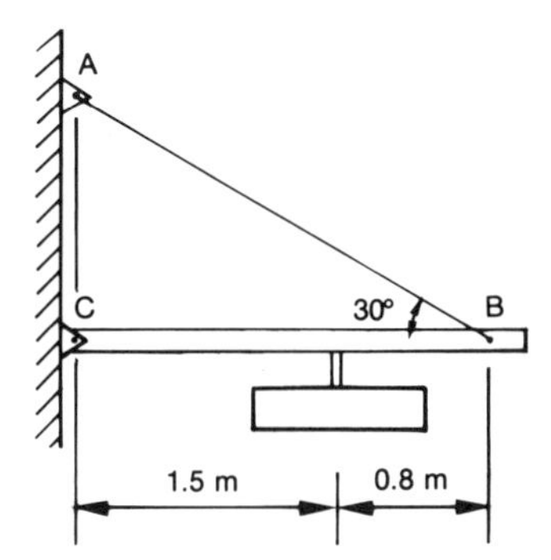

Fig. 4.19

4.14 Determine the force F required to start to tip the cabinet shown (Fig. 4.20) about *A*, assuming it is not going to slide. Determine also the reaction at *A*. The mass of the cabinet is 50 kg.

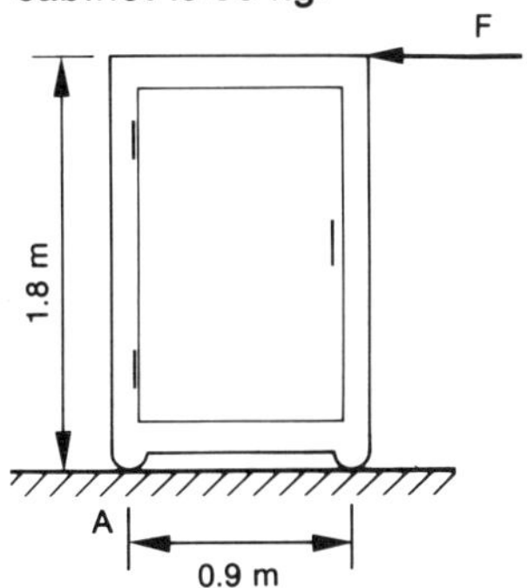

Fig. 4.20

4.15 A pin-jointed frame supporting a score board is subjected to a horizontal wind load as shown in Figure 4.21. Determine reactions at *A* and *B*.

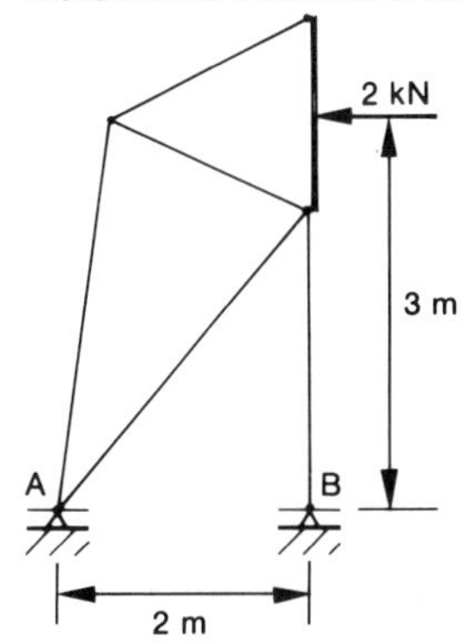

Fig. 4.21

4.16 A ladder rests on a rough floor surface and against a smooth wall as shown in Figure 4.22. Neglecting the mass of the ladder, determine the reactions at *A* and *B* when a man of mass 85 kg climbs to a point on the ladder as shown.

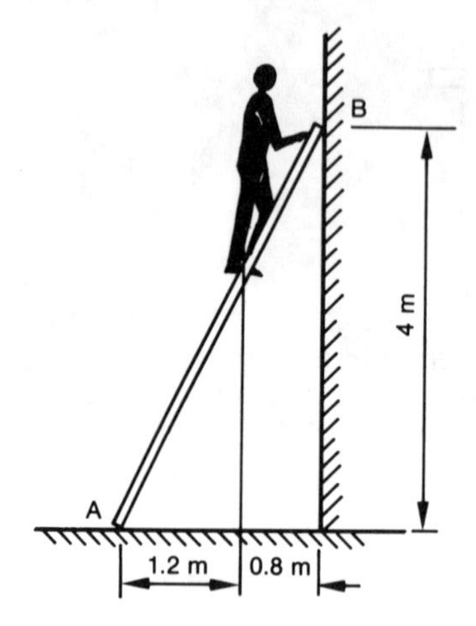

Fig. 4.22

4.17 Determine the force F required to pull a 500 mm diameter roller, having a mass of 100 kg, over a step 60 mm high, and the reaction at the point of contact (see Fig. 4.23).

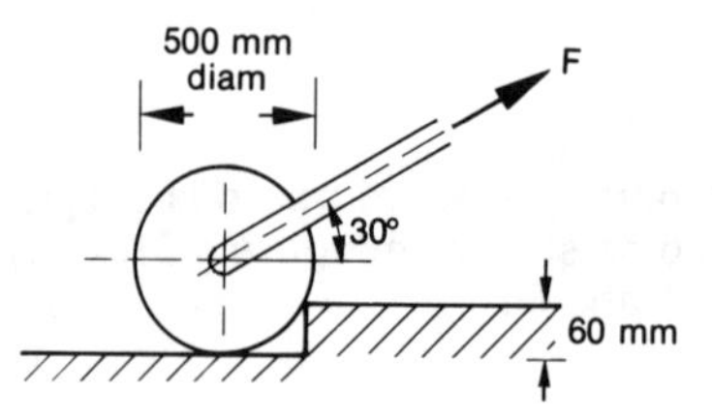

Fig. 4.23

4.18 Find the force in cable *AB* and reaction at *C* when a load of 500 kg is supported by a flexible rope passing over a sheave as shown in Figure 4.24.

4.19 The door in Figure 4.25 is held in closed position by tension in the spring equal to 70 N. Determine the reaction at the hinge.

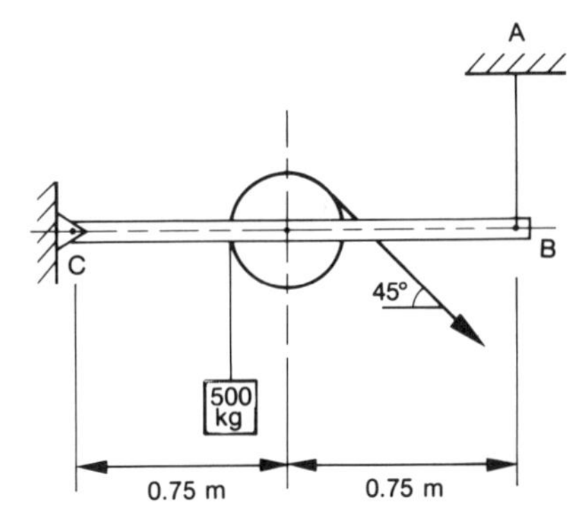

Fig. 4.24

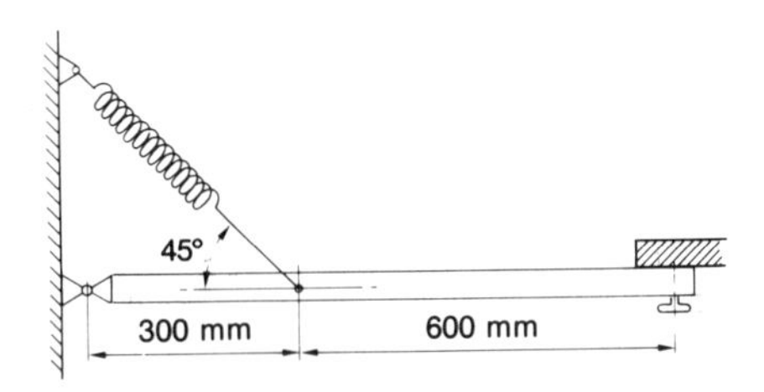

Fig. 4.25

4.20 For each metre of its length, the retaining wall shown in Figure 4.26 has a mass of 1.8 tonnes and the pressure of the Earth behind it is equivalent to a force of 7.57 kN. The point at which the reaction at the base can be regarded as applying is point *A*. Determine the reaction in magnitude and direction and the position *x* of point *A*.

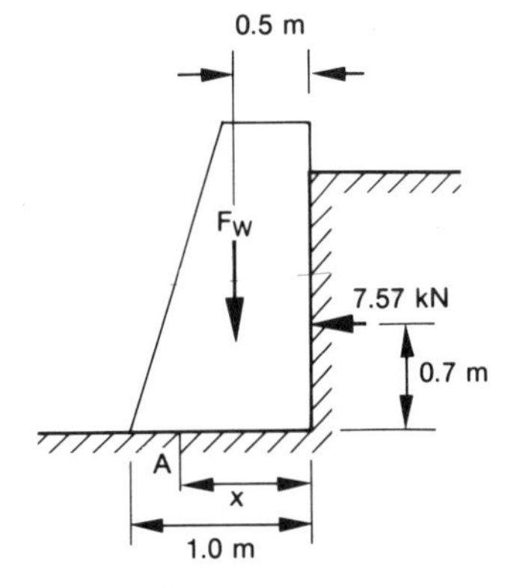

Fig. 4.26

Review questions

4.1 What is meant by equilibrium of forces?
4.2 Define concurrent forces.
4.3 What is the condition for equilibrium of a concurrent force system?
4.4 State two equations commonly used to express the conditions of equilibrium.
4.5 What is a free-body diagram?
4.6 Define equilibrant force.
4.7 What is a three-force body?
4.8 State the three-force principle.
4.9 What is the only exception from the three-force principle?
4.10 Explain the difference between smooth and rough surface contact.

5

Moment and torque

Every time we drive a car, tighten up a nut, ride a bicycle or open a door, we make use of the turning effect of a force or forces, applied at some distance from the axis or fulcrum about which turning takes place. In this chapter we consider the turning effect of a force about a point and the concept of torque.

5.1 Moment of a force

At the very dawn of physical science, Archimedes is reputed to have declared, "Give me a place to stand, and I can move the Earth". This was not just an idle boast, but a poetical assertion of the law of the lever, which Archimedes understood and developed mathematically. The lever is a simple device, one of the earliest used by mankind, by means of which we have been able to multiply the force available in our own hands to move very large loads. Today, the lever, in various forms and combinations, is the most common machine element.

The lever is only one example of the application of the principle of moments which is the subject of this chapter. **The moment of a force about a point is defined as the product of the force and the perpendicular distance of its line of action from the point.**

$$M = F.d$$

The moment is a measure of the turning effect of the force acting on a body, relative to the specified point. As a product of force and distance, moment of a force is measured in units derived from those of force and length. The SI unit of moment of force is the product of the newton and the metre, called "newton metre", with a compound symbol N.m. The multiples and submultiples of the newton metre are formed by using decimal prefixes before the new unit, e.g. kilonewton metre (kN.m) equal to 1000 N.m.

Example 5.1

If a force of 65 N is applied to the lever shown in Figure 5.1, and the length of the moment arm is 0.75 m, determine the moment of the force about the pivot point.

Fig. 5.1

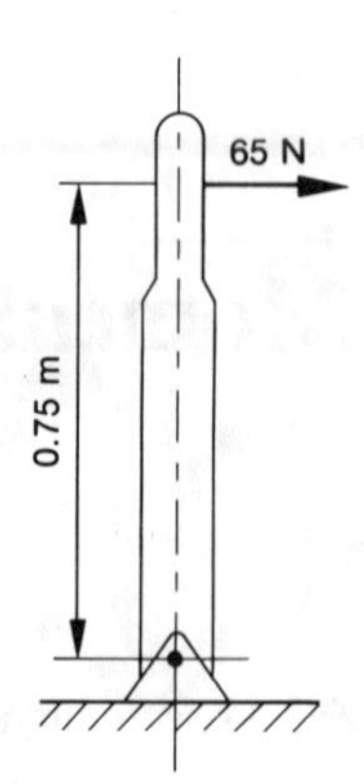

Solution

Moment is the product of the force and the perpendicular distance called the moment arm.

$M = F \times d = 65 \text{ N} \times 0.75 \text{ m} = 48.75 \text{ N.m clockwise.}$

Note that the answer also indicates the directional sense of the moment, clockwise in this case. In this example, the actual length of the lever is in fact the moment arm at right angles to the force. In many problems, care must be exercised in using correct perpendicular distance with each force, as is illustrated by the following example.

Example 5.2

Determine the magnitude and sense of the moments of forces F_1, F_2 and F_3 about point *A* in Figure 5.2.

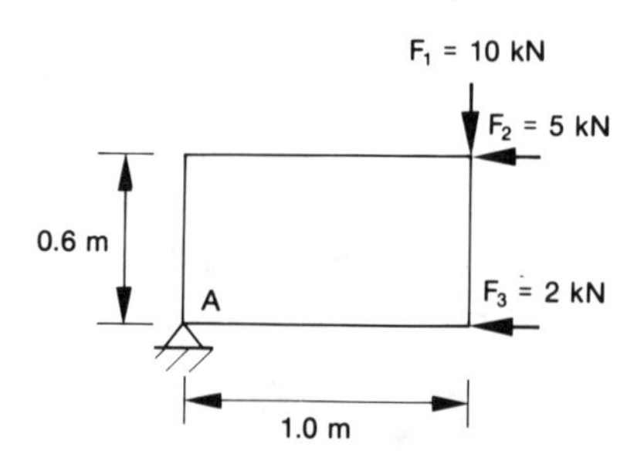

Fig. 5.2

Solution

Moment of force F_1

$M_1 = F_1 \times d_1 = 10 \text{ kN} \times 1.0 \text{ m} = 10 \text{ kN.m clockwise}$

Moment of force F_2

$M_2 = F_2 \times d_2 = 5 \text{ kN} \times 0.6 \text{ m} = 3 \text{ kN.m anticlockwise}$

Moment of force F_3

$M_3 = F_3 \times d_3 = 2 \text{ kN} \times 0 = 0$

Note that force F_3 does not have a moment about point *A*, due to the fact that its line of action passes through the point, i.e. its perpendicular distance from *A* is equal to zero. Students should study this example very carefully in order to understand which distance is in fact the moment arm for each of the forces in question. Diagrams in Figure 5.3 may be helpful in this regard.

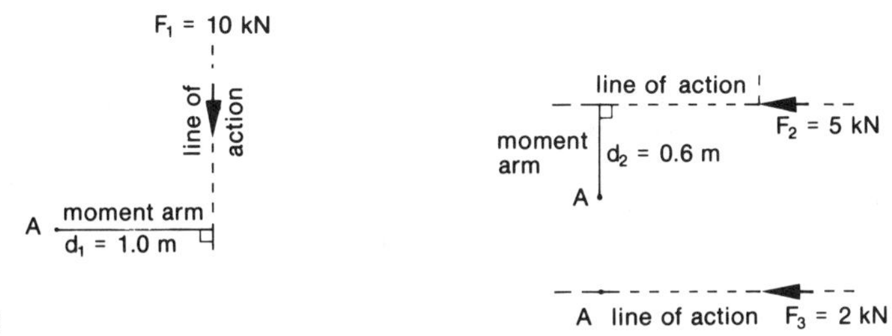

Fig. 5.3

If a force is inclined to the convenient principal directions, such as the horizontal and vertical directions, the correct perpendicular distance must be determined graphically or trigonometrically, as in the following example.

Example 5.3

Determine the moment of force $F = 50$ N about point A, if the distance AB is 800 mm. (See Figure 5.4.)

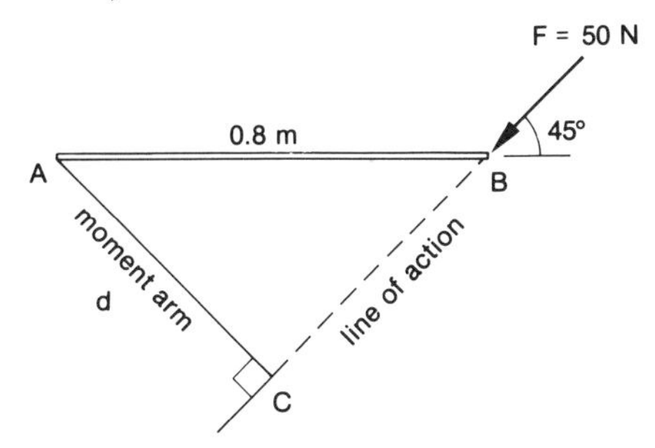

Fig. 5.4

Solution

In this case, the given distance AB is not the moment arm, because AB is not perpendicular to the line of action of the force. To solve the problem, we need distance AC. This can be found graphically by drawing the diagram to scale and measuring the required distance AC. Alternatively, from the right-angle triangle ABC,
the moment arm $d = AB \sin 45° = 0.8 \times 0.707 = 0.566$ m.

The moment of the force is therefore

$$M = F \times d = 50 \text{ N} \times 0.566 \text{ m} = 28.3 \text{ N.m clockwise.}$$

5.2 Addition of moments

If more than one force is acting on a body, as in Example 5.2, there is a corresponding number of moments of force, each tending to produce a turning effect about the point. The total turning effect, or resultant moment, is the algebraic sum of the moments of all the forces acting on the body. Algebraic sum means that the different sense of the various moments must be taken into account when the moments are being added. A sign convention usually used is that clockwise moments are taken to be positive, and anticlockwise moments, negative.

Returning to Example 5.2, the total moment about point A of all the forces is

$$M = M_1 + M_2 + M_3 = +10 \text{ kN.m} - 3 \text{ kN.m} + 0 = 7 \text{ kN.m clockwise}$$

An extension of the principle of addition of moments is known as the Varignon Theorem, originally proposed by the French mathematician, Varignon (1654–1722). It states that the moment of a force about any axis is equal to the sum of the moments of its components about that axis. This theorem can be used to solve Example 5.3 mathematically.

Mathematical solution of Example 5.3

To solve this problem mathematically, resolve force F into its horizontal and vertical components, as shown in Figure 5.5.

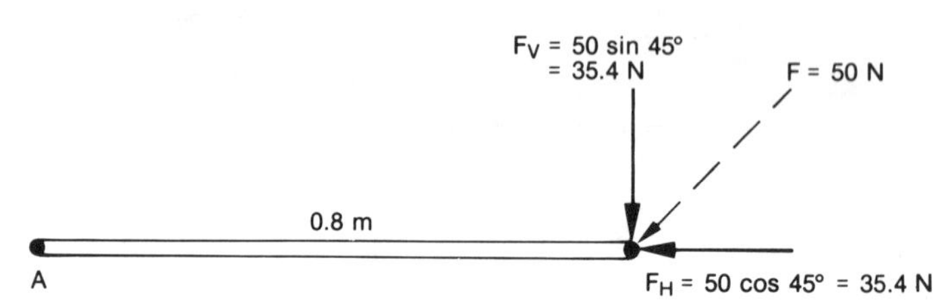

Fig. 5.5.

As the moment of the horizontal component is zero, due to the line of action passing through point A, the total moment about A is

$$M = F_V \times d + F_H \times 0 = 35.4 \text{ N} \times 0.8 + 0 = 28.3 \text{ N.m}$$

It is important to realise that for a given system of forces, there is no single answer for the moment of forces, unless the reference point is specified. A moment must always be calculated with respect to a particular reference point. It must also be understood that in many problems there is no rotation actually taking place. The moment represents only the tendency for rotation under the influence of a force or forces, and not actual rotation. The following example should help to illustrate these points.

Example 5.4

A horizontal beam rests on two supports A and B, and supports three forces as shown in Figure 5.6. Calculate the total moment due to the applied forces (a) about the left-hand support A and (b) about the right-hand support B.

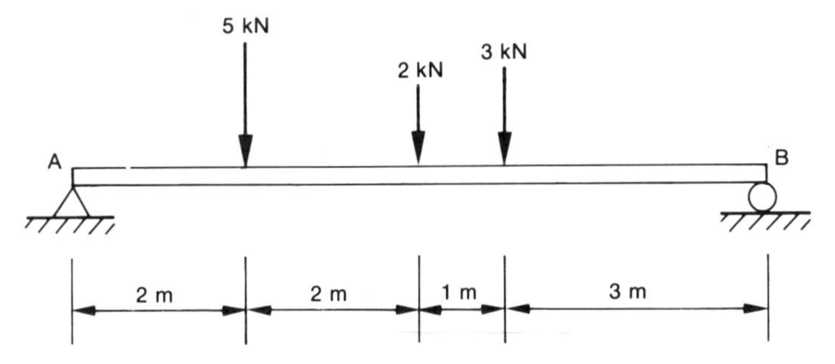

Fig. 5.6

Solution

(a) The total moment about A is the sum of the moments of all forces about A. (Note: All distances are measured from point A).

$M_A = \Sigma(F \times d) = 5 \times 2 + 2 \times 4 + 3 \times 5 = 33$ kN.m clockwise

(b) The total moment about point B is (distances measured from B)

$M_B = \Sigma(F \times d) = 5 \times 6 + 2 \times 4 + 3 \times 3 = 47$ kN.m anticlockwise

It is obvious that the moments about A and B are different, not only in magnitude but also in sense. There is no actual rotation taking place. However, if the right-hand support B were suddenly removed, there would be rotation produced by the turning moment M_A about A in the clockwise direction. If, instead, the left-hand support A were removed, the beam would turn anticlockwise about B under the influence of the moment M_B.

5.3 Equilibrium of moments

When a structure or a machine component is in equilibrium, the moments, as well as forces, must be in a state of balance, otherwise the unbalanced resultant moment would cause rotation of the body.

Thus for a body to be in rotational equilibrium, the resultant moment about any point must be equal to zero, i.e. there must be no resultant turning effect.

Mathematically this can be stated as

$$\Sigma M = 0$$

Alternatively this principle can be expressed in terms of equivalence of the sum of the clockwise moments about any point and the sum of anticlockwise moments about the same point. That is, for equilibrium:

sum of clockwise moments = sum of anticlockwise moments.

To prove that a given system of forces is in rotational equilibrium, we must demonstrate that the algebraic sum of all moments about any point is zero.

Example 5.5

A bridge structure is subjected to forces as shown in Figure 5.7. Prove that the structure is in rotational equilibrium.

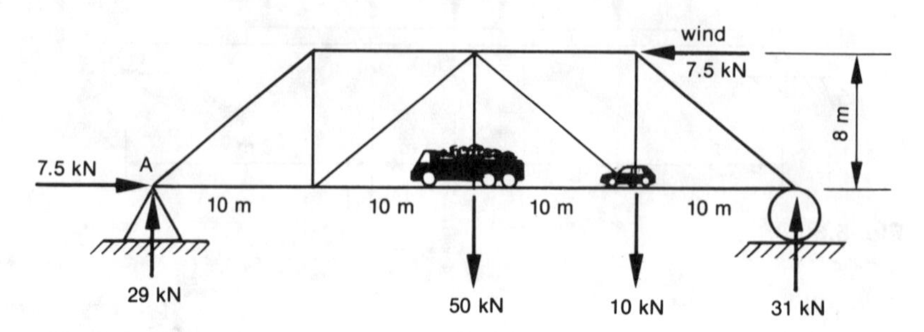

Fig. 5.7

Solution

Take the algebraic sum of all moments about the left-hand support A.

$\Sigma M_A = 50 \times 20 + 10 \times 30 - 7.5 \times 8 - 31 \times 40 = 0$

$\therefore \Sigma M_A = 0$

The moments are balanced about point A.

Take the algebraic sum of all moments about the right-hand support B.

$\Sigma M_B = 29 \times 40 - 50 \times 20 - 10 \times 10 - 7.5 \times 8 = 0$

$\therefore \Sigma M_B = 0$

The moments are balanced about point B.

5.4 Torque

The examples considered so far in this chapter involved static forces, such as those acting on a beam resting on its supports, or forces producing a small amount of rotation, often less than one revolution and of discontinuous nature, such as that of the lever in a previous example. There are many engineering applications in which rotation is continuous, as in gear boxes, electric motors, fly-wheels, etc.

Furthermore, the turning effort which produces and maintains rotation is not always readily identifiable as a product of a single force located at a fixed distance from the axis of rotation. For example, in an electric motor there are distributed forces acting on the rotor, but to an outside observer the effort produced by the output shaft of the motor appears to be a "pure turning effort".

We are going to describe this pure turning effort by the term **torque**.* As a pure turning effort, akin in its effect to the moment of a force, torque is

* The word torque derives its origin from a Latin word for twisting, or twisted necklace, as worn by the ancient Gauls.

measured in the same units as those for turning moments,* i.e. newton metres, or their multiples and submultiples, such as kilonewton metres. Dynamic torque, i.e. torque transmitted during continuous rotation of a component, can be measured directly by transmission dynamometers, which are essentially torsion bars, acting as a coupling between a power source and a rotating load, measuring twist while transmitting rotational motion, and translating it into torque.

It should be pointed out that there is always a close relationship between torque and forces acting on various rotating components, such as gears, sprockets, pulleys, as the following example will illustrate.

Example 5.6

A gearbox consists of two spur gears in mesh. The larger gear has a 200 mm pitch-circle diameter, and the pinion, i.e. the smaller gear, is 50 mm pitch-circle diameter.** (Note: Mating gears have their pitch circles tangent.) The input shaft transmits a torque through the pinion of 160 N.m. Determine the tangential force between the two gears, and the value of output torque.

Solution

Input torque $T_{in} = 160$ N.m

This produces a force between the two gears, such that the moment of the force about the centreline of the pinion is equal to the transmitted torque,

$$(\text{force}) \times (\text{radius}) = (\text{torque})$$

$$F \times 0.025 \text{ m} = 160 \text{ N.m}$$

hence

$$F = \frac{160 \text{ N.m}}{0.025 \text{ m}} = 6400 \text{ N} = 6.4 \text{ kN}$$

But the same force or, to be precise, its equal and opposite reaction force, is also applied to the larger gear at a distance of 0.1 m from its centreline, producing a turning moment of

$$M = F \times d = 6400 \text{ N} \times 0.1 \text{ m} = 640 \text{ N.m}$$

This is transmitted as torque through the output shaft, i.e.

$$T_{out} = 640 \text{ N.m}$$

Unlike moments of a force, which can be calculated in relation to any arbitrarily selected points of reference, torque is usually understood to act about the axis of rotation of a component such as a shaft, gear or flywheel, irrespective of whether actual rotation is taking place or not.

This chapter is limited to those aspects of torque, such as equilibrium, which can readily be studied by the methods of statics. The effects of unbalanced torque on rotating bodies, e.g. acceleration, work and power, are the province of dynamics and will be discussed in Chapters 9 and 10.

---

* It is necessary to be aware that the distinction between torque and moment adopted in this book is by no means universal. Elsewhere the term "torque" is sometimes used instead of "moment" to mean a turning effort of any kind, as for example in the expression "torque wrench".

** The pitch-circle diameter is the diameter of the circle which by a pure rolling action would transmit the same motion as the actual gear wheel.

Problems

5.1 A force at the end of a spanner applied 300 mm from the centre of the nut is 45 N. What is the maximum turning moment produced by this force? What should be the direction of the force to achieve the maximum turning effect?

5.2 A winding hoist has a drum of 300 mm diameter. Determine the moment about its centreline produced by tension in the cable equal to 0.5 kN.

5.3 A horizontal beam 2 m long is supported at its ends. Determine the moments about each of the supports due to the following loads:

(a) A downward force of 3 kN at mid-point.

(b) A downward force of 3 kN at a point 0.5 m from the left support.

(c) Three downward forces 1 kN each located at 0.5 m, 1 m and 1.5 m from the left support.

(d) A downward force of 1 kN and a downward force of 2 kN located at 0.5 m and 1.5 m from the left support respectively.

(e) A force of 3 kN applied at mid-point and inclined at 60° to the horizontal.

5.4 The beam in Figure 5.8 is built into the wall and carries a load of 10 kN as shown.

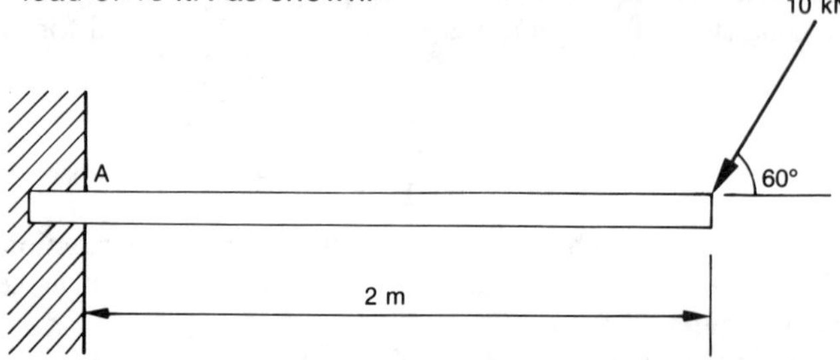

Fig. 5.8

Determine the moment of the force about the support *A*,

(a) by using perpendicular distance to the force,

(b) by using rectangular components of the force.

5.5 The truss shown in Figure 5.9 is subjected to three loads as indicated. Determine the total moment due to the applied forces about

(a) left-hand support *A*,

(b) right-hand support *B*.

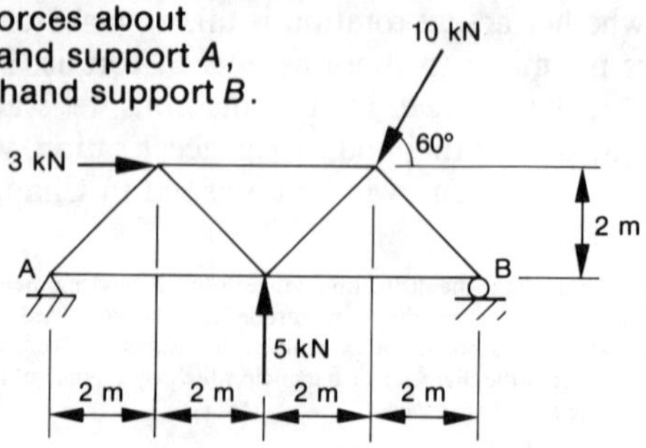

Fig. 5.9

5.6 A car has 750 mm diameter wheels and requires a force of 0.5 kN between each driving wheel and the road surface. What torque must be applied to each of the driving wheels to supply this force?

5.7 A pull in a bicycle chain of 250 N applied to a sprocket of 80 mm diameter provides the necessary torque to the rear wheel. If the wheel is 600 mm diameter, what is the force between the wheel and the ground?

5.8 A 300 mm diameter pulley is subjected to belt tensions equal to 500 N and 200 N. Determine the torque acting on the pulley.

5.9 The arm shown in Figure 5.10 is keyed to a 200 mm diameter shaft. Assuming that the turning effort is transmitted by the key only, determine the force on the key if the load at the end of the arm is 250 N.

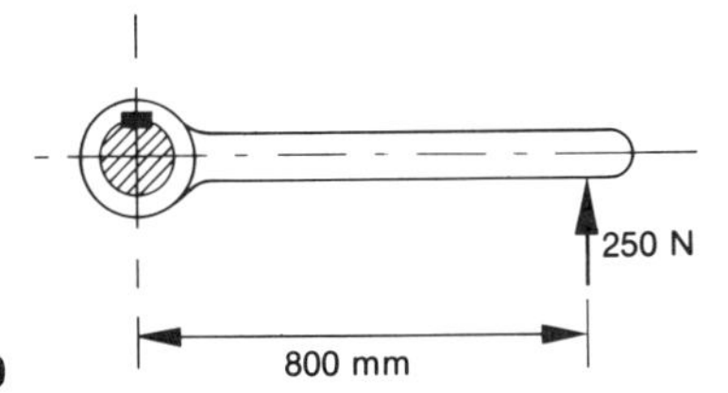

Fig. 5.10

5.10 Determine the force that must be applied to the foot-pedal shown in Figure 5.11 to produce a force in the connecting link of 200 N.

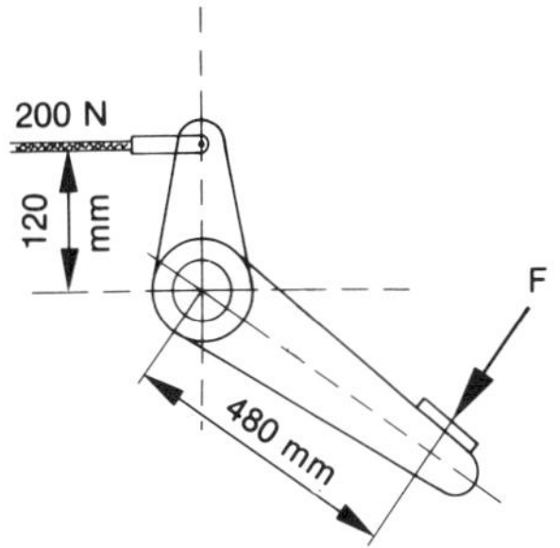

Fig. 5.11

5.11 Determine the force F required to start to tip the cabinet shown in Figure 5.12, if the mass of the cabinet is 50 kg.

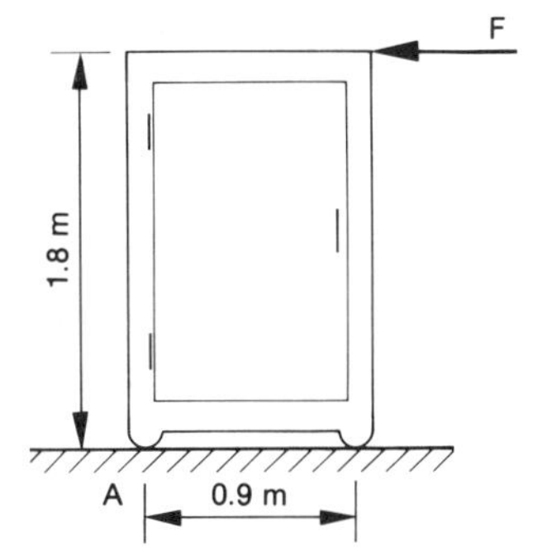

Fig. 5.12

5.12 Two shafts are to be connected by a coupling having six bolts equally spaced around a 300 mm diameter pitch circle as shown in Figure 5.13. Assuming the bolts are equally loaded, determine the force transmitted by each bolt if the torque transmitted is 5000 N.m.

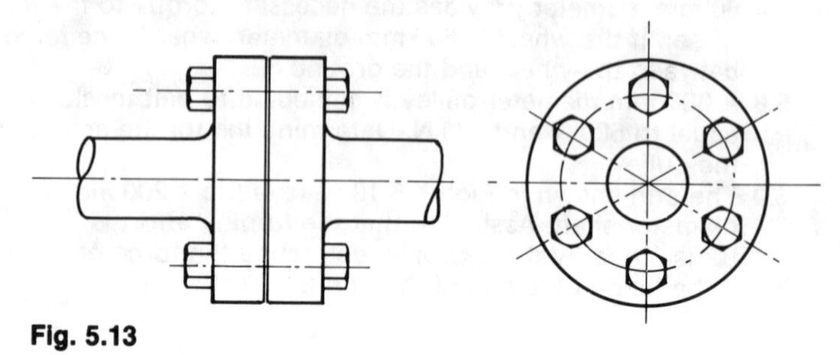

Fig. 5.13

5.5 Moment of a couple

So far we have discussed turning effects of single forces and the pure turning effort or torque. We now turn our attention to a very special combination of forces called a **couple. A couple consists of two forces having the same magnitude, parallel lines of action, and opposite sense.**

When your hands are on the steering wheel of a car, at two opposite points of the wheel, one hand pushing up and the other pulling down, with equal but opposite forces, the result is a couple (see Fig. 5.14).

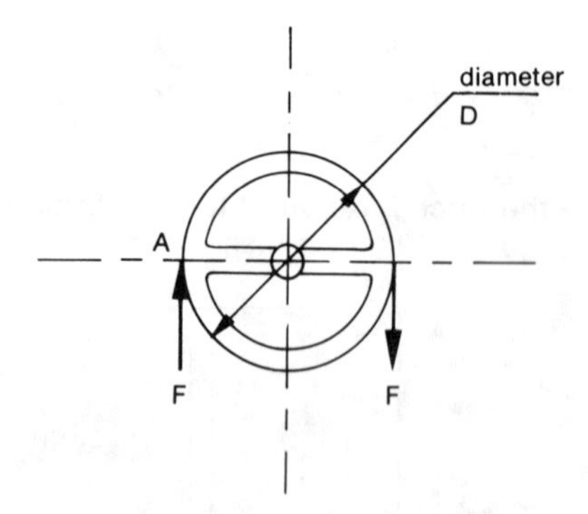

Fig. 5.14

Clearly the algebraic sum of the two forces is equal to zero, i.e. there is no net push or pull in any direction. However, there is a turning effect, which can be calculated relative to any point. The obvious point of reference for calculating the moments is the centre point of the wheel, which is the axis of its rotation. The total moment is

$$M_o = F \times \frac{D}{2} + F \times \frac{D}{2} = F\left(\frac{D}{2} + \frac{D}{2}\right) = F \times D$$

where F is the magnitude of each of the forces in the couple, and
D is the distance between them, in this case the wheel diameter.
The sense of the total moment is clockwise for the forces shown.

The product of one of the forces and the distance between them is called the **moment of the couple**. One important characteristic of a couple is that its moment, $M = F \times D$, does not depend on the choice of the reference point. For example, the moment about point A is

$M_A = F \times D + 0 = F \times D$, clockwise as before.

This can be checked by repeating the same calculation about any other point on, or even outside, the wheel.

The independence of the magnitude of the moment of a couple from the choice of the reference point about which it is calculated (i.e. the moment of a couple is the same about any point) often leads to the statement that a couple has no point of application, meaning that it does not matter where the two forces forming the couple act, the only thing that counts is the moment of the couple. While this is theoretically correct, and one couple can be replaced by an equivalent couple applied elsewhere on the body, in practice it is convenient to regard a couple as a pair of actual forces applied to the body at specified points. Furthermore, wherever possible the turning effect of a couple should be visualised with respect to a real point about which the tendency for rotation is in fact real, e.g. a hinge, an axle.

Example 5.7

Determine the moment of the couple acting on the steering wheel in Figure 5.14 if the wheel diameter is 350 mm and the forces applied as shown are 5 N each. Determine also the forces required to produce the same moment if the hands are held on the spokes, half-way between the axis and the rim of the wheel.

Solution

Moment of the couple

$$M = F \times D = 5\text{ N} \times 0.35\text{ m} = 1.75\text{ N.m clockwise}$$

To apply the same moment when $D = \dfrac{0.35\text{ m}}{2} = 0.175$ m, the forces required can be calculated from $M = F \times D$,

$$1.75\text{ N.m} = F \times 0.175\text{ m}$$

$$\therefore F = \frac{1.75\text{ N.m}}{0.175\text{ m}} = 10\text{ N each}$$

This example illustrates the idea of equivalent couples, i.e. couples which involve different forces but produce equal turning moments. Next time you drive a car, turn off a tap or turn a key in a lock, think of the two forces applied by your hands or fingers, producing a turning moment.

5.6 Equivalent force-moment system

When a force F is applied to a rigid body, such as the spanner shown in Figure 5.15, at a point A, it is sometimes useful to consider the effect of the force viewed from another point, B.

It can be shown that the given force at A can be replaced by an equal force acting at another point, B, and a turning moment, equal to the moment of F, in its original position, about point B. The moment can be represented by the symbol ↻ placed at point B.

The turning moment at B can be interpreted as a pure moment, i.e. torque, or as a couple, as can be seen from the following examples.

Example 5.8

A bolt is being tightened by a force of 60 N applied to a spanner at a distance of 300 mm as shown in Figure 5.15(a). Replace the force by a force-moment system at B.

Solution

The equivalent force-moment system at B comprises a force of 60 N and a turning moment equal to

$$M = 60 \text{ N} \times 0.3 \text{ m} = 18 \text{ N.m clockwise,}$$

as shown in Figure 5.15(b).

This implies that the effect of the force applied at A combines the tendency to push, or shear, the bolt with a force of 60 N and at the same time to turn it clockwise with a turning moment of 18 N.m.

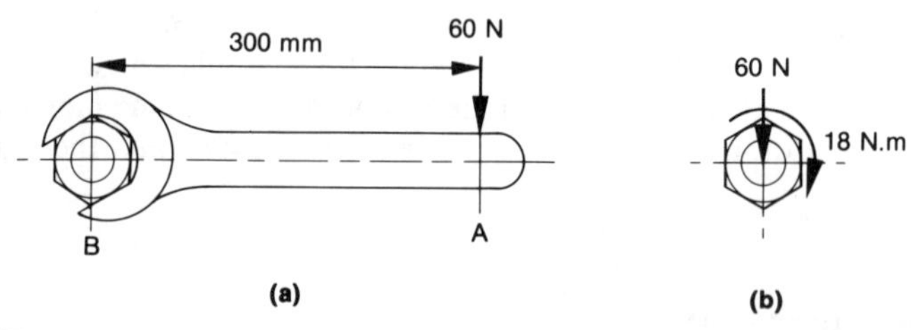

Fig. 5.15

Example 5.9

For the hook shown in Figure 5.16(a), replace the force of 200 N acting in plane A by an equivalent force-moment system at B.

Solution

The equivalent force-moment system at B consists of a force of 200 N and a clockwise turning moment equal to

$$M = 200 \text{ N} \times 0.075 \text{ m} = 15 \text{ N.m,}$$

as shown in Figure 5.16(b).

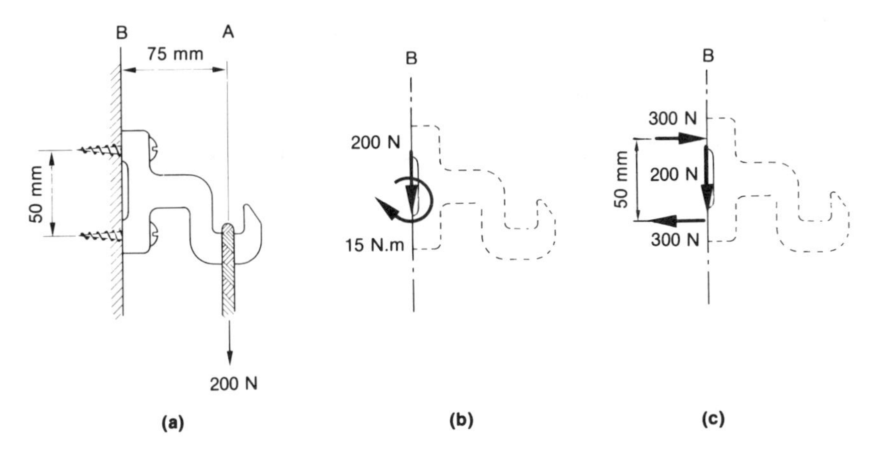

Fig. 5.16

In this case, where the moment is taken by the supports, it may be meaningful to interpret the moment as an equivalent couple with a distance of 50 mm between forces, i.e.

$$M = F \times d$$
$$15 \text{ N.m} = F \times 0.05 \text{ m}$$
$$\text{Hence } F = \frac{15}{0.05} = 300 \text{ N, as in Figure 5.16(c).}$$

Problems

5.13 A screw jack is operated by means of 40 N forces applied at each end of a double arm, at right angles to it. Determine the magnitude of the turning moment if the total length of the double arm is 700 mm.

5.14 A 400 mm × 300 mm plate is subjected to forces as shown in Figure 5.17. Show that the turning effects of the forces add up to zero at any point.

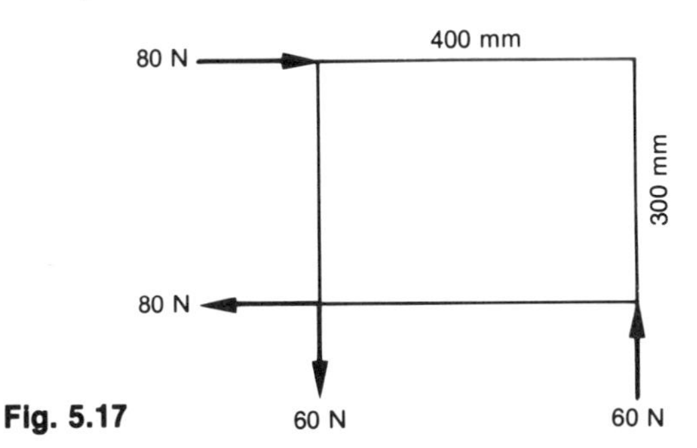

Fig. 5.17

5.15 A horizontal bar is subjected to a number of forces as shown in Figure 5.18. Prove that the bar is in rotational equilibrium:

(a) by adding the moments of all forces about any point on the bar and then repeating for one other point;

(b) by recognising pairs of forces as couples and adding their moments.

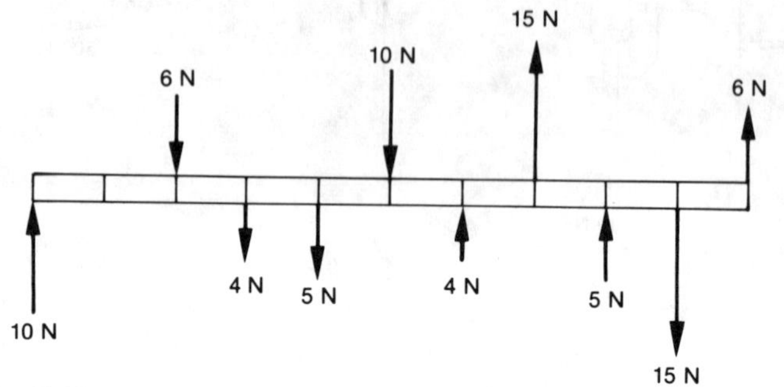

Fig. 5.18

5.16 Determine the force F at each end of the bar AB (Fig. 5.19) required to maintain equilibrium, if the force in the string which is passed around the pulleys is 160 N.

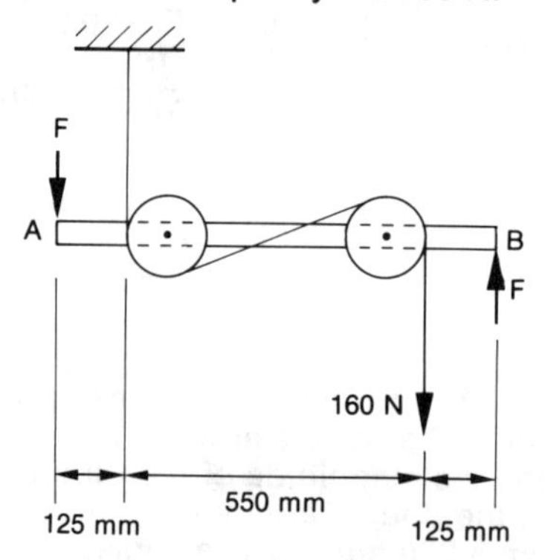

Fig. 5.19

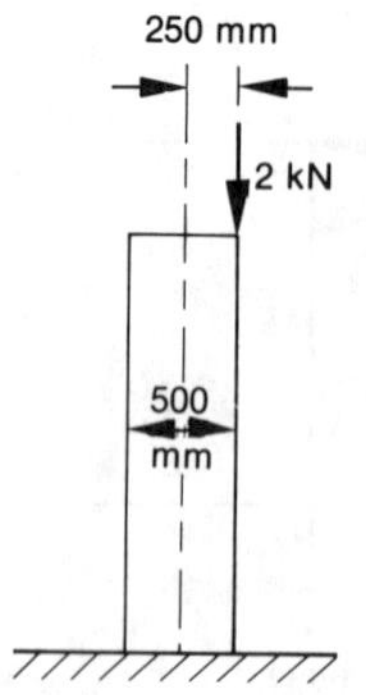

Fig. 5.20

5.17 A column 0.5 m wide carries a 2 kN force as shown in Figure 5.20. Reduce the force to an equivalent axial load and a moment.

5.18 Determine the eccentricity d of a load which can be represented as an axial load of 1 kN and a moment of 15 N.m acting on a column.

5.19 Determine the equivalent force-moment system at the built-in end of a cantilever beam shown in Figure 5.21.

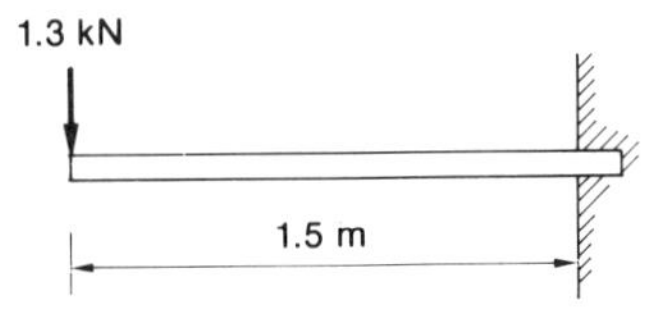

Fig. 5.21

5.20 If the mass supported by the bracket in Figure 5.22 is 5 kg, interpret the load as a force and a couple acting through the bolts.

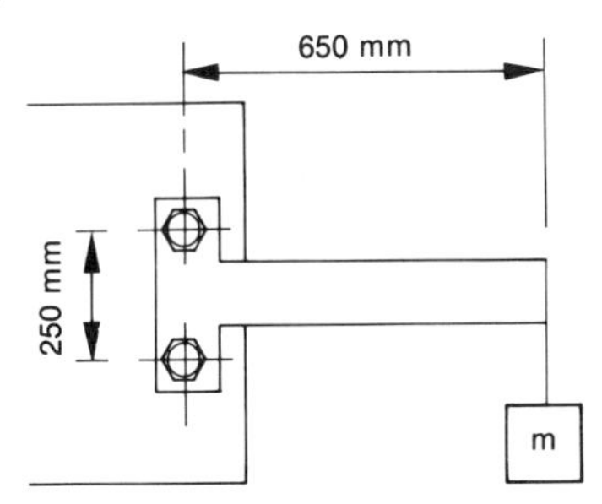

Fig. 5.22

Review questions

1. Define moment of a force.
2. What is the SI unit of moment of a force?
3. Does moment of a force have directional sense?
4. What is the principle known as the Varignon Theorem?
5. What is the condition of rotational equilibrium of a system of forces?
6. Define torque and state the SI unit of torque.
7. Define moment of a couple.
8. What is an equivalent couple?
9. What is meant by an equivalent force-moment system?

6 Equilibrium of non-concurrent forces

This chapter is concerned with the equilibrium of rigid bodies under the action of non-concurrent force systems. A system of forces is called **non-concurrent** if there is no single point of concurrency, i.e. the lines of action of the forces do not all meet at a common point.

6.1 Conditions of equilibrium

It has already been established that for a system of forces to be in equilibrium, the resultant force, i.e. the result of vectorial summation of all forces, must be equal to zero. That is, there must be no resultant push or pull.

This statement is often interpreted mathematically in terms of the perpendicular, usually horizontal and vertical, components of the resultant force, and can be stated simply as

$$\Sigma F_x = 0 \text{ and } \Sigma F_y = 0$$

Only when both of these conditions are satisfied is the force equilibrium established.

These two conditions were found sufficient for the study of equilibrium of concurrent force systems, as discussed in Chapter 4. However, we have since discovered that, while there is no resultant force under certain conditions, e.g. a couple of parallel, equal but opposite forces, there could be a resultant moment and rotation may occur. Clearly, under these circumstances the absence of a resultant force does not by itself ensure equilibrium.

An additional condition for equilibrium of a system of non-concurrent forces must, therefore, ensure the absence of a turning moment, stated as

$$\Sigma M = 0$$

Collectively, the three conditions of equilibrium are known as the **three equations of statics**:

1. The sum of x components of all forces must equal zero.

$$\boxed{\Sigma F_x = 0}$$

2. The sum of y components of all forces must equal zero.

$$\Sigma F_y = 0$$

3. The sum of moments of all forces about any point must be zero.

$$\Sigma M = 0$$

These equations may be used to prove that a particular structure is in static equilibrium under the combined action of a system of non-concurrent forces.

Example 6.1

The structure in Figure 6.1 is subjected to forces as shown. Prove that the structure is in equilibrium.

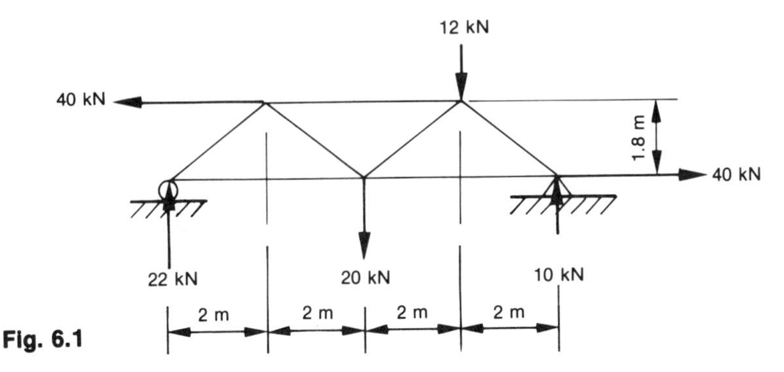

Fig. 6.1

Solution

Sum of horizontal forces:

$$\Sigma F_x = +\ 40\ \text{kN} - 40\ \text{kN} = 0$$

Sum of vertical forces:

$$\Sigma F_y = 22\ \text{kN} + 10\ \text{kN} - 20\ \text{kN} - 12\ \text{kN} = 0$$

Sum of moments about left-hand support:*

$$\Sigma M = 20 \times 4 - 10 \times 8 + 12 \times 6 - 40 \times 1.8 = 0$$

All three equations are satisfied. The conclusion is that the structure is in static equilibrium.

---

* It is suggested that the student should try calculating moments about some alternative point, e.g. right-hand support, to be satisfied that the sum of the moments is equal to zero irrespective of the position of the reference point.

6.2 Calculation of beam reactions

Beams are structural members used to support loads applied at various points along the member. They are usually long straight members made of wood, steel, light alloy, reinforced or prestressed concrete. In most cases, beams are supported in a horizontal position with the loads vertical, i.e. perpendicular to the axis of the beam.

A beam may be subjected to concentrated or point loads, to distributed loads or to a combination of both. Only concentrated vertical loads are treated in this book. In the majority of cases the loads are weights, i.e. forces of gravity acting on various masses supported by a beam. In mechanical engineering there are also many beam-like machine components, such as levers and line shafts, which are subjected to transverse loads due to various causes, e.g. belt tension, interaction of gear wheels, which are not weights.

There are two types of beams considered in this book, according to the method of support. The first, known as **simply supported beam**, is a beam which rests freely on two supports, one at each end. It is assumed that the reaction force at each support acts at a knife-edge point. With all loads vertical, both reaction forces are also vertical.* The distance between supports is called the **span** which, in a simply supported beam, corresponds to the length of the beam.

The second type, called the **cantilever beam**, is a beam which is built-in to a wall or otherwise securely fixed at one end, and hangs freely at the other.

This chapter deals with the problem of determining the magnitude of reactions which must exist at beam supports. Later in the book (Chap. 19), we will discuss the effects of the applied loads on the beam and the relationship between the cross-section of the beam and the internal resistance in the material of the beam.

Reactions: Simply supported beam

To calculate unknown reactions at the knife-edge supports of a simply supported beam, an obvious assumption of static equilibrium is necessary, i.e. the three equations of statics must be applied.

In the absence of any horizontal forces, the first equation, $\Sigma F_x = 0$, is only a formality and not very helpful. However, since there are only two unknown forces, one at each support, the other two equations are sufficient, i.e. $\Sigma F_y = 0$ and $\Sigma M = 0$.

The method illustrated by the following example consists of two steps. First, take moments about one of the supports and equate the algebraic sum of all moments to zero. This gives an equation in which the reaction force at the opposite support is the only unknown. Solve for the unknown.

---

* When not all loads are vertical, they produce axial forces in the beam which must be taken by at least one of the supports. This requires a distinction to be made between a fixed hinged support and a movable support on rollers. However, this lies outside the scope of this book.

Second, equate the algebraic sum of all vertical forces, including reactions, to zero, and solve for the second unknown reaction force.

Example 6.2

Calculate the reactions for the simply supported beam shown in Figure 6.2.

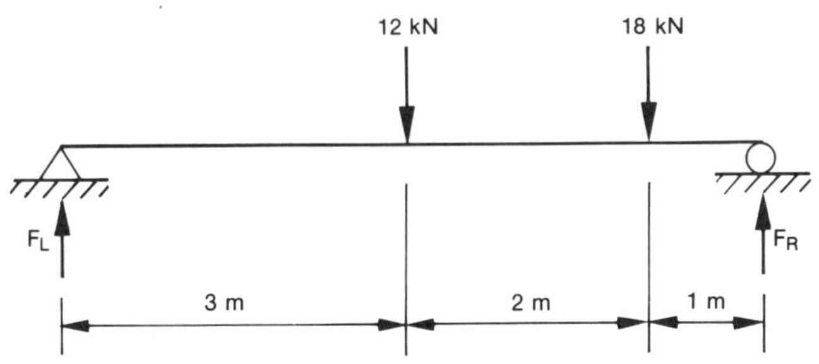

Fig. 6.2

Solution

Let F_L and F_R stand for the reaction forces at the left-hand and right-hand support respectively.

Taking moments about the left-hand support yields

$$\Sigma M_L = 12 \text{ kN} \times 3 \text{ m} + 18 \text{ kN} \times 5 \text{ m} - F_R \times 6 \text{ m} = 0$$

$$36 + 90 - F_R \times 6 = 0$$

$$\text{Hence } F_R = 21 \text{ kN}$$

Note that the moment of F_L about the left-hand support is zero, because the force passes through the point there. Remember also that by convention clockwise moments are taken to be positive and counterclockwise moments, negative.

Now, summation of vertical forces yields:

$$\Sigma F = F_L - 12 \text{ kN} - 18 \text{ kN} + F_R = 0$$

but $F_R = 21$ kN

Therefore, $F_L - 12 \text{ kN} - 18 \text{ kN} + 21 \text{ kN} = 0$

$$\text{Hence } F_L = 9 \text{ kN}$$

Here again a sign convention is observed—upward forces are positive and downward forces are negative.

Alternatively, F_L could also be found by taking moments about the right-hand support, as follows:

$$\Sigma M_R = -12 \text{ kN} \times 3 \text{ m} - 18 \text{ kN} \times 1 \text{ m} + F_L \times 6 \text{ m} = 0$$

$$\text{Hence } F_L = 9 \text{ kN}$$

In this case the result may be checked by the summation of vertical forces:

$$\begin{aligned} \Sigma F &= F_L - 12 \text{ kN} - 18 \text{ kN} + F_R \\ &= 9 \text{ kN} - 12 \text{ kN} - 18 \text{ kN} + 21 \text{ kN} \\ &= 0 \end{aligned}$$

Reactions: Cantilever beam

In the case of a cantilever beam, there is only one support at the fixed end, where the reaction is a combination of a vertical reaction force (F) and a moment (M) exerted by the fixing.

The method for finding the magnitude of the reaction force and the reaction moment again involves the use of the equations $\Sigma M = 0$ and $\Sigma F_y = 0$.

Example 6.3

Calculate the reactions for the cantilever beam shown in Figure 6.3.

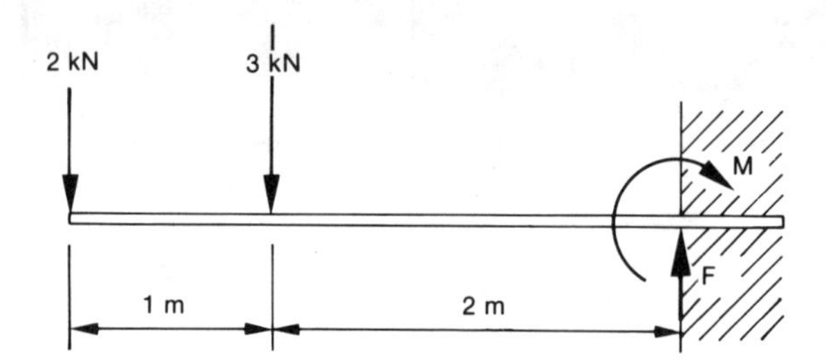

Fig. 6.3

Solution

The moments equation about the support yields

$$\Sigma M = M - 3 \text{ kN} \times 2 \text{ m} - 2 \text{ kN} \times 3 \text{ m} = 0$$

Note carefully that moment M acting at the support must be included in the equation.

Hence reaction moment $M = 12$ kN.m.

The summation of forces gives

$$\Sigma F = F - 3 \text{ kN} - 2 \text{ kN} = 0$$

Hence, reaction force $F = 5$ kN.

Problem

6.1 For each of the beams in Figure 6.4, calculate reactions at the supports.

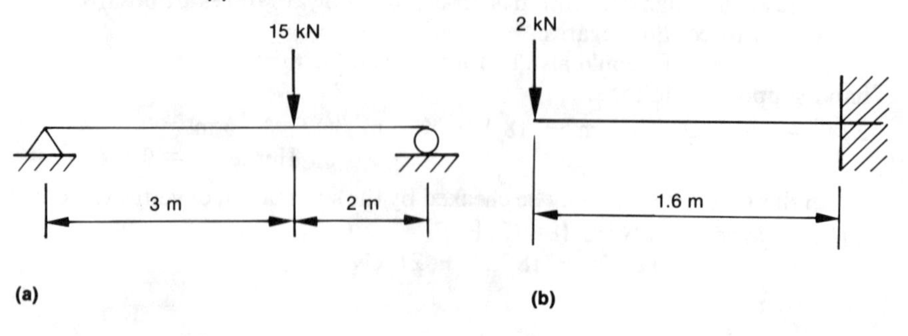

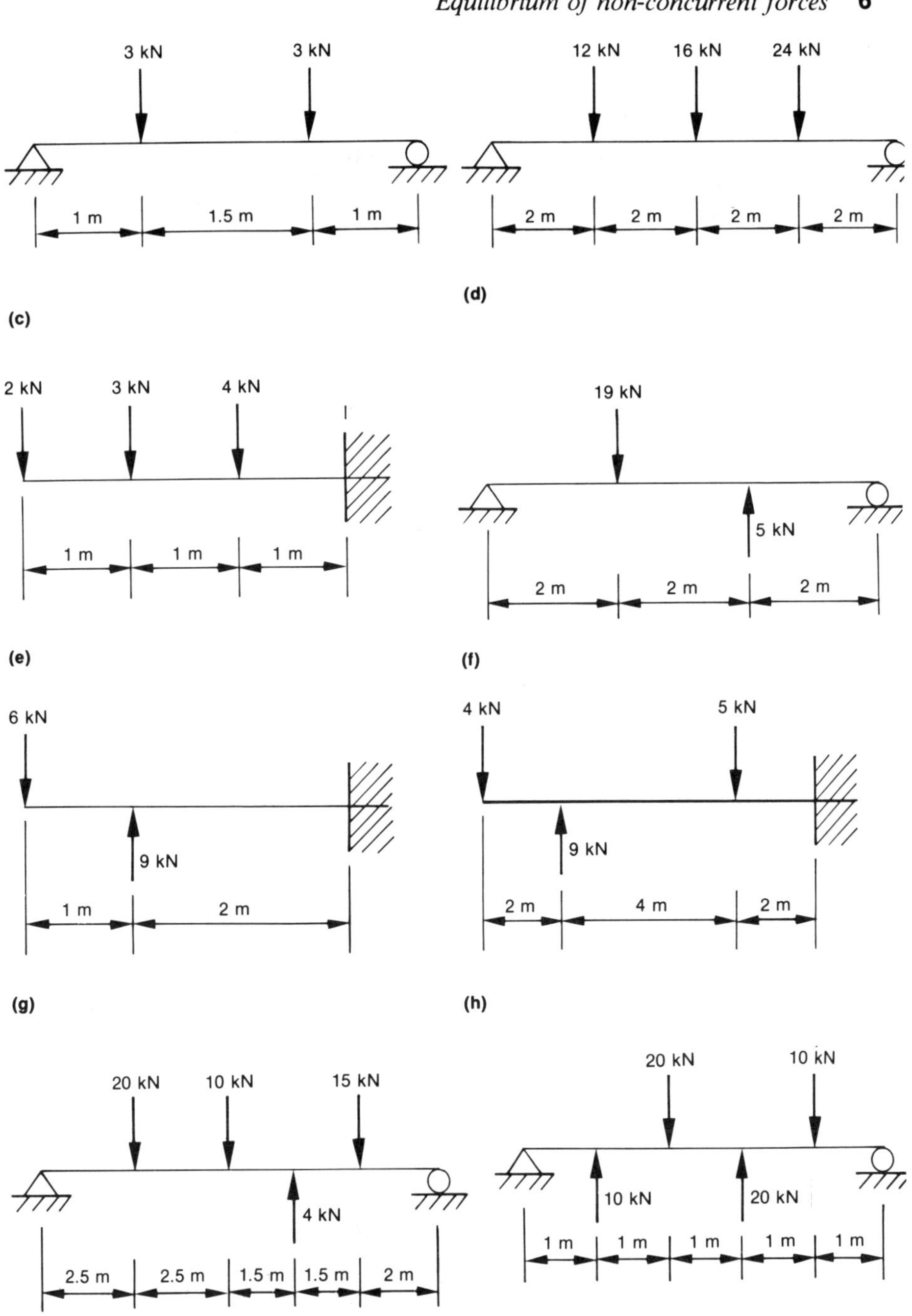
3 kN
3 kN
1 m
1.5 m
1 m
(c)
12 kN
16 kN
24 kN
2 m
2 m
2 m
2 m
(d)
2 kN
3 kN
4 kN
1 m
1 m
1 m
(e)
19 kN
5 kN
2 m
2 m
2 m
(f)
6 kN
9 kN
1 m
2 m
(g)
4 kN
5 kN
9 kN
2 m
4 m
2 m
(h)
20 kN
10 kN
15 kN
4 kN
2.5 m
2.5 m
1.5 m
1.5 m
2 m
(i)
20 kN
10 kN
10 kN
20 kN
1 m
1 m
1 m
1 m
1 m
(j)

Fig. 6.4

6.3 Resultant of non-concurrent forces

We have previously explained the resultant force as that single force which has exactly the same effect as a given system of forces, and learned how to determine its magnitude and direction by the force polygon method or by mathematical addition of forces.

So far this is applicable to the case of non-concurrent forces acting on a rigid body, except that each force in such a system will also tend to rotate the body upon which it acts, about some axis. It is therefore necessary to consider the rotational effect of the force system in addition to the linear push-pull action of the forces.

When solving for the resultant of a non-concurrent force system, the problem usually consists of two steps:

1. finding the magnitude and direction of the resultant force, by mathematical or graphical addition of all forces, and
2. finding the location of the resultant relative to an arbitrary reference point, usually by applying the principle of moments.

Example 6.4

Determine the magnitude, direction and location of the resultant of the system of forces in Figure 6.5.

Solution

The magnitude of the resultant force is

$$F = \Sigma F = -10 + 15 - 19 - 8 = -22 \text{ kN}$$

i.e. 22 kN down.

The location is found by taking moments about an arbitrary point, such as point A.

$$\Sigma M = 10 \times 0 - 15 \times 1 + 19 \times 2 + 8 \times 4 = 55 \text{ kN.m}$$

This sum must be equal to the moment of the resultant about the same point.

$$F \times d = 55 \text{ kN.m}$$
$$22 \text{ kN} \times d = 55 \text{ kN.m}$$

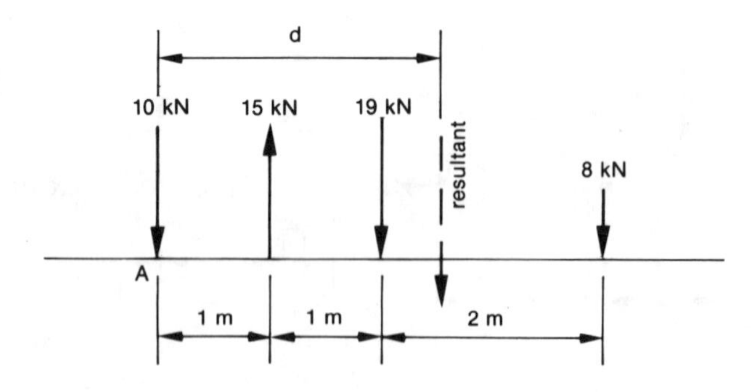

Fig. 6.5

Hence,

$$d = \frac{55 \text{ kN.m}}{22 \text{ kN}} = 2.5 \text{ m}$$

It is therefore possible to replace the given system of forces by a single downward force of 22 kN, located at 2.5 m from point A, as shown.

Example 6.5

A concrete foundation for a brick wall must support two vertical loads and a horizontal load due to soil pressure on one side, as shown in Figure 6.6(a). Its own weight is also shown. Determine the resultant and check if it passes through the base of the foundation, as required for stability.

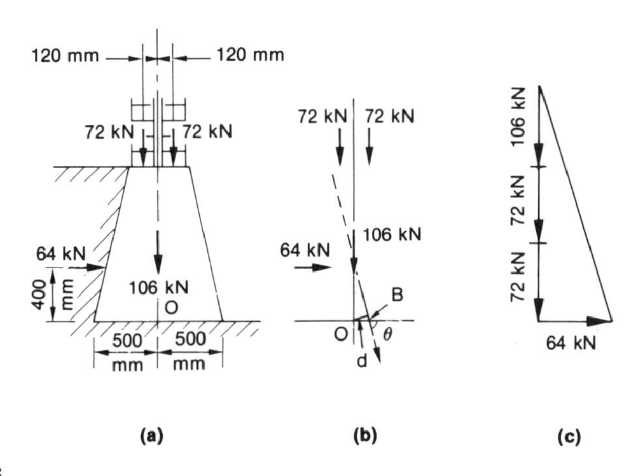

Fig. 6.6

Solution

Sum of the vertical forces is:

$$\Sigma F_V = 106 + 72 + 72 = 250 \text{ kN down}$$

Sum of the horizontal forces is:

$$\Sigma F_H = 64 \text{ kN to the right}$$

The resultant is therefore:

$$F = \sqrt{250^2 + 64^2} = 258.1 \text{ kN}$$

The angle to the horizontal is:

$$\theta = \tan^{-1}\left(\frac{250}{64}\right) = 75.6°$$

To locate the resultant, take moments about a convenient point, such as mid-point of the base.

$$\Sigma M = 64 \times 0.4 - 72 \times 0.12 + 72 \times 0.12 = 25.6 \text{ kN.m}$$

Distance of the resultant from the mid-point is:

$$d = \frac{\Sigma M}{F} = \frac{25.6 \text{ kN.m}}{258.1 \text{ kN}} = 0.0992 \text{ m} = 99.2 \text{ mm}$$

The point at which the resultant intersects with the base can be calculated from:

$$OB = \frac{d}{\sin 75.6°} = \frac{99.2}{\sin 75.6°} = 102.4 \text{ mm}$$

Parts of this solution can also be done graphically. For example, the magnitude and direction of the resultant can very conveniently be found by constructing the polygon of forces as in Figure 6.6(c).

However, the principle of moments, for determining the distance d, is best dealt with mathematically.*

The last part of the question can also be answered by a diagram such as Figure 6.6(b) if drawn to scale.

Problems

6.2 For the systems of forces in Figure 6.7 determine and locate the resultant force, stating the distance along horizontal line from the first force on the left.

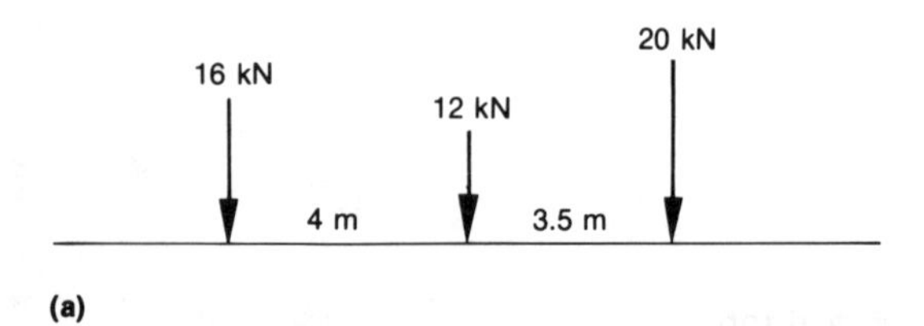

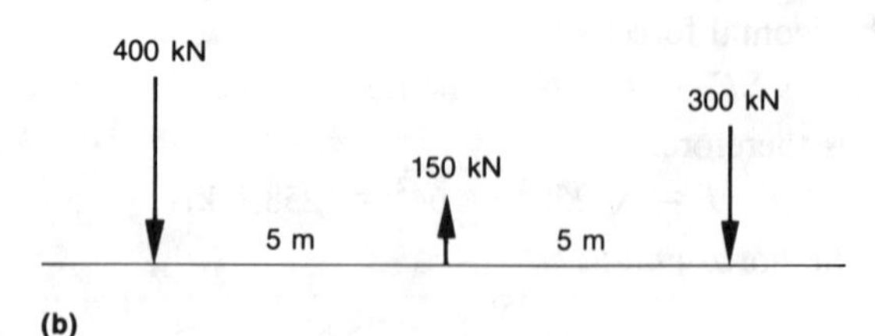

* The alternative graphical method, known as the **funicular** (from *funis* = rope) **polygon**, involves a special construction procedure which locates a point on the line of action of the resultant force, thus locating the force itself.

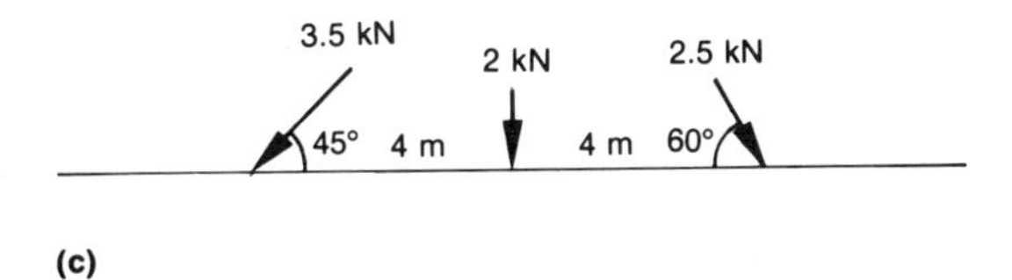

(c)

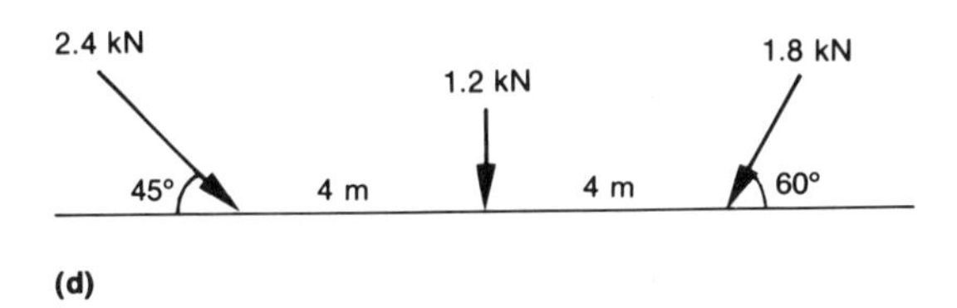

(d)

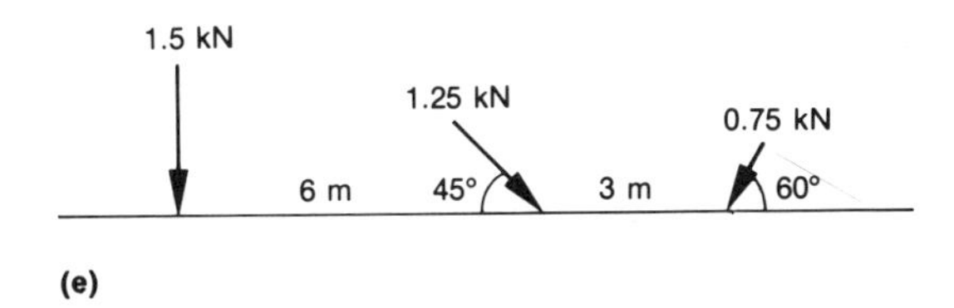

(e)

Fig. 6.7

6.3 Determine and locate the resultant of the three forces acting on the ladder as shown in Figure 6.8.

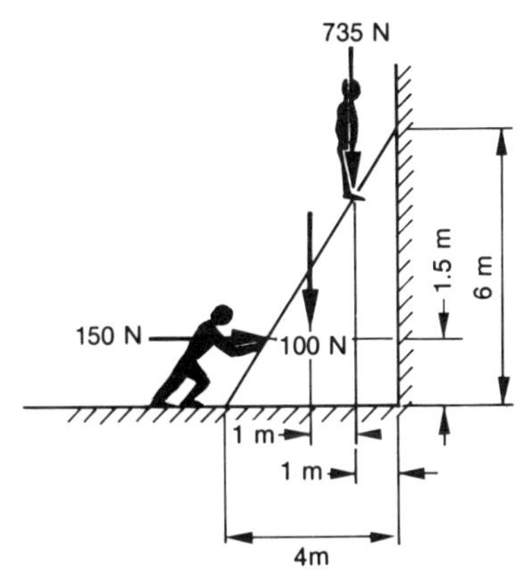

Fig. 6.8

6.4 Three forces acting on a 300 mm diameter pulley are as shown in Figure 6.9. These include belt tensions and the weight of the pulley. Determine the resultant in magnitude, direction and distance from the centreline of the pulley.

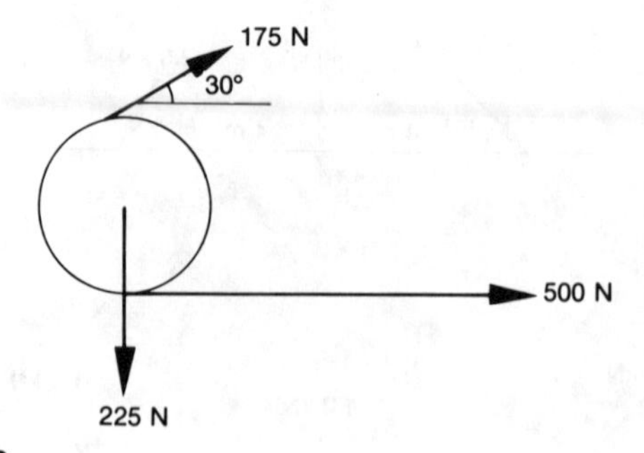

Fig. 6.9

6.5 Determine the resultant of the loads acting on the structure as shown in Figure 6.10 and locate it relative to the left-hand support.

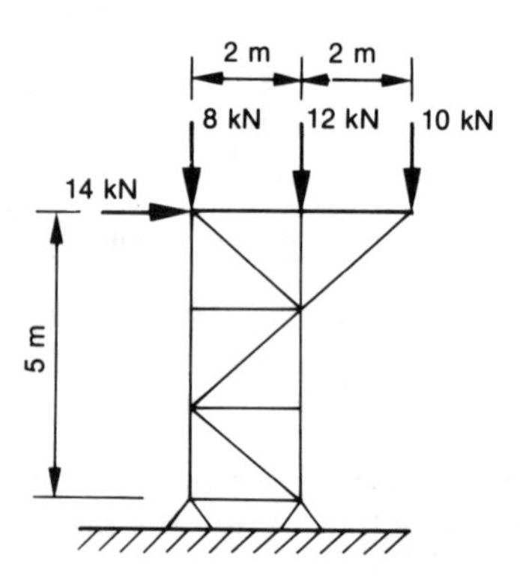

Fig. 6.10

Review questions

1. Define non-concurrent force system.
2. What are the conditions of equilibrium of non-concurrent forces?
3. What is meant by beam reactions?
4. Describe a simply supported beam.
5. Describe a cantilever beam.
6. What kind of reactions apply to
   (a) simply supported beam?
   (b) cantilever beam?
7. Explain the method of determining the magnitude and direction of the resultant of a non-concurrent force system.
8. How can the line of action be located?

7

Friction

It is common experience when one object is pushed or pulled along the surface of another object that there is resistance to such motion which must be overcome by applying an external force. The property of the surface that causes this resistance is called **friction**. More precisely, friction is a force that resists the sliding of one solid object over another.

The phenomenon of friction is one of the most fundamental and most common encountered by the engineer. It has been studied as a branch of mechanics, where the methods of statics are best suited for estimating the magnitude and effects of friction.

A new science of tribology is now emerging which studies friction and its related subjects of wear and lubrication.

7.1 Frictional resistance

Friction is not a simple phenomenon. Its laws have been studied for hundreds of years and reasonably satisfactory methods for calculating friction in simple cases have been known for at least two centuries. However, the exact mechanism by which frictional forces are formed is not yet completely understood.

Friction is believed to be caused by the adhesion of or interference between microscopic irregularities of the surfaces in the areas of contact. It depends on the nature of the surfaces in contact, the force pressing one of the surfaces on to the other, and to some extent on whether the surfaces are already in relative motion or at rest.

Some frictional effects are beneficial, such as traction needed to walk without slipping. The operation of brakes, clutches and power transmission belts depends on the presence of friction. Friction also makes the wheels of a locomotive grip the rails. On the other hand, much of the power used to drive machines of all kinds is consumed in overcoming friction between moving parts. Friction also tends to produce heat, which causes the surfaces to wear, necessitating costly repairs and replacements. Lubrication tends to reduce frictional resistance and wear between moving machinery parts and is therefore widely used.

Another form of resistance similar to friction, known as **rolling resistance**, is produced when a deformable tyre rolls over a surface. The main reason for rolling friction is deformation which occurs when the tyre is slightly flattened and the road surface is somewhat indented at the point of contact. The resistance to motion due to rolling friction is generally considerably less than that of sliding friction. A common engineering application which combines the effects of rolling friction and lubrication is the ball or roller bearings used to reduce frictional resistance to rotational motion of shafts and wheels.

Resistance of air to the motion of vehicles and projectiles through it also represents a friction-like force. However, mathematical treatment of air resistance is quite complex and different from that of sliding friction. Except for the laws of dry sliding friction discussed in some detail in this chapter, we will simply regard all forms of friction as forces resisting motion.

7.2 The laws of dry sliding friction

The first research on the laws governing sliding friction was carried out by Leonardo da Vinci in the fifteenth century. However, the credit for discovering the laws of friction is given to Guillaume Amontons, French physicist and inventor of scientific instruments, who published his investigations on friction in 1699. Another early worker was Charles Coulomb, also a French scientist and military engineer best known for his work in electrostatics, who conducted the first complete investigation of dry friction in 1781. The theory of dry friction often bears the name of Coulomb friction.

Sliding friction is characterised by two simple experimental facts. First, it is nearly independent of the area in contact. Secondly, friction is proportional to the normal force that presses the surfaces together.

Consider a solid block, such as a brick, resting on a horizontal table surface (Fig. 7.1). The weight (F_w) acting down and the normal reaction (F_n) from the table, acting up, are equal and opposite. If a small push (F_p) is applied to the brick and the brick does not move, there must be a friction force (F_f), equal and opposite to the push, at the surface of contact between the brick and the table. If the push is increased gradually, it will eventually cause the brick to slide on the table. There is, therefore, a limiting value of the force of friction beyond which it cannot increase.

Experiments show that the limiting value of the frictional force is the same whether the brick is lying down or standing on its end, i.e. independent of the area in contact. Furthermore, if a pile of three bricks is pushed along the table, the friction is three times greater than when one brick is pushed, i.e. friction is proportional to the normal force.

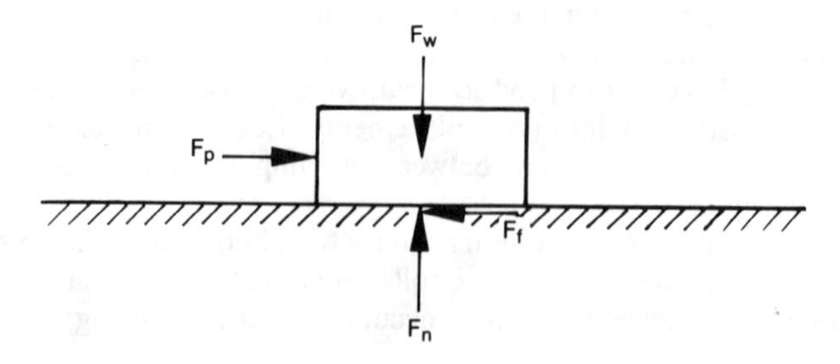

Fig. 7.1

Table 7.1 Approximate values of the coefficient of static friction

Surfaces	*Typical value*	*Usual range*
Metal on metal (greasy)	0.1	0.08—0.2
Hardwood on metal (greasy)	0.2	0.15—0.3
Metal on metal (dry)	0.2	0.15—0.35
Wire rope on metal pulley	0.2	0.15—0.4
Hemp rope on metal pulley	0.3	0.2 —0.5
Wood on wood	0.35	0.25—0.5
Hardwood on metal (dry)	0.35	0.2 —0.6
Rubber or leather on metal	0.4	0.3 —0.60
Brake lining on metal	0.4	0.3 —0.7
Metal on stone	0.4	0.3 —0.7
Wood on stone	0.4	0.3 —0.7
Masonry on brickwork	0.6	0.55—0.7
Rubber tyre on concrete	0.8	0.6 —1.00

At the moment of impending motion, the value of F_f is always a fixed ratio of the normal force F_n, which depends on the materials and the roughness of the contacting surfaces. The constant value of this ratio is called the **coefficient of static friction** and is usually symbolised by the Greek letter μ (mu).

$$\mu = \frac{F_f}{F_n}$$

Because both friction and normal forces are measured in units of force, i.e. newtons, the coefficient of static friction is dimensionless.

Owing to the variation in friction with the condition of the rubbing surfaces, it is impossible to specify exact values for each pair of materials in contact. Typical clean, unlubricated surfaces have friction coefficients in the range of 0.2 to 0.4. Table 7.1 gives generally accepted approximate values.

The presence of a liquid lubricant entirely alters the character of friction and often has a profound effect on the friction coefficient. Good lubricants are able to reduce the coefficient of friction to as low as 0.05. A typical range for lubricated polished surfaces is 0.08 to 0.15.

The friction thus far described arises between surfaces at rest with respect to each other. At the moment of impending motion, the smallest force (F_p) required to overcome static friction and to start motion is

$$F_p = F_f = \mu F_n$$

Once the motion begins, the force required to continue the motion, i.e. to maintain motion against continuous frictional resistance called kinetic friction, is usually less than static friction, by about 25 per cent.

From these conclusions, the laws of friction can be summarised as follows:

1. Friction always acts in a direction opposite to impending or actual motion.
2. Static friction has a limiting value beyond which it cannot increase.

3. The limiting value of static friction is given by

$$F_f = \mu F_n$$

4. The value of coefficient of static friction (μ) depends on the nature and condition of the surfaces in contact, but is independent of the areas in contact.
5. In general, kinetic friction is less than the limiting static friction.

Example 7.1

A body of mass 5 kg rests on a horizontal surface and the coefficient of friction between the two surfaces is 0.33. What horizontal force will be required to start the body moving?

Solution (Refer to Fig. 7.1)

Weight of body: $F_w = m.g = 5\text{ kg} \times 9.81\frac{\text{N}}{\text{kg}} = 49.05\text{ N}$

Normal force: $F_n = F_w = 49.05\text{ N}$

Limiting friction: $F_f = \mu F_n = 0.33 \times 49.05\text{ N} = 16.2\text{ N}$

Force required to just start the body moving is $F_p = F_f = 16.2\text{ N}$.

Example 7.2

In an experiment to determine the coefficient of static friction, a horizontal force of 50 N was required to start a 10 kg block moving on a horizontal surface. What was the value of the coefficient?

Solution (Refer to Fig. 7.1)

$$F_n = F_w = \text{m.g} = 10\text{ kg} \times 9.81\frac{\text{N}}{\text{kg}} = 98.1\text{ N}$$

$$F_f = F_p = 50\text{ N}$$

$$\mu = \frac{F_f}{F_n} = \frac{50\text{ N}}{98.1\text{ N}} = 0.51$$

Example 7.3

A 100 kg block rests on a plate as shown in Figure 7.2. The coefficient of friction between all surfaces is 0.2. Determine the force required to pull the plate from under the block.

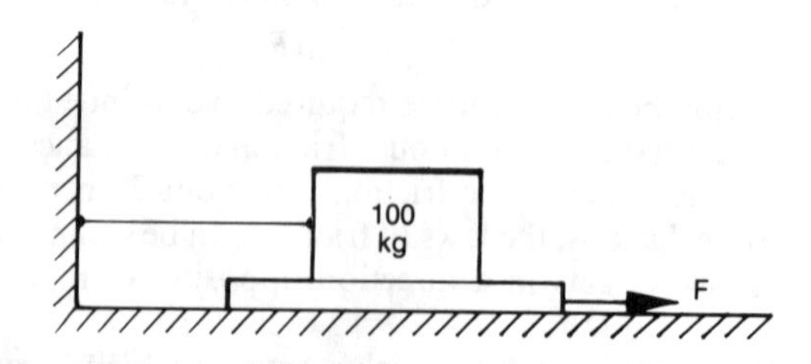

Fig. 7.2

Solution

Normal force:

$$F_n = F_w = m.g = 100 \text{ kg} \times 9.81\frac{\text{N}}{\text{kg}} = 981 \text{ N}$$

Friction force:

$$F_f = \mu F_n = 0.2 \times 981 \text{ N} = 196.2 \text{ N}$$

The applied force F_p must overcome friction between two pairs of surfaces in contact.

Therefore, $F_p = 2F_f = 2 \times 196.2 \text{ N} = 392.4 \text{ N}$

Problems

7.1 A block of mass 2 kg rests on a horizontal table and the coefficient of static friction between the surfaces is 0.28. What horizontal force will be required to start the block moving?

7.2 A block of wood having a mass of 5 kg rests on a horizontal table. A horizontal force of 12 N is just sufficient to cause it to slide. What is the coefficient of static friction between the surfaces?

7.3 A 2 tonne girder is pulled along a horizontal floor by a winch. The coefficient of friction between the girder and the floor is 0.35. What is the tension in the horizontal rope between the girder and the winch?

7.4 A 20 kg block rests on a horizontal surface and is attached to a mass of 3.5 kg by a cable as shown in Figure 7.3. The pulley is frictionless and the coefficient of static friction is 0.2.

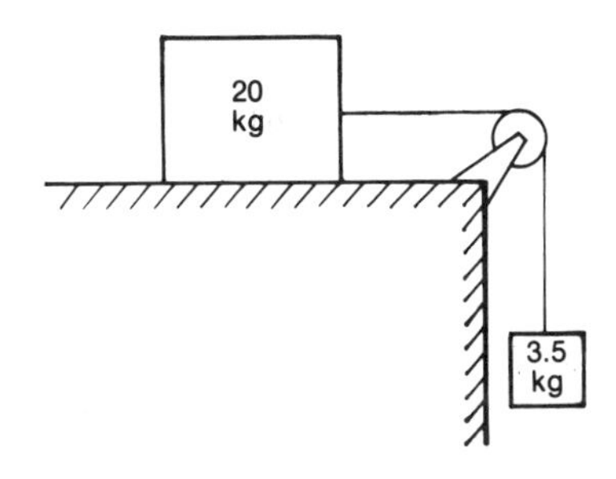

Fig. 7.3

Prove that the mass is not sufficient to start the block moving. What is the additional mass that would be required to start motion?

7.5 Each of the two blocks in Figure 7.4 has a mass of 10 kg and the coefficient of friction between all surfaces is 0.3. Determine the force F_p required to pull one block from under the other.

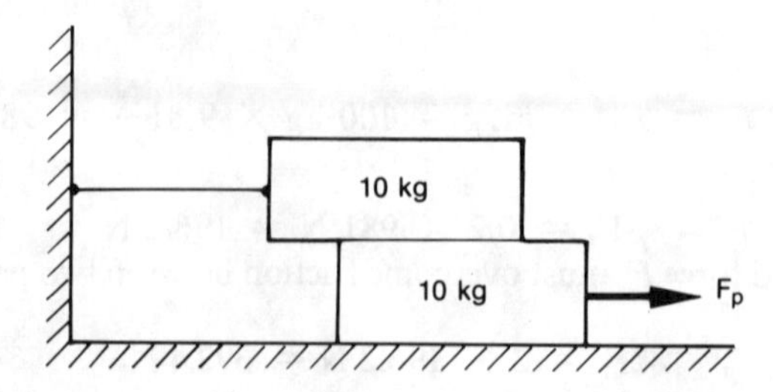

Fig. 7.4

7.6 In an automatic materials handling operation metal blocks 1.5 kg each are pushed one at a time from the bottom of a stack six blocks high as shown in Figure 7.5. If the coefficient of friction is 0.25, determine the horizontal force required.

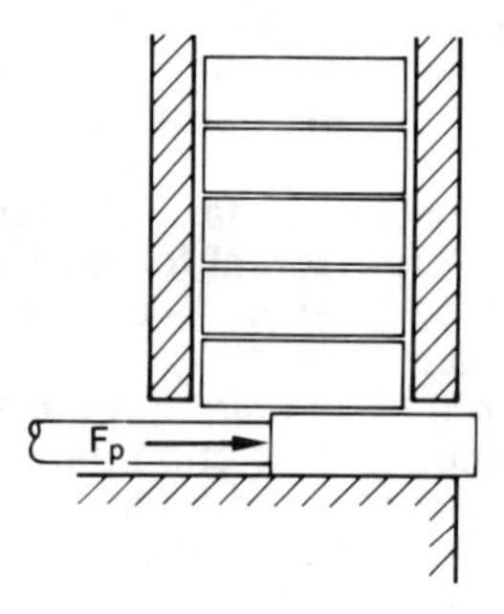

Fig. 7.5

7.7 If the steel straps shown in Figure 7.6 rely on friction to transmit a load of 3 kN, what should the force be in each of the four bolts? Take $\mu = 0.2$.

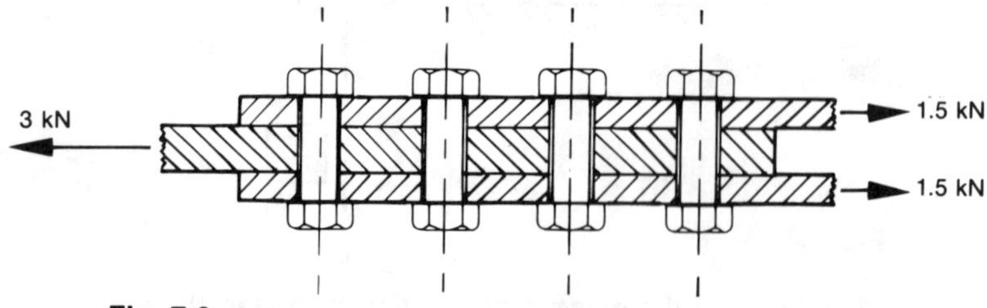

Fig. 7.6

7.8 What is the braking torque applied to the brake drum of 300 mm diameter (Fig. 7.7) if the brake shoes are pressed to the drum with a force of 800 N each? The coefficient of friction is 0.6.

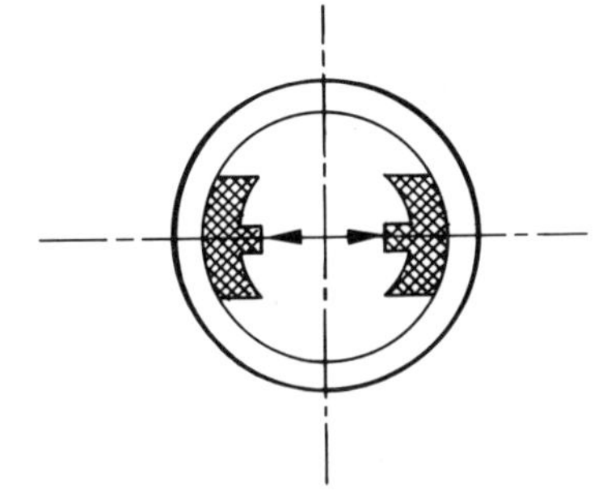

Fig. 7.7

7.9 In the brake shown in Figure 7.8, the coefficient of friction between the brake shoe and the drum is 0.45. Find the smallest value of force F required to prevent rotation of the drum against an applied torque of 75 N.m.

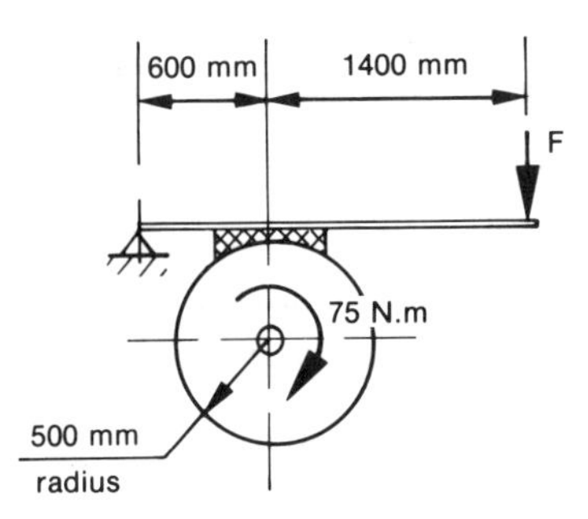

Fig. 7.8

7.10 A 35 kg cabinet, 0.75 m × 0.75 m × 1.8 m high, is pushed by a gradually increasing horizontal force applied 1.5 m above floor level. If the coefficient of friction between the cabinet and the floor is 0.4, determine if the cabinet will slide or tip.

7.3 The angle of friction

Let us now consider another approach to the analysis of friction on a horizontal plane. Referring to Figure 7.9, it is possible to combine the force of friction (F_f) and the normal reaction (F_n) into a single resultant force (F_r), representing a total reaction at the surface of contact to the action of weight (F_w) and applied force (F_p).

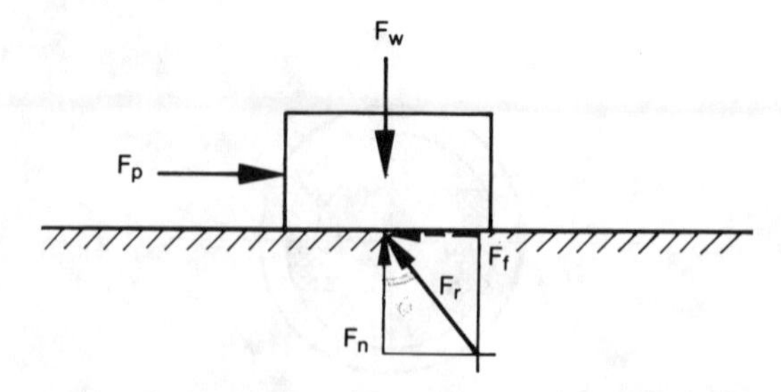

Fig. 7.9

The resultant force (F_r) will be inclined at an angle to the normal direction. Using the rules for finding resultants of two mutually perpendicular forces yields

$$F_r = \sqrt{F_f^2 + F_n^2}$$

and

$$\tan \phi = \frac{F_f}{F_n}$$

But we know that at the moment of impending motion, i.e. when sliding motion is about to begin, the ratio of F_f to F_n is equal to the coefficient of static friction (μ). Therefore, for limiting friction

$$\boxed{\tan \phi = \mu}$$

Under these circumstances the angle ϕ, known as the **angle of friction** is constant and depends only on the nature and condition of the surfaces in contact. The values of the angle of friction for each corresponding value of μ can easily be calculated and are summarised in Table 7.2.

Table 7.2 The angle and coefficient of static friction

μ	0.1	0.15	0.2	0.25	0.3	0.35	0.4	0.45	0.5
ϕ	5.71°	8.53°	11.31°	14.04°	16.70°	19.29°	21.80°	24.23°	26.57°

The use of the angle of friction enables problems involving sliding friction to be solved graphically as illustrated by the following example.

Example 7.4

A body of mass 3 kg rests on a horizontal surface and the coefficient of friction between the two surfaces is 0.3. What horizontal force is required to start the body moving?

Solution

The weight of the body is

$$F_w = \text{m.g} = 3\ \text{kg} \times 9.81\frac{\text{N}}{\text{kg}} = 29.43\ \text{N}$$

The angle of friction is $\phi = \tan^{-1} 0.3 = 16.7°$

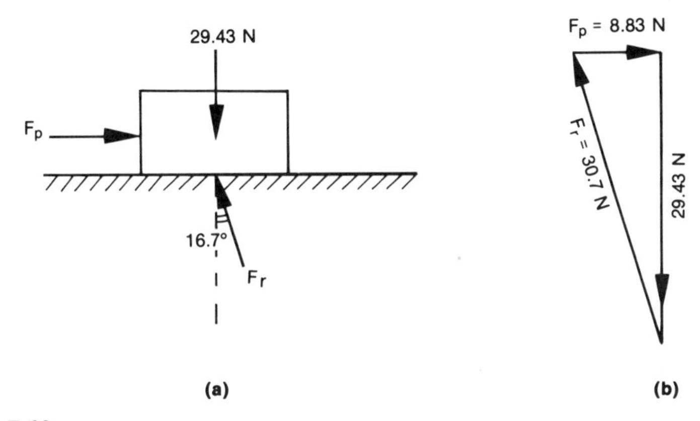

Fig. 7.10

The three forces in equilibrium are as shown in Figure 7.10(a). The answer is found by constructing a triangle of forces, Figure 7.10(b).

The force required is $F_p = 8.83$ N.

7.4 The angle of repose

An easy way of observing the laws of friction and, in particular, of measuring the coefficient of friction between two surfaces, is by means of a plane tilted slowly through an angle θ to the horizontal.

Take a book or a board and place on it a matchbox or a similar object. Tilt the book slowly through a small angle θ. The weight of the box can now be resolved into components, one along $F_w \sin \theta$ and the other perpendicular to the surface of the book $F_w \cos \theta$. (See Fig. 7.11.)

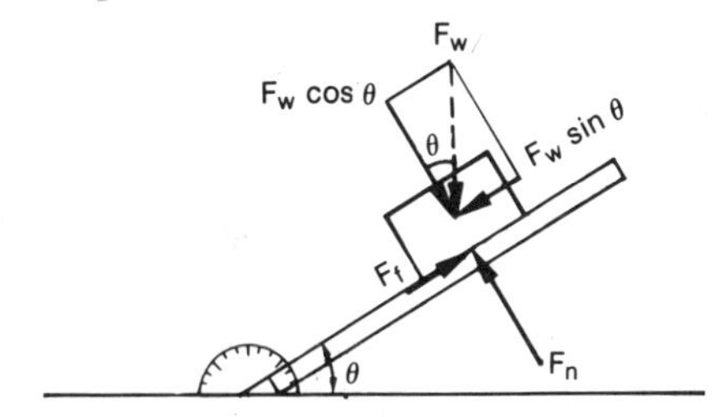

Fig. 7.11

If the angle of tilting is small, the box will not slide down the book but will remain at rest, the force of friction being sufficient for equilibrium.

As the book is tilted more and more, the friction will reach its limiting value beyond which it cannot increase. At this moment the box is on the point of slipping and all forces are exactly balanced:

$$F_n = F_w \cos \theta$$

and

$$F_f = F_w \sin \theta$$

Remembering that $\mu = \dfrac{F_f}{F_n}$, we write

$$\mu = \frac{F_f}{F_n} = \frac{F_w \sin\theta}{F_w \cos\theta} = \tan\theta$$

$$\text{but } \mu = \tan\phi$$

$$\text{therefore } \tan\theta = \tan\phi$$

$$\text{or } \theta = \phi$$

The value of the angle of inclination corresponding to impending motion is called the **angle of repose**. Clearly, the angle of repose is equal to the angle of static friction.

Understanding the concept of the angle of repose is important in the design of bins and hoppers for the storage and handling of granular materials, and for calculating the steepest angle to the horizontal which can be made by the inclined surface of a heap of loose material or an embankment.

Example 7.5

What is the steepest ramp on which a car can stand without slipping down, if the coefficient of friction between the tyres and the ramp surface is 0.8?

Solution

$$\begin{aligned}\text{Angle of repose} &= \text{Angle of friction } \phi \\ &= \tan^{-1}\mu \\ &= \tan^{-1} 0.8 \\ &= 38.7^\circ\end{aligned}$$

Problems

7.11 Determine the angle of friction corresponding to a coefficient of static friction of 0.6.

7.12 If the normal and frictional forces between two surfaces which are about to slip are 100 N and 35 N respectively, determine the coefficient of static friction, the angle of friction and the magnitude of the resultant force between the surfaces.

7.13 If the normal reaction between two surfaces is 120 N and the coefficient of friction is 0.25, determine the magnitude and direction of the total reaction for the case of limiting friction.

7.14 A body of 2 kg mass rests on a board which is gradually tilted until, at an angle of 27° to the horizontal, the body begins to move down the plane. Determine the coefficient of friction and the magnitude of the normal and frictional forces when the body begins to slip.

7.15 Solve problem 7.1 graphically.

7.16 Solve problem 7.2 graphically.

7.17 Solve problem 7.3 graphically.

7.18 Solve problem 7.4 graphically.
7.19 Solve problem 7.5 graphically.
7.20 Solve problem 7.6 graphically.

7.5 The inclined plane

The inclined plane is one of the earliest mechanical devices used by people in their efforts to influence their environment. It was used as a simple machine for moving and lifting heavy stones for the construction of buildings and pyramids in the ancient world. Today it is still used as an element of modern machinery in the form of the power screw. Many other mechanical devices, such as cams, wedges and knife-like metal cutting tools, also operate on the inclined plane principle. Friction is a prominent and unavoidable feature of the operation of the inclined plane.

Problems involving friction on an inclined plane can be solved mathematically or graphically. The mathematical solution is best suited for problems in which the applied force is acting parallel to the plane, while the graphical method is useful for solving forces inclined to the plane at any angle. To some extent the choice of a method is a matter of personal preference.

Regardless of the method used, it is very important to remember that friction always opposes motion and to show all active forces accordingly.

Example 7.6

A 200 kg block rests on a 25° incline as shown in Figure 7.12(a). The coefficient of friction between the block and the plane is 0.2. Determine the magnitude of force F_p acting along the inclined plane required.

(a) to start the block moving up the inclined plane, and
(b) to prevent it slipping down the inclined plane.

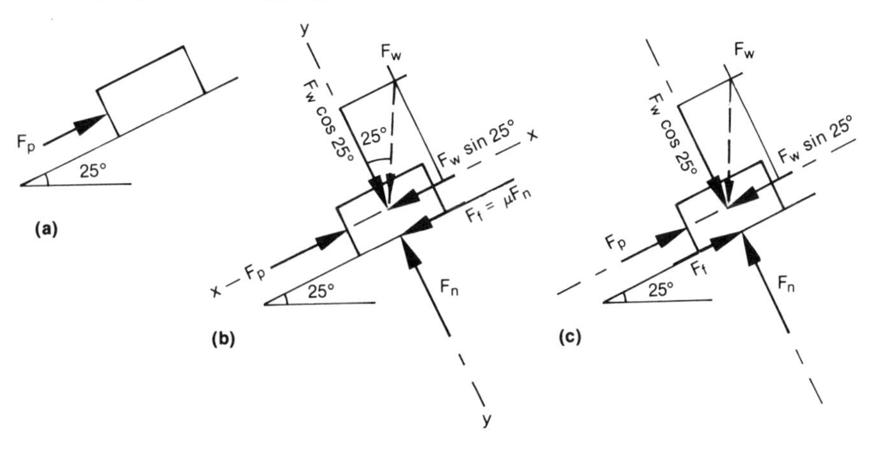

Fig. 7.12

Solution

(a) In the first case, while motion is about to begin up the plane, the force of friction (F_f) will act down the plane in opposition to the impending motion, as shown in Figure 7.12(b).

The weight of the block can be resolved into components along and perpendicular to the plane:

$$(F_w)_x = F_w \sin 25° = 200 \times 9.81 \times \sin 25° = 829.2 \text{ N}$$
$$(F_w)_y = F_w \sin 25° = 200 \times 9.81 \times \cos 25° = 1778 \text{ N}$$

Summation of forces in the direction not containing the unknown force (F_p) yields

$$\Sigma F_y = 0$$
$$F_n = F_w \cos 25°$$
$$F_n = 1778 \text{ N}$$

The law of friction can now be used to find the frictional force $F_f = \mu F_n$.

$$F_f = 0.2 \times 1778 = 355.6 \text{ N}$$

Summation of forces in the other direction now gives

$$\Sigma F_x = 0$$
$$F_p = F_w \sin 25° + F_f$$
$$F_p = 829.2 + 355.6 = 1184.8 \text{ N}$$

(b) In the second case, the tendency for motion is down the plane, and the direction of friction will be reversed as in Figure 7.12(c).

The solution follows the same basic steps:

$$(F_w)_x = F_w . \sin 25° = 829.2 \text{ N}$$
$$(F_w)_y = F_w . \cos 25° = 1778 \text{ N}$$
$$\Sigma F_y = 0 \qquad F_n = 1778 \text{ N}$$
$$F_f = \mu F_n \qquad F_f = 0.2 \times 1778 = 355.6 \text{ N}$$

The summation of forces in the direction parallel to the plane has friction acting with the applied force, i.e. up the plane, to balance the component of weight acting along the plane.

$$\Sigma F_x = 0 \qquad F_p + F_f = F_w . \sin 25°$$
$$F_p + 355.6 = 829.2$$
$$F_p = 473.5 \text{ N}$$

Comparing the answers to parts (a) and (b), it can be seen that there is a range of magnitudes of force F_p between 473.5 N and 1184.8 N within which the block remains stationary on the plane. However, if the force is 1184.8 N or greater, the block will move up the plane. On the other hand, if the force is equal to or less than 473.5 N the block will slip down the plane.

It is also worth noting that while the weight of the block is 1962 N, it can be pushed up the plane with a force of only 1184.8 N. It is suggested that the student should repeat this solution a few times using different values of the coefficient of friction and the angle of inclination to see what effect these have on the magnitude of the force required.

Example 7.7

A 100 kg block resting on a 30° incline is acted upon by a force F_p at 20° to the plane as shown in Figure 7.13(a). Determine the magnitude of the force required to start the motion up the plane, if the coefficient of static friction is 0.2.

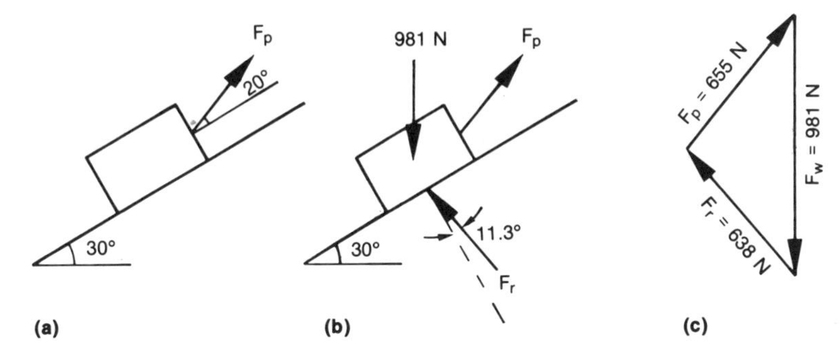

Fig. 7.13

Solution

The weight of the block is $F_w = 100 \times 9.81 = 981$ N.

The resultant of normal and frictional forces is inclined at the angle of friction from the normal direction, the angle of friction being

$$\phi = \tan^{-1} \mu = \tan^{-1} 0.2 = 11.3°$$

One must be careful in deciding which way the friction acts, in order to draw the correct direction of the resultant. See Figure 7.13(b).

A triangle of forces is then constructed (Fig. 7.13(c)) from which the magnitude of the applied force F_p is scaled off.

The answer, in this case, is 655 N.

It is strongly recommended that, as a learning exercise, graphical solutions should be verified by mathematical solutions, and vice versa.

Problems

7.21 A 150 kg box is resting on a 15° incline with a coefficient of friction of 0.35. What force acting along the plane is required to start the box up the plane?

7.22 What force parallel to the slipway is required to lift a 1.5 tonne boat up an incline of 20°, if the coefficient of friction is 0.25?

7.23 A load of 20 kg resting on an inclined plane begins to slip when the plane is tilted to an angle of 25° to the horizontal. What force parallel to the plane will be necessary to keep the load from slipping down the plane, if the angle of the slope is increased to 35°?

7.24 If the applied force in the previous problem was horizontal, what magnitude would be required?

7.25 In problem 7.23, what horizontal force must be applied to the load to just start it up the plane inclined at 25°? At 35°?

7.26 What is the mass of the largest stone that can be dragged uphill at an angle of 20° by a tractor capable of exerting a pull of 2 kN, if the coefficient of friction is 0.6?

7.27 A casting of mass 100 kg is pulled along a horizontal floor by a rope at 30° to the floor. The tension in the rope just sufficient to overcome friction is 300 N. Determine the coefficient of friction.

7.28 Determine the maximum and minimum values of the mass *m* between which static equilibrium will be maintained for the system shown in Figure 7.14. Take $\mu = 0.3$.

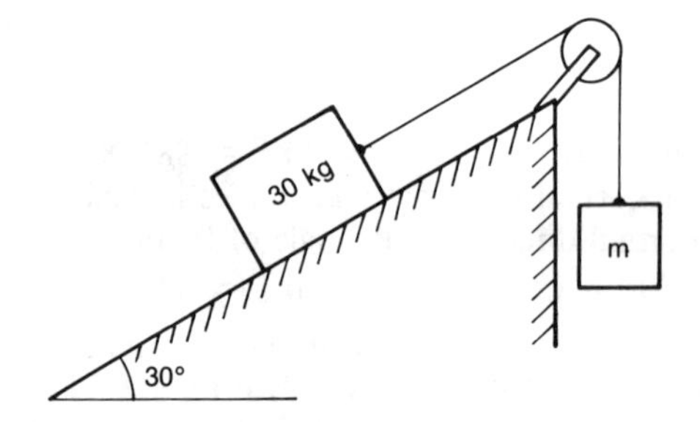

Fig. 7.14

7.29 Determine the range of values of mass *m* within which static equilibrium will be maintained for the system shown in Figure 7.15. Take $\mu = 0.25$.

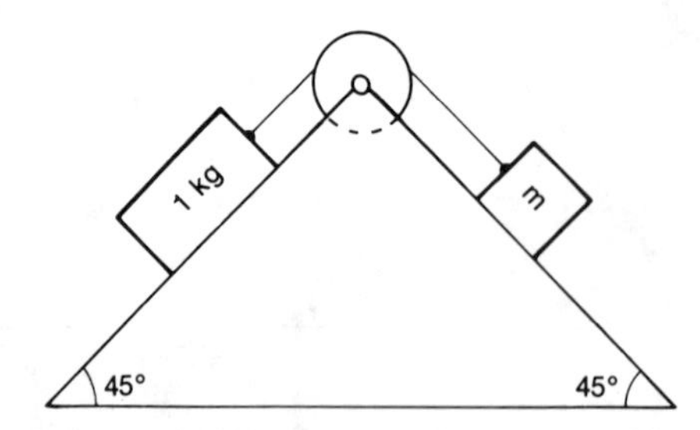

Fig. 7.15

Review questions

1. What is meant by friction?
2. Are frictional effects beneficial or harmful? State some examples.
3. What are the main two characteristics of sliding friction?
4. What is the ratio of the frictional and normal forces called?
5. What is the unit of the coefficient of static friction?
6. What is the effect of liquid lubricants on the magnitude of friction?
7. What is the difference between static and kinetic friction?
8. State the laws of dry sliding friction.
9. Define the angle of friction. How is it related to the coefficient of static friction?
10. What is meant by the angle of repose? How is it related to the angle of friction?

PART THREE

Dynamics

Now, to have access to the science of motion, one does not need to scale the heights of mathematics first; on the contrary, nature itself reveals its beauty in full splendour, and even a person with meagre abilities can see a multitude of things that have hitherto been hidden from the greatest minds.

Robert Mayer,
on the law of conservation of energy

PART THREE

8

Linear motion

One of the primary concerns of mechanical engineering is with motion. Motion exists where there is a change in position or orientation of an object with reference to some other object or objects. Aeroplanes, trains and automobiles are part of everyday life. A reciprocating piston in an internal combustion engine, a rotating flywheel, water flowing inside a pipe, and steam driving the rotor of a turbine are typical examples of different kinds of motion with which the science and practice of mechanical engineering are concerned.

Historically, mechanics is the oldest of the physical sciences and it provided the foundation for the growth and development of all other areas of physics. The study of motion in mechanics is called **dynamics**. Unlike statics, which goes back to Archimedes and the Greek philosophers, serious study of dynamics only began with experimental work of Galileo (1564–1642) which led to the mathematical formulation of the fundamental laws of motion by Newton in his famous publication, known as the "Principia".*

The part of dynamics that describes the geometry of motion in terms of two elementary concepts, those of position and time, is called **kinematics**. The part that relates the motion to the forces causing it is called **kinetics**.

The problems of dynamics can be classified and studied according to the type of motion that exists. In this chapter, we discuss motion along a straight line, known as **rectilinear translation**. We shall simply refer to it as linear motion. With some limitations our discussion also applies to motion along a curved path.

Chapter 9 is devoted to **rotation**, which is motion in a circular path around a fixed axis. In the following chapters a number of important concepts related to motion will be introduced. These will include work, power, energy and momentum.

8.1 Displacement, velocity and acceleration

Let us now consider how linear motion of an object, such as a motor-car, can be described in terms of its position and time. More specifically, motion is usually described in terms of linear displacement, linear velocity and linear acceleration—the concepts which are defined below.

---

* *Philosophiae Naturalis Principia Mathematica*, or *Mathematical Principles of Science.*

Later, in this and the following chapters, if it is clear from the context that motion is in a straight line, the word "linear" will be omitted from the terms linear displacement, linear velocity and linear acceleration. Furthermore, provided that displacement, velocity and acceleration are measured along the path of the motion at every point, e.g. road distances and not as the crow flies, the definitions and the formulae that follow are also applicable to curved paths.

Linear displacement

Linear motion can be generally described as a change in the position of an object, or as the passage of an object from one place to another along its path. **Linear displacement**, S, of an object is a measure of its change of position along the path of its motion with respect to an arbitrary fixed point.

Displacement is a vector quantity, i.e. it has direction as well as magnitude. For our purposes, it is usually convenient to measure displacement from the initial position of the object, in the direction of motion which is assumed to be positive, in which case displacement is also a measure of the distance travelled by the object.*

The magnitude of linear displacement can be measured in metres, kilometres or millimetres, depending on the actual distance measured. However, when relating displacement to velocity, acceleration, and especially to force, work and power, it is usually necessary to convert to base units, i.e. metres, before any further calculations are performed.

Linear velocity

Motion is a change in the position of an object which occurs in time. The time rate of change in the position of an object is known as its **linear velocity**, v. As such, velocity is a vector quantity, implying a magnitude and direction.

For motion which occurs in one direction only, as is the case in the majority of problems treated in this book, the directional sense of velocity coincides with that of displacement and is taken to be positive. If reversal in the direction of motion occurs, such as when an object is thrown upwards, reaches a maximum altitude and starts falling, problems can easily be solved by considering the different stages of motion separately.**

For linear motion in one direction, velocity at any point can best be understood as the distance travelled by an object per unit time in a specified direction along its linear path. The base unit of velocity is metre per second (m/s). However, the original unit of measurement commonly used to express vehicle speeds is kilometre per hour (km/h), which is not decimally related to

---

* One should be aware that if reversal in the direction of motion occurs, e.g. in reciprocating or oscillatory motion, the displacement from a fixed origin as defined above is not equivalent to the total distance travelled by the object.

** In more advanced problems, involving oscillatory and curvilinear motion, it is usually necessary to treat velocities as vectors observing appropriate sign convention and the rules of vector mathematics.

the metre per second $\left(1 \text{ km/h} = 1000 \text{ m}/3600 \text{ s} = \frac{1}{3.6} \text{ m/s}\right)$. It is usually necessary to convert velocity from kilometres per hour to metres per second, particularly if calculations involve other related concepts, such as acceleration, power or momentum.

Note that speed is not a synonym of velocity, but a scalar quantity which expresses the magnitude of velocity only without any reference to its direction. In your car, the speedometer reading tells you how fast you are travelling at any instant of time without any regard to the direction of your travel. Although in many situations the meaning is obvious, one should not regard the terms speed and velocity as equivalent.

If distances travelled in successive intervals of time are the same, the speed is said to be constant. Otherwise average speed can be calculated using total distance and the time taken to cover the distance.

Example 8.1

In the last Grand Prix race of the series on his way to the World Drivers' Championship title in 1980, Alan Jones of Australia established a new lap record for the 6 km road course at Watkins Glen, NY, by covering the distance in 1 minute and 43.8 seconds.

What was his speed in metres per second? What was it in kilometres per hour?

Solution

Distance $S = 6 \text{ km} = 6000 \text{ m}$

Time $t = 103.8 \text{ s}$

Speed $\frac{S}{t} = \frac{6000 \text{ m}}{103.8 \text{ s}} = 57.8 \text{ m/s}$

Converting to km/h,

Speed $= 57.8 \text{ m/s} \left[\frac{3600}{1000}\right] = 208.1 \text{ km/h}$

Linear acceleration

If the velocity is not constant, but increasing gradually at a uniform rate, an object is said to be moving with a uniformly accelerated motion.* The rate at which linear velocity is changing with time is called **linear acceleration,** a.

If over a period of time equal to t seconds, velocity of an object changes from its initial value v_0 to a final value v, it follows from the definition that acceleration is the quotient of the increment of velocity $(v - v_0)$ and the time t:

$$a = \frac{v - v_0}{t}$$

* The more general case of non-uniform acceleration is outside the scope of this book.

It can readily be seen that the unit of acceleration must be the unit of velocity, m/s, divided by the unit of time, s, i.e. metre per second squared, m/s^2.

The relationship between initial and final velocities, time and acceleration for uniformly accelerated motion is usually stated as a formula in which final velocity is the subject

$$v = v_0 + at$$

where v is final velocity, in m/s
v_0 is initial velocity, in m/s
a is acceleration, in m/s^2
t is time taken, in s

Example 8.2

A car starts from rest and accelerates at the rate of 1.2 m/s^2 for 15 seconds. Determine the velocity reached after 15 seconds.

Solution

Initial velocity $v_0 = 0$
Time $t = 15$ s
Acceleration $a = 1.2$ m/s^2
Substitute into $v = v_0 + at$
$v = 0 + 1.2 \text{ m/s}^2 \times 15 \text{ s}$

Hence, velocity after 15 seconds is

$$v = 18 \text{ m/s } (= 64.8 \text{ km/h})$$

If instead of increasing velocity is gradually decreasing, the motion is said to be uniformly decelerated. Deceleration, or retardation, can be regarded as negative acceleration, i.e. acceleration acting in the direction opposite to velocity.

Example 8.3

If after travelling for some distance at constant velocity of 18 m/s, brakes are applied to the car producing a retardation of 2 m/s^2, determine the time taken to reduce its velocity to 10 m/s (36 km/h).

Solution

Initial velocity $v_0 = 18$ m/s
Final velocity $v = 10$ m/s
Acceleration $a = -2$ m/s^2
Substitute into $v = v_0 + at$
$10 = 18 - 2t$

Hence, time taken $t = \dfrac{10 - 18}{-2} = 4$ s

8.2 Equations of linear motion

In the case of uniformly accelerated linear motion, the distance travelled from the starting point is the product of time and average velocity:

$$S = t \times v_{av}$$

where simple arithmetic averaging of velocities gives:

$$v_{av} = \frac{v_0 + v}{2}$$

When these equations are combined, we have:

$$S = t\left(\frac{v_0 + v}{2}\right)$$

which is an additional independent equation to:

$$v = v_0 + at$$

Eliminating final velocity v from these equations yields:

$$S = v_0 t + \frac{at^2}{2}$$

Similarly, if time t is eliminated we get:

$$2aS = v^2 - v_0^2$$

Given any three of the five variables, i.e. displacement, time, acceleration, initial and final velocities, any problem involving uniformly accelerated linear motion can be solved using these equations.

Example 8.4

Find the total emergency stopping distance of a car and the total time taken from the point where the driver sights danger, if the driver's reaction time before applying the brakes is 0.9 s, the initial velocity is 60 km/h and retardation due to the brakes is 7.5 m/s^2. (See Fig. 8.1.)

Solution

There are two stages in this problem, (a) motion with uniform velocity before the brakes are applied and (b) uniformly decelerated motion that brings the car to rest. Let us consider them separately.

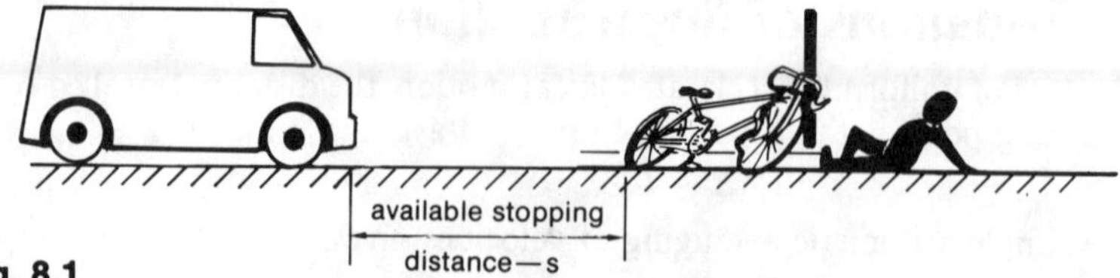

Fig. 8.1

(a) Uniform motion

Velocity (constant) $v = 60 \text{ km/h} \frac{1000}{3600} = 16.67 \text{ m/s}$

Time $t = 0.9 \text{ s}$

Displacement $S = t \times v_{av} = 0.9 \times 16.67 \text{ m/s} = 15 \text{ m}$

(b) Decelerated motion

Initial velocity $v_0 = 16.67 \text{ m/s}$

Final velocity $v = 0$

Acceleration $a = -7.5 \text{ m/s}^2$

(N.B. Retardation is negative acceleration.)

Substitute into $v = v_0 + at$ to find time taken to bring the car to rest:

$$0 = 16.67 + (-7.5) \times t$$

$$t = 2.22 \text{ s}$$

Substitute into $S = t \dfrac{v_0 + v}{2}$ to find the displacement during the period of decelerated motion:

$$S = 2.22 \times \left(\frac{16.67 + 0}{2}\right) = 18.5 \text{ m}$$

Combining the answers obtained in (a) and (b) yields:

Total stopping distance = 15 + 18.5 = 33.5 m

Total time taken = 0.9 + 2.22 = 3.12 s*

8.3 Freely falling bodies

In ancient times, ignorant of the effects of air resistance, an early Greek scientist-philosopher, Aristotle, had observed a leaf and a stone fall to the ground and had come to the general conclusion that a light body falls more slowly than a heavy one. Such was the appeal of the apparently irrefutable observation and the extraordinary influence of this outstanding thinker, that it took almost two thousand years and the perseverance of Galileo to dispel this erroneous principle.

* There must be a lesson in these results not only for the student of mechanics but also for the road safety conscious as well.

Galileo suspected that, if falling bodies were heavy enough to neglect air resistance, Aristotle's idea was demonstrably wrong. He undertook a series of experiments which led him to the correct conclusions that the maximum speed attained by a falling body, in the absence of air resistance, is proportional to the time taken and that the distance travelled from rest is proportional to the square of the time. This is consistent with the equations of linear motion if we take the initial velocity to be zero, $v_0 = 0$, and the acceleration due to gravity as a_g.

$$v = v_0 + at \quad \text{gives} \quad v = a_g t$$

$$\text{and } S = v_0 t + \frac{at^2}{2} \quad \text{gives} \quad S = \frac{a_g t^2}{2}$$

Experiments conducted since the time of Galileo, combined with the principles expounded by Newton, set the value of gravitational acceleration at:

$$\boxed{a_g = 9.81 \text{ m/s}^2}$$

This is more or less constant anywhere on or near the surface of the Earth.*

Example 8.5

A stone is dropped from the deck of a bridge and strikes the water below after 3.4 seconds of free fall. Neglecting air resistance, calculate the height of the bridge above water and the velocity with which the stone strikes the water.

Solution

Since the initial velocity is zero, $v_0 = 0$ and $a_g = 9.81 \text{ m/s}^2$, the distance of free fall is found from:

$$\begin{aligned} S &= v_0 t + \frac{at^2}{2} \\ &= 0 + \frac{9.81 \times 3.4^2}{2} \\ &= 56.7 \text{ m} \end{aligned}$$

The final velocity is found from:

$$\begin{aligned} v &= v_0 + at \\ &= 0 + 9.81 \times 3.4 \\ &= 33.4 \text{ m/s} \end{aligned}$$

If a body such as a stone is projected vertically upwards with an initial velocity v_0, the acceleration due to gravity will be negative, i.e. it will be a retardation causing a decrease in velocity at the rate of 9.81 m/s^2.

---

* The exact numerical value of gravitational acceleration is equal to the value of the gravitational constant and is, therefore, subject to local variations to the same extent (see Ch. 3). The relationship between gravitational acceleration and gravitational constant is illustrated in Example 8.8.

Example 8.6

If a stone is thrown upwards from the edge of the bridge in the previous example with an initial velocity of 15.6 m/s, determine the maximum height reached above the deck of the bridge, the velocity with which it strikes the water below and the total time taken.

Solution

Refer to Figure 8.2.

Fig. 8.2

(a) Consider the upward motion first.

Initial velocity $v_0 = 15.6$ m/s

Acceleration $a_g = -9.81$ m/s^2

Final velocity at point A $v = 0$

Substitute into $2aS = v^2 - v_0^2$

$$2 \times (-9.81) \times S = 0 - 15.6^2$$

$$S = 12.4 \text{ m}$$

Therefore the maximum height above the deck is 12.4 m.

Now using $v = v_0 + at$,

$$0 = 15.6 + (-9.81)\, t$$

Therefore, time taken to reach this height is:

$$t = 1.59 \text{ s}$$

(b) Consider free fall from the maximum height into the water below. Measured from point A as the origin,

Displacement $S = 56.7 \text{ m} + 12.4 \text{ m} = 69.1 \text{ m}$

Initial velocity $v_0 = 0$

Acceleration $a_g = 9.81$ m/s^2

Substitute into $S = v_0 t + \dfrac{at^2}{2}$

$$69.1 = 0 + \frac{9.81 t^2}{2}$$

Therefore, time taken in free fall is:

$$t = \sqrt{\frac{69.1 \times 2}{9.81}} = 3.75 \text{ s}$$

Now using $v = v_0 + at$, velocity when striking water is equal to:

$$v = 0 + 9.81 \times 3.75 = 36.8 \text{ m/s}$$

Combining the time taken for each part of this motion, total time from the moment the stone is thrown upwards to the moment it splashes into water is:

$$t_{tot} = 1.59 \text{ s} + 3.75 \text{ s} = 5.34 \text{ seconds}$$

It should be noted that since several alternative equations are available (see Section 8.2), there is usually more than one way in which a particular problem can be solved. The choice of a suitable sequence of calculations depends not only on the given information, but also on the selection of particular equations, which to some extent is a matter of personal preference.

Problems

8.1 When Herb Elliott of Australia established an Olympic record in the 1500 m event in 1960 in Rome, his time for the distance was 3 minutes and 35.6 seconds. What was his average speed?

8.2 What average speed should you travel at to cover a distance from Brisbane to Sydney, equal to 1020 km, in 12 hours? Express your answer in kilometres per hour and in metres per second.

8.3 On a forward journey a distance of 120 km was covered at an average speed of 100 km/h and on the return journey at an average speed of 60 km/h. What was the average speed for the total journey?

8.4 Determine the acceleration required to increase the velocity of a motorcycle from 60 km/h to 100 km/h in 5 seconds.

8.5 A rocket is launched from rest with a constant upward acceleration of 18 m/s^2. Determine its velocity after 25 seconds.

8.6 Calculate the altitude reached by the rocket in the previous problem after 25 seconds.

8.7 A car travelling at 47 km/h accelerates to 97 km/h in a distance of 260 metres. Calculate the acceleration and the time taken.

8.8 A train is moving at 56 km/h. If it is to pull up in 200 m, what must be the retardation and the time taken?

8.9 A launch approaches a wharf at a speed of 10.5 knots* (= 5.4 m/s). The engines are put in reverse when the launch is 10 metres from the wharf, the retardation being 1.25 m/s^2. Will the launch come to rest before hitting the wharf and if not, with what velocity will it strike the wharf?

---

* A knot, originally defined as one nautical mile per hour, is a non-metric unit of speed, equal to 1.852 km/h, which is allowable in Australia for restricted application in marine and aerial navigation.

8.10 A train accelerates from rest at station A at a rate of 0.8 m/s² for 25 seconds, then travels at constant velocity for 77.5 seconds before it comes to rest at station B after a period of retardation lasting 20 seconds. What is the distance between stations A and B and the average velocity of the train between the stations?

8.11 One train starts from rest and has an acceleration of 0.8 m/s². A second train has an initial velocity of 14.3 m/s and a retardation of 0.5 m/s². After how many seconds are they moving with the same velocity?

8.12 If the first train in the previous problem starts at the instant the second train passes it on parallel track and they move in the same direction, after how many seconds will the first train overtake the second train and at what distance from the original point?

8.13 Determine the time taken and the final velocity when a hammer dropped accidentally from a height of 49 m on a construction site hits the ground.

8.14 A ball is kicked vertically upwards with a velocity of 35 m/s. Determine the greatest height and the time taken to reach it. Neglect air resistance.*

8.15 A rocket fired vertically burns its fuel for 25 seconds, which provides a thrust producing a constant upward acceleration of 8 m/s². Determine the altitude reached at the instant the rocket runs out of fuel and the maximum altitude reached by the rocket. Determine also the total time, ground to ground, including free fall. Neglect air resistance.

8.4 Force, mass and acceleration

The fundamental relationship between force, mass and acceleration is the central concept of dynamics. It is embodied in the system of mechanics summarised by Isaac Newton in his three Laws of Motion.

Slightly reworded, these laws are as follows:

1. *First Law:* If there is no unbalanced force acting on a body, the body will remain at rest or continue to move in a straight line with a constant linear velocity.
2. *Second Law:* The acceleration of a body is proportional to the resultant force acting on it, inversely proportional to the mass of the body, and is in the direction of the force.
3. *Third Law:* The forces of action and reaction are equal in magnitude and opposite in direction.

* In reality, motion of a relatively light object, such as a soccer ball, could be retarded quite significantly by air resistance. However, at this stage, the influence of air resistance will be ignored.

Taken in a different order these laws take us through the basic principles of kinetics as follows.

The Third Law suggests that forces acting on a body always originate in other bodies, i.e. a push or pull experienced by an object is always a result of interaction with some other object. Single isolated forces do not exist. The First Law introduces the property of matter, often called inertia, that determines its resistance to a change in its state of rest or uniform motion. Mass of a body is then regarded as the quantitative measure of its inertia, which only a force can overcome. The Second Law gives mathematical expression to the relationship between the mass of a body, the unbalanced force acting on it and the acceleration produced by the force:

$$F = m \times a$$

where F is resultant of all forces acting on a body
m is mass of the body
a is acceleration

It is very important to understand that the International System (SI) of units is a coherent system in which the product of any two unit quantities in the system is the unit of the resultant quantity. This applies to Newton's Second Law of motion for which the product of the SI units of mass and acceleration, i.e. kilogram and metre per second squared, is the SI unit of force, $\mathrm{kg.m/s^2}$, which is given the special name **newton** (symbol, N). Thus, the Second Law provides the definition of the unit of force:

$$F = m.a$$

$$1 \text{ N} = 1 \text{ kg} \times 1 \text{ m/s}^2$$

$$\text{N} = \frac{\text{kg.m}}{\text{s}^2}$$

Let us now consider a few examples of the applications of the Second Law of motion.

Example 8.7

Determine the net force required to give a body of 300 kg mass a horizontal acceleration of 2.5 $\mathrm{m/s^2}$.

Solution

$$F = m \times a = 300 \text{ kg} \times 2.5 \text{ m/s}^2 = 750 \text{ kg.m/s}^2 = 750 \text{ N}$$

Example 8.8

Determine the acceleration of a body of 25 kg mass due entirely to its own weight.

Solution

The weight of the body, i.e. the force of gravity acting on it, is:

$$F_w = m.g = 25 \text{ kg} \times 9.81\frac{\text{N}}{\text{kg}} = 245.3 \text{ N}$$

Therefore, from $F = m \times a$, acceleration due to gravity is:

$$a = \frac{F}{m} = \frac{245.3 \text{ N}}{25 \text{ kg}} = \frac{245.3 \text{ kg.m/s}^2}{25 \text{ kg}} = 9.81 \text{ m/s}^2$$

Obviously, this only proves what we already knew about acceleration of a body under the action of gravity only, i.e. freely falling bodies. Furthermore, it is apparent that acceleration due to gravity is independent of mass.

Example 8.9

Determine the acceleration of a body sliding down a smooth surface inclined to the horizontal at 35°.

Solution

It is necessary to consider components of weight acting along and at right angles to the surface. Let the weight be $F_w = m.g$. The perpendicular component of weight is balanced by the normal reaction at the surface, as shown in Figure 8.3.

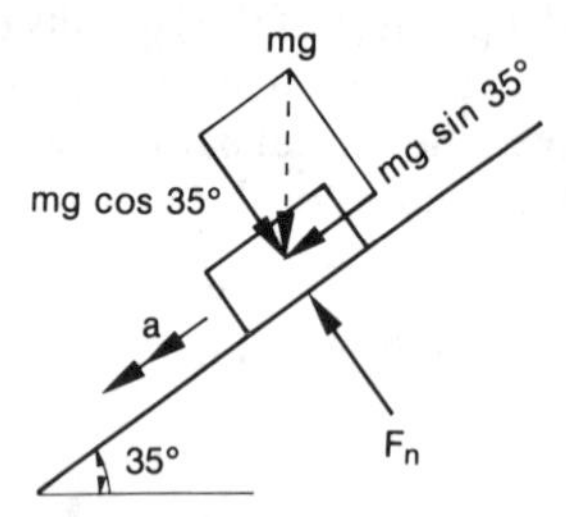

Fig. 8.3

The parallel component equal to *mg* sin 35° is not balanced by any other force, assuming no frictional resistance, i.e. a smooth surface.

Therefore, from $F = m \times a$

$$a = \frac{F}{m} = \frac{mg \sin 35°}{m} = 9.81 \times \sin 35° = 5.63 \text{m/s}^2$$

The acceleration is in the direction of the net unbalanced force, i.e. parallel to the plane as shown.

Example 8.10

Determine the force required to accelerate a vehicle of 1.5 tonne mass from rest to 60 km/h in 12 seconds.

Solution

The acceleration required is found from $v = v_0 + at$, where $v = 60 \text{ km/h} = 16.67 \text{ m/s}$.

$$16.67 = 0 + a \times 12$$
$$a = 1.389 \text{ m/s}^2$$

Applying Newton's Law,

$$F = m.a = 1500 \text{ kg} \times 1.389 \text{ m/s}^2 = 2083 \text{ N} = 2.08 \text{ kN}$$

8.5 Acceleration against resistance

It has already been stated that, in the equation $F = ma$, the force is net accelerating force, i.e. the resultant of all forces applied to the body. The resultant unbalanced force, i.e. net accelerating force, is the difference between the applied push or pull F_p and the resistance force F_r.

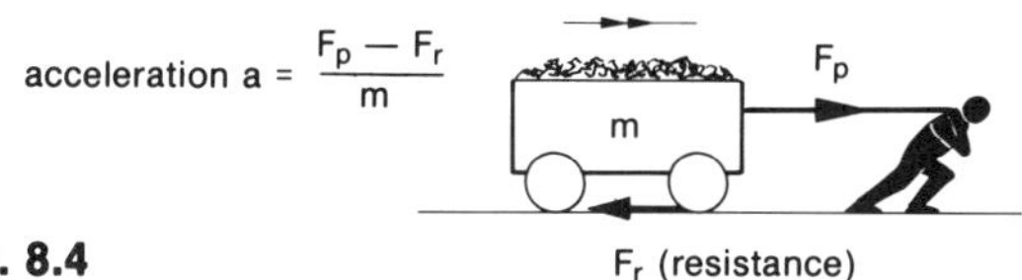

Fig. 8.4

$$F = F_p - F_r$$

The resistance is usually due to friction of some kind, which may include sliding friction, friction in bearings and air friction.

Acceleration produced by the resultant force is found from Newton's Law (see Fig. 8.4).

$$a = \frac{F}{m} = \frac{F_p - F_r}{m}$$

It should be clear from this equation that if the applied force F_p is equal to the resistance force F_r, there will be no acceleration, i.e. the motion, if any, will continue at constant velocity. Thus a vehicle moving on level road at constant speed requires a force equal to the tractive resistance to maintain uniform motion. Any force in excess of tractive resistance will accelerate the vehicle. On the other hand, a tractive effort which is less than tractive resistance will cause retardation.

Example 8.11

A train of total mass 120 tonnes is travelling at 60 km/h on level track. The tractive resistance is 80 newtons per tonne. Calculate the tractive effort required to accelerate the train to 100 km/h in 30 seconds.

Solution

Acceleration must be calculated using $v = v_0 + at$, where $v = 100$ km/h $= 27.78$ m/s and $v_0 = 60$ km/h $= 16.67$ m/s.

$$27.78 = 16.67 + a \times 30$$
$$a = 0.37 \text{ m/s}^2$$

The net accelerating force required to accelerate the mass is:

$$F = m.a = 120\ 000 \text{ kg} \times 0.37 \text{ m/s}^2 = 44\ 400 \text{ N} = 44.4 \text{ kN}$$

Tractive resistance, which is often stated as resistance per unit mass of the vehicle, is equal to:

$$F_r = 80 \frac{\text{N}}{t} \times 120\ t = 9600 \text{ N} = 9.6 \text{ kN}$$

Therefore, from the balance of forces,

$$F = F_p - F_r$$

$$44.4 \text{ kN} = F_p - 9.6 \text{ kN}$$

the required tractive effort F_p is:

$$F_p = 44.4 + 9.6 = 54 \text{ kN}$$

Example 8.12

Determine the time taken and the distance travelled by the train in the previous example when it is brought to rest from 100 km/h by a braking force of 72.7 kN.

Solution

If tractive resistance F_r is unchanged:

$$F_r = 9.6 \text{ kN}$$

In this case the resistance will assist the braking effort F_b in slowing the train down. Therefore the total decelerating force will be:

$$F = F_r + F_b = 9.6 + 72.7 = 82.3 \text{ kN}$$

Deceleration is therefore equal to:

$$a = \frac{F}{m} = \frac{82\ 300 \text{ N}}{120\ 000 \text{ kg}} = 0.686 \text{ m/s}^2$$

Remembering that deceleration is negative acceleration, we can substitute into $2aS = v^2 - v_0^2$ and solve for distance travelled during the period of retardation.

$$2aS = v^2 - v_0^2$$

$$2 \times (-0.686)\ S = 0 - 27.78^2$$

$$S = 562 \text{ m}$$

The time taken can be found from:

$$S = t\left(\frac{v + v_0}{2}\right)$$

$$562 = t\left(\frac{0 + 27.78}{2}\right)$$

$$t = 40.5 \text{ seconds}$$

8.6 Acceleration against gravity

When motion occurs in a vertical direction, such as that of a lift, the force required to produce acceleration is influenced by the force of gravity, i.e. by the weight, F_w, of the object.

The resultant unbalanced force in this case is the difference between the applied force, F_p, which is often provided by the pull in a cable, and the force of gravity F_w (see Fig. 8.5).

$$F = F_p - F_w$$

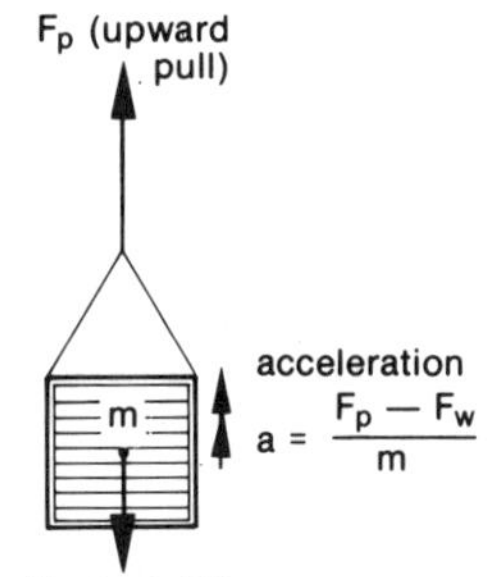

Fig. 8.5 Fw (weight)

Gravity in this case can be regarded as the resistance to upward acceleration. The acceleration produced by the resultant force is found from Newton's Law.

$$a = \frac{F}{m} = \frac{F_p - F_w}{m}, \text{ where } F_w = m.g$$

It follows that, if the applied pull F_p is equal to the weight F_w, there will be no acceleration, i.e. the object will remain in a state of rest or, if moving, will continue to move with constant velocity.

A force F_p greater than the weight will produce an upward acceleration, while a force less than the weight will allow downward acceleration.

Example 8.13

A loaded lift has a total mass of 1500 kg. Determine the force in the cables when:

(a) the acceleration of 2 m/s^2 is upwards.
(b) the acceleration of 2 m/s^2 is downwards.

Solution

Net accelerating force is:

$$F = m.a = 1500 \text{ kg} \times 2 \text{ m/s}^2 = 3000 \text{ N} = 3 \text{ kN}$$

The force of gravity (weight) of the lift is:

$$F_w = m.g = 1500 \text{ kg} \times 9.81 \frac{\text{N}}{\text{kg}} = 14\,715 \text{ N} = 14.7 \text{ kN}$$

When acceleration is upwards

$$F = F_p - F_w$$
$$3 \text{ kN} = F_p - 14.7 \text{ kN}$$

Hence, force in the cable $F_p = 17.7$ kN.

When acceleration is downwards, weight is greater than the force in the cable:

$$F = F_w - F_p$$
$$3 \text{ kN} = 14.7 \text{ kN} - F_p$$

Hence, force in the cable $F_p = 14.7 - 3 = 11.7$ kN.

Problems

8.16 Calculate the acceleration of a car of mass 1.2 tonne on level road if the net accelerating force is 6 kN.

8.17 A boy and his bicycle have a mass of 45 kg. When travelling at 20 km/h on a level road, he ceases to pedal and finds his speed reduced to 16 km/h in 10 seconds. What is the total force resisting motion?

8.18 A planing machine table has a mass of 450 kg and must be accelerated from rest to a velocity of 0.35 m/s in 100 mm. Determine the net accelerating force required.

8.19 When a ship of 6000 tonnes is launched, it leaves the slip with a velocity of 2.5 m/s and is then stopped within a distance of 150 m from the slip. What is the magnitude of the force that brings her to rest?

8.20 A trolley of 100 kg mass is to be pulled up a smooth incline at 10° to the horizontal with an acceleration of 2.5 m/s^2. What force parallel to the incline is required if friction at the wheels is negligible?

8.21 A locomotive exerts a pull of 30 kN on a train whose total mass is 200 tonnes. If total tractive resistance is 8 kN on level track, how long will it take to reach a velocity of 50 km/h, starting from rest?

8.22 The train in the previous problem reaches an upward slope of 4° to the horizontal. Determine the pull that must be exerted by the locomotive to maintain a steady velocity of 50 km/h. Assume that tractive resistance remains the same.

8.23 A train of mass 500 tonnes is drawn by a locomotive of mass 50 tonnes, which exerts a tractive effort of 75 kN, while tractive resistance is 60 N/t.

Determine the time and distance required to reach a velocity of 80 km/h from rest, on level track.

8.24 A lift cage, together with its load, has a mass of 2000 kg and is raised vertically by a cable. Determine the pull exerted by the cable if the lift reaches a velocity of 5 m/s after rising 5 m from rest.

8.25 A lift has a mass of 1000 kg. Calculate the force in the lifting cable when the lift is:
(a) moving at constant velocity.
(b) moving upwards with an acceleration of 1.6 m/s^2.
(c) moving upwards with a retardation of 1.4 m/s^2.
(d) moving downwards with an acceleration of 1.0 m/s^2.

8.26 A hoist lowers a load of 3 tonnes at a uniform velocity of 2.4 m/s by means of a wire rope. Calculate the force in the rope required to bring the load to rest in 3 metres.

8.27 A man in a lift holds a spring balance graduated to read in newtons, with a 2 kg mass hanging from it. What will the readings of the spring balance be when the lift:
(a) accelerates upwards at 2 m/s^2?
(b) moves upwards with a constant velocity?
(c) moves upwards with a retardation of 2 m/s^2?

8.7 Systems of bodies in motion

Mechanical devices and simple machines usually consist of two or more components constrained to move together, a situation that can be described as a system of bodies in motion.

Such systems often contain cords or cables and pulleys used to connect the components and to transmit or change the direction of force and motion. It is common practice to neglect the weights and frictional resistance of cords and pulleys, unless specific information is available to describe their effect.

We will limit our discussion to systems involving two separate masses connected in a way which forces them to move simultaneously and imposes a fixed ratio on the distances moved by each mass.

The method of solving problems of this kind consists of first determining the relationship between aspects of motion of two masses, which depends on how they are interconnected, and then considering each mass as a separate free body. This produces a system of simultaneous equations, which can readily be solved.

Example 8.14

For the system of bodies shown in Figure 8.6(a), determine the acceleration and the force in the cord.

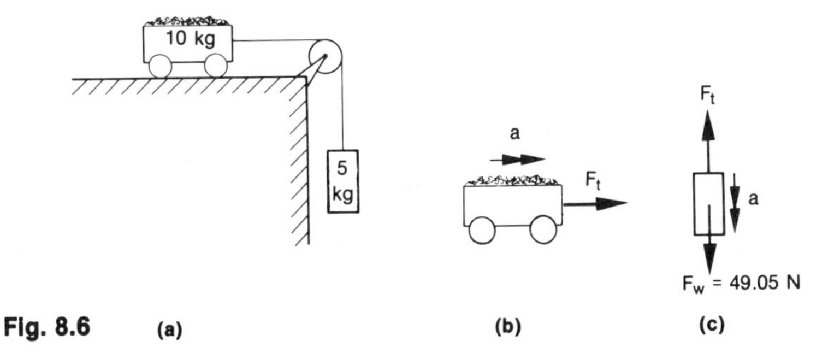

Fig. 8.6 **(a)** **(b)** **(c)**

Solution

In this system, when the 5 kg mass accelerates downwards, the 10 kg mass will accelerate to the right at the same rate because they are simply connected by the cord, running over a pulley. Let us call this acceleration, a. Let us also call the tension in the cord, F_t.

We are now ready to consider each mass as a separate free body.

For the 10 kg mass, we can write:

$$F = ma$$

$$\text{or } F_t = 10 \text{ kg} \times a \qquad (1)$$

For the 5 kg mass, the net accelerating force is the difference between its weight and tension in the cord. Therefore, $F = ma$ becomes

$$F_w - F_t = ma$$

$$\text{or } 5 \times 9.81 - F_t = 5 \times a \qquad (2)$$

Solving Equations 1 and 2 yields:

$$a = 3.27 \text{ m/s}^2 \quad \text{and} \quad F_t = 32.7 \text{ N}$$

Example 8.15

Determine the accelerations of bodies A and B and the force of tension in the cord, for the system in Figure 8.7(a).

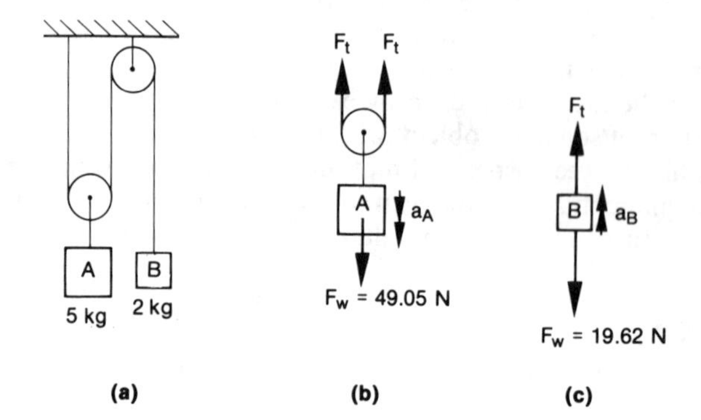

Fig. 8.7 (a) (b) (c)

Solution

In this case it is not immediately obvious in which direction motion will occur. This is decided by considering which body has a greater mass per fall of cord supporting it.* Body A is supported by two falls of cord and has a mass of 5 kg, i.e. 2.5 kg per fall. Body B has 2 kg per fall. Therefore, body A will accelerate downwards and body B upwards.

It is not hard to see that the displacement of A will be one half that of B. Acceleration will be in the same ratio:

$$a_A = \frac{a_B}{2}$$

---

* This rule is based on the principle that the motion, if any, will take place in the direction of the external pull that produces the greatest static tension in the cord, when all but one body in turn are held immovable while the static tension in the connecting cord is determined.

Considering free body diagrams yields:

for body B

$$F = m \times a$$

$$(F_t - F_{wB}) = m_B \times a_B$$

$$F_t - 2 \times 9.81 = 2 \times a_B \quad (1)$$

for body A

$$F = m \times a$$

$$(F_{wA} - 2F_t) = m_A \times a_A$$

$$5 \times 9.81 - 2F_t = 5 \times a_A$$

Substitute

$$a_A = \frac{a_B}{2}$$

$$5 \times 9.81 - 2F_t = \frac{5 \times a_B}{2} \quad (2)$$

After Equations 1 and 2 are simplified, we have

$$F_t - 19.62 = 2a_B$$

$$49.05 - 2F_t = 2.5a_B$$

Multiplying the first equation by 2 and then adding the left and right hand sides respectively eliminates F_t as follows:

$$2F_t - 39.24 = 4a_B$$

$$+$$

$$49.05 - 2F_t = 2.5a_B$$

$$9.81 = 6.5a_B$$

Hence,

$$a_B = \frac{9.81}{6.5} = 1.51 \text{ m/s}^2$$

It follows that:

$$a_A = \frac{a_B}{2} = \frac{1.51}{2} = 0.755 \text{ m/s}^2$$

and $F_t = 2a_B + 19.62 = 2 \times 1.51 + 19.62 = 22.6$ N

Problems

8.28 Determine the horizontal pull required to accelerate two wagons of mass 400 kg each along a level surface at the rate of 0.4 m/s², and the force in the bar connecting the two wagons (Fig. 8.8).

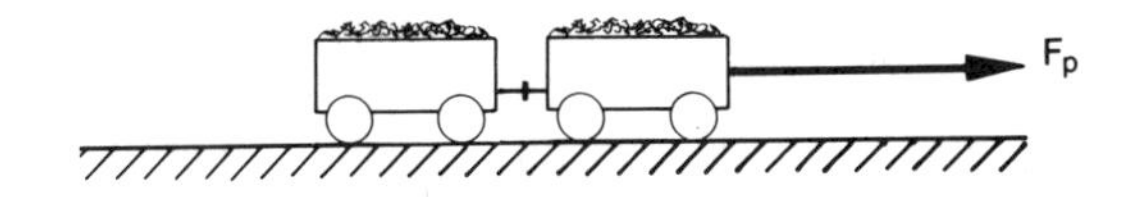

Fig. 8.8

8.29 A mass of 5 kg resting on a smooth horizontal table is connected by a fine string, passing over a smooth pulley on the edge of the table, with a mass of 100 g hanging freely. How far will the mass of 100 g descend in 3 seconds, starting from rest?

8.30 The trolley *A* shown in Figure 8.9 has a mass of 25 kg and rolls with negligible resistance. Determine the acceleration produced by mass *B* equal to 1 kg and the tension in the cord.

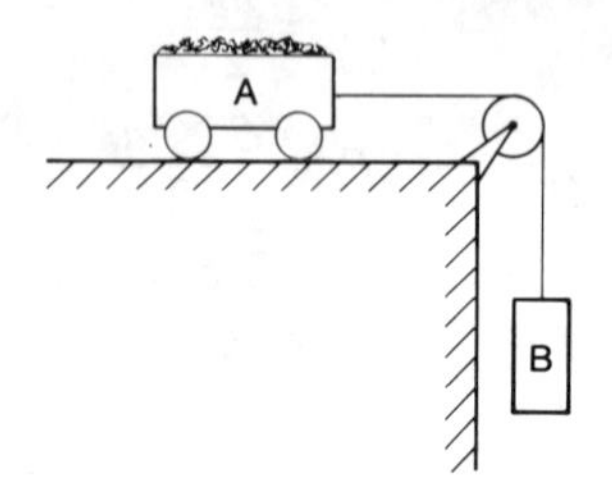

Fig. 8.9

8.31 If friction in the wheels of the trolley in problem 8.30 increases from lack of lubrication to produce a total tractive resistance of 3 N, what additional mass should be placed on block *B* to maintain the same rate of acceleration as in the previous problem?

8.32 In Figure 8.10, block *A* has a mass of 15 kg and the coefficient of friction between the block and the horizontal surface is 0.3. What is the mass of block *B* required to accelerate the system 2.5 m/s²?

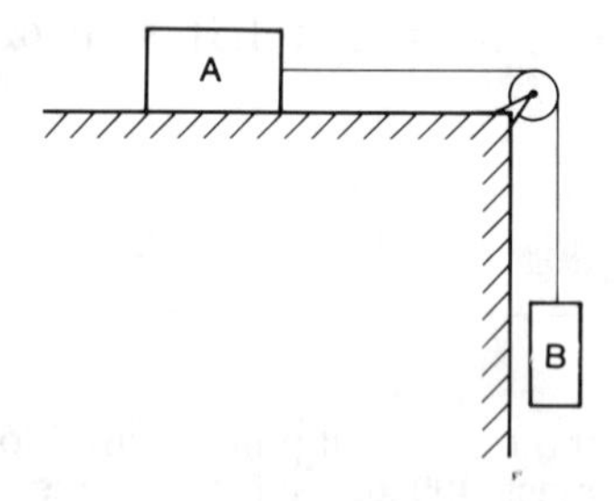

Fig. 8.10

8.33 Two masses of 1.0 kg and 1.1 kg are suspended by a fine string passing over a smooth pulley. With what acceleration will the masses move?

8.34 A 70 kg man hoists himself on a bosun's chair as shown in Figure 8.11. If the pull exerted by the man on the rope is 300 N, what is the acceleration?

Fig. 8.11

8.35 In the arrangement shown in Figure 8.12, the loaded trolley, having a total mass of 100 kg, can remain at rest or move without acceleration. Calculate the mass of the counterweight attached to the rope. If the mass of the trolley when empty is 40 kg, how long will it take to reach a velocity of 4 m/s starting from rest? Neglect frictional resistance.

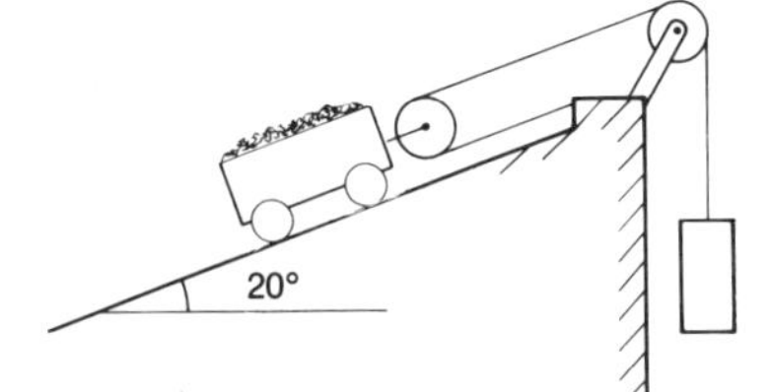

Fig. 8.12

Review questions

1. What is meant by
   (a) dynamics? (b) kinematics? (c) kinetics?
2. Define linear displacement, velocity and acceleration and state their symbols and units used.
3. What is implied by the term "uniformly accelerated motion"?
4. List four equations of linear motion.
5. What is the acceleration of a body in free fall?
6. State Newton's Laws of Motion.
7. What is the relationship between force acting on a body, the mass of the body and acceleration produced by the force?
8. What is the relationship between the newton and the kilogram?
9. Describe the method of solving problems involving a system of two bodies in motion.

9

Rotation

Up to this point we have considered linear motion, which could be described as motion in which the body moves from one place to another. Let us now discuss the case of **rotation**, or rotary motion, during which a body turns around a fixed axis in such a way that every particle of the body, except the axis, travels along a circular path. In rotary motion, the body as a whole does not move from one place to another. Common examples of rotation are flywheels, pulleys, shafts, gears, turbine rotors, and many others.

The most general type of motion is that where linear motion and rotation occur simultaneously, such as the motion of a bicycle wheel along a road. In most cases combined rotation and translation can be dealt with by separating the rotation from the linear motion and considering each separately. At the level for which this book is intended, problems involving combined rotation and translation will not be encountered. It will therefore help always to regard rotation as taking place about a fixed axis.

You may find that this chapter is in some ways very similar to the previous chapter dealing with linear motion. This is because there is a definite analogy between linear and rotational motion, so that previously discussed concepts of linear displacement, velocity and acceleration have their rotational equivalents, which can be described by relations similar to equations and laws of linear motion.

9.1 Angular displacement, velocity and acceleration

Let us now consider how rotational motion of an object, such as a flywheel, can be described in terms of its orientation in space and time. More specifically, rotation is usually described in terms of angular displacement, angular velocity and angular acceleration—the concepts which are defined below.

Angular displacement

When a body undergoes rotational motion its orientation in space changes. **Angular displacement** of a rotating object is a measure of its change of orientation with respect to a fixed radius as an arbitrary datum. This is usually denoted by the Greek letter θ (theta).

Angular displacement is a rotational vector quantity, i.e. it has direction as well as magnitude. For our purposes, it is usually convenient to measure angular displacement from the initial position of the object, in the direction of rotation which is assumed to be positive, in which case angular displacement is simply the angle through which the object turns.

The magnitude of angular displacement can be measured in revolutions, degrees or radians. Revolutions are the most convenient units for measuring angular displacement of mechanical components such as shafts and flywheels. However, when relating angular displacement to torque, work and power, it is necessary to convert to base units, i.e. radians, before any further calculations are performed. Remember that one revolution contains 360 degrees or 2π radians.

$$1 \text{ rev} = 360^\circ = 2\pi \text{ rad}$$

Angular velocity

The rate at which a body changes its angular position is called its **angular velocity**, usually denoted by the symbol ω.* Angular velocity is also a vector quantity, implying a magnitude and direction.

For rotational motion in a particular direction, the directional sense of angular velocity coincides with that of displacement and can be taken to be positive. The magnitude of angular velocity at any instant of time can be described as the angle turned through per unit time. The base unit of angular velocity is radian per second (rad/s). However, the most common practical unit of measurement used to express rotational speeds of mechanical components is revolution per minute (r.p.m.), which is not decimally related to the radian per second. It is normally required to convert angular velocity from revolutions per minute to radians per second,** particularly if calculations involve other related concepts, such as angular acceleration, energy or power.

If the angle turned through in each successive interval of time is the same, the angular velocity is said to be constant. Otherwise, average angular velocity can be calculated using total angular displacement and the time in which it occurs.

Example 9.1

A cam in a mechanism makes 500 revolutions in 2 minutes 37 seconds. What is its average angular velocity in revolutions per minute and in radians per second?

---

* Omega—the last letter of the Greek alphabet.

** To convert, multiply speed in r.p.m. by $2\pi/60$.

For example, 955 r.p.m. $= 955 \times \frac{2\pi}{60} = 100$ rad/s.

Solution

Total angular displacement $\theta = 500$ rev

Time $t = 157$ s $= 2.617$ min

Average angular velocity $\omega = \dfrac{\theta}{t} = \dfrac{500 \text{ rev}}{2.617 \text{ min}} = 191$ r.p.m.

Also equal to $\omega = \dfrac{500 \times 2\pi \text{ rad}}{157 \text{ s}} = 20$ rad/s

Angular acceleration

If angular velocity is not constant, but is increasing gradually at a uniform rate, an object is said to be rotating with a uniformly accelerated motion.* The rate at which angular velocity is changing with time is called **angular acceleration,** α.**

If over a period of time equal to t seconds, angular velocity of an object changes from its initial value ω_0 to a final value ω, it follows from the definition that angular acceleration is the quotient of the increment of angular velocity $(\omega - \omega_0)$ and the time, t.

$$\alpha = \frac{\omega - \omega_0}{t}$$

It can easily be seen that the unit of angular acceleration must be the unit of angular velocity, rad/s, divided by the unit of time, s, i.e. radian per second squared, rad/s^2.

The relationship between the initial and final angular velocities, time and angular acceleration for uniformly accelerated rotational motion is usually stated as a formula in which final angular velocity is the subject

$$\omega = \omega_0 + \alpha t$$

where ω is final angular velocity in rad/s
ω_0 is initial angular velocity in rad/s
α is angular acceleration in rad/s^2
t is time taken in seconds.

Note the similarity between this relationship and that for linear velocity and acceleration, $v = v_0 + at$.

Example 9.2

A flywheel starts from rest and is accelerated at the rate of 2.4 rad/s^2 for 30 seconds. Determine the angular velocity reached after 30 seconds.

Solution

Initial angular velocity $\omega_0 = 0$

Time $t = 30$ s

---

* The more general case of non-uniform angular acceleration is not covered in this book.
** This is alpha—the first letter of the Greek alphabet.

Angular acceleration $\alpha = 2.4 \text{ rad/s}^2$

Substitute into $\omega = \omega_0 + \alpha t$

$$\omega = 0 + 2.4 \text{ rad/s}^2 \times 30 \text{ s}$$

Hence, angular velocity after 30 seconds is

$$72 \text{ rad/s} (= 687.5 \text{ r.p.m.})$$

If instead of increasing angular velocity is gradually decreasing, the rotation is said to be uniformly decelerated. Angular deceleration, or retardation, is regarded as negative acceleration, i.e. angular acceleration acting in the direction opposite to angular velocity.

Example 9.3

If after rotating for some time at constant angular velocity of 72 rad/s, brakes are applied to the flywheel producing a retardation of 4 rad/s^2, determine the time taken to reduce its angular velocity to 40 rad/s (382 r.p.m.).

Solution

Initial angular velocity $\omega_0 = 72 \text{ rad/s}$

Final angular velocity $\omega = 40 \text{ rad/s}$

Angular acceleration $\alpha = -4 \text{ rad/s}^2$

Substitute into $\omega = \omega_0 + \alpha t$

$$40 = 72 - 4t$$

Hence, time taken $t = \dfrac{72 - 40}{4} = 8 \text{ s}$

9.2 Equations of rotational motion

In the case of uniformly accelerated rotational motion, the angular displacement from the initial position is the product of time and the average velocity:

$$\theta = t \times \omega_{av}$$

where simple arithmetic averaging of angular velocities gives:

$$\omega_{av} = \frac{\omega_0 + \omega}{2}$$

When these equations are combined, we have

$$\boxed{\theta = t\left(\frac{\omega_0 + \omega}{2}\right)}$$

which is an additional independent equation to:

$$\boxed{\omega = \omega_0 + \alpha t}$$

Eliminating final velocity, ω, from these equations yields:

$$\theta = \omega_0 t + \frac{\alpha t^2}{2}$$

Similarly, if time t is eliminated we get:

$$2\alpha\theta = \omega^2 - \omega_0^2$$

Note very carefully how these equations compare with those for linear motion in Section 8.2.

Any problem involving kinematics of rotational motion can be solved using these equations. However, great care must be exercised in use of appropriate units. It is usually better to make all necessary unit conversions before the substitution into appropriate equations.

Example 9.4

A flywheel turns at a constant angular velocity of 150 revolutions per minute for 45 seconds before a brake is used to bring it to rest with a retardation of 0.5 radians per second squared.

Determine the total time and the total angular displacement of the wheel.

Solution

There are two stages in this problem, motion with uniform angular velocity before the brake is applied and uniformly decelerated motion that brings the flywheel to rest. Each stage can be considered separately and the results combined.

(a) Uniform motion

Angular velocity (constant) $\omega = 150$ r.p.m.

$$= 150 \times \frac{2\pi}{60}$$

$$= 15.71 \text{ rad/s}$$

Time $t = 45$ s

Angular displacement $\theta = t \times \omega$

$$= 45 \text{ s} \times 15.71 \text{ rad/s} = 706.9 \text{ rad}$$

Also equal to 112.5 revolutions.

(b) Decelerated motion

Initial angular velocity $\omega_0 = 15.71$ rad/s

Final angular velocity $\omega = 0$

Angular acceleration $\alpha = -0.5$ rad/s^2

(Note the negative sign)

Substitute into $\omega = \omega_0 + \alpha t$ to find the time taken to bring the flywheel to rest:

$$0 = 15.71 - 0.5t$$
$$t = 31.4 \text{ s}$$

Substitute into $\theta = t\left(\frac{\omega_0 + \omega}{2}\right)$ to find the angular displacement during the period of decelerated motion:

$$\theta = 31.4\left(\frac{15.71 + 0}{2}\right) = 246.6 \text{ rad}$$

Also equal to 39.3 revolutions.
Combining the answers obtained in (a) and (b) yields:
Total angular displacement = 706.9 + 246.6
= 953.5 rad or 112.5 + 39.3 = 151.8 revolutions.

Total time taken = 45 s + 31.4 s = 76.4 s.

9.3 Relation between rotation and circular motion

It is useful to note the subtle difference between rotation of an object about an axis and linear motion of a point travelling in a circular path. In rotary motion, the object simply spins, usually around its own geometrical axis. In circular motion, a point or an object moves along a circumference of a circle at a fixed distance, r, from the centre of its circular path, such as a car driven around a curve of known radius.

There is, however, a relationship between angular terms which describe the rotary motion, i.e. angular displacement, velocity and acceleration, and their linear counterparts measured along the circumference. This relationship stems from the definition of a radian as the angle subtended at the centre of a circle by an arc equal in length to the radius. If we consider circular motion of a point P on a rotating disc at a radius r from the centre, as in Figure 9.1, it can be seen that:

$$S = r \times \theta$$

where S is linear displacement along the circumference
r is radius, in the same units as displacement
θ is corresponding angular displacement in radians.

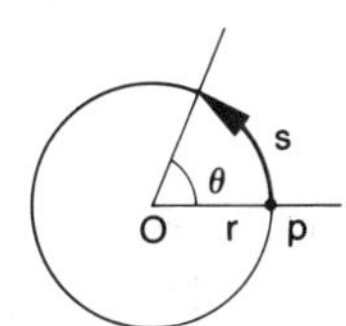

Fig. 9.1

Dividing by time, $\frac{S}{t} = r \times \frac{\theta}{t}$, produces the equation relating instantaneous angular velocity, ω, to linear velocity along the circular path, v:

$$v = r \times \omega$$

Likewise, from $\frac{v}{t} = r \times \frac{\omega}{t}$, it follows that accelerations are similarly related:

$$a = r \times \alpha$$

Example 9.5

On the former British passenger liner *Queen Mary* there were four gearboxes transmitting power from turbines to a propeller shaft rotating at 180 revolutions per minute at normal cruising speed. Each large driven gear was approximately 4 metres in pitch circle diameter.

For a point on the pitch circle of the gear, determine the linear velocity and the distance travelled along the circumference per revolution.

Solution

The radius of circular motion, $r = 2$ m.

Angular velocity is $\qquad \omega = 180 \times \frac{2\pi}{60} = 18.85$ rad/s

Angular displacement $\qquad \theta = 1$ revolution $= 2\pi$ rad

Substitute and solve:

Linear circumferential velocity

$$v = r \times \omega = 2 \text{ m} \times 18.85 \text{ rad/s} = 37.7 \text{ m/s}^*$$

Linear displacement per revolution is:

$$S = r \times \theta = 2 \text{ m} \times 2\pi \text{ rad} = 12.57 \text{ m}$$

The comparison between linear motion of a vehicle and the rotation of its wheels also depends on the above-mentioned relationships. We know that when a wheel rolls on a road surface, its axis is actually moving forward relative to the road. One can also visualise this situation from the driver's point of view in terms of a "fixed" axis of rotation, in relation to which the road surface is "moving backwards" under the wheel. In either case, the point of contact between the wheel and the road has momentarily a linear velocity v and an angular velocity ω, which are related by $v = r \times \omega$.

* Note that since the radian is dimensionless (see Chap. 2, Section 2.4) it may be used to compose units involving angular measure, and left out when transition to linear units is involved, as above, without disturbing dimensional homogeneity of an equation.

Example 9.6

Determine the speed of a car in km/h when its wheels, which are 650 mm in diameter, are rotating at 440 r.p.m. (see Fig. 9.2).

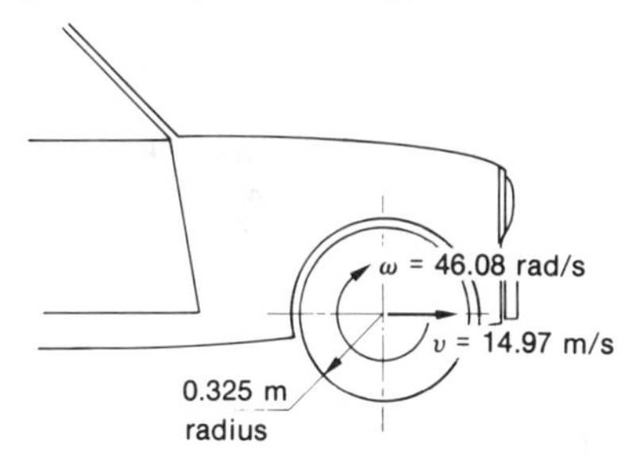

Fig. 9.2

Solution

Angular velocity $\qquad \omega = 440 \times \frac{2\pi}{60} = 46.08 \text{ rad/s}$

Radius of rotation $\qquad r = \frac{0.65}{2} = 0.325 \text{ m}$

Linear velocity $\qquad v = r \times \omega = 0.325 \text{ m} \times 46.08 \text{ rad/s}$

$$= 14.97 \text{ m/s}$$

Converting to kilometres per hour:

$$v = 14.97 \times \frac{3600}{1000} = 53.9 \text{ km/h}$$

Problems

9.1 What is the angular speed of each of the three hands of a clock expressed in revolutions per minute, radians per second?

9.2 A flywheel with an initial angular velocity of 35 rad/s is accelerated at 12 rad/s^2 to a final angular velocity of 275 rad/s. Determine the time taken and the angular displacement in radians during that time.

9.3 A rotor of a steam turbine revolving at 8000 r.p.m. slows down to 2000 r.p.m. in 15.7 seconds after steam supply has been adjusted. Determine the angular deceleration and the number of revolutions made by the rotor in that time.

9.4 The output shaft of an electric motor accelerates from rest, at the rate of 10 rad/s^2, to its final angular velocity in 15.2 seconds. What is the final angular velocity in rad/s and r.p.m.? Determine also the total number of revolutions made by the shaft in that period.

9.5 A mass at the end of a 700 mm pendulum moves through an arc of 20 degrees. Determine the angular displacement in radians and the corresponding linear displacement along the arc.

9.6 If at a particular instant the velocity of the mass in problem 9.5 is 0.3 m/s^2, and deceleration is 0.15 m/s^2, determine the corresponding angular velocity and deceleration in rad/s and rad/s^2 respectively.

9.7 A fan impeller is 800 mm in diameter and rotates at a constant angular velocity of 600 r.p.m. Determine the tip speed, i.e. circumferential velocity, of its blades.

9.8 A centrifugal pump is to be driven through a V-belt drive at a speed of 580 r.p.m. by an electric motor rotating at 1450 r.p.m. and the driver pulley is 100 mm diameter. Determine the linear velocity of the belts and the diameter of the driven pulley.

9.9 A car accelerates from rest at the rate of 1.2 m/s^2. Determine the angular acceleration, angular velocity and angular displacement of its wheels after 15 seconds, if the wheel diameter is 650 mm.

9.10 During a test on a large boring mill, its circular work table revolving in a horizontal plane was observed to accelerate from rest at the rate of 0.2 rad/s^2 for 18 seconds, then rotate at constant speed for 90 seconds before coming to rest after a period of retardation lasting 24 seconds. Determine the total number of revolutions made by the work table during the test.

9.4 Torque and rotational motion

The analogy between rotation and rectilinear motion can usefully be extended to the study of kinetics of rotational motion.

Just as it is with rectilinear motion, rotation, once started, will tend to continue at constant angular velocity unless an outside influence acts to increase or decrease the velocity. Thus it can be concluded that every physical body possesses a property of rotational inertia, which determines its resistance to a change in its state of rest or uniform rotational motion. The correct technical term to describe this property is **mass moment of inertia**, with a symbol, I.* Mass moment of inertia is a quantitative measure of the rotational inertia of a body, and as such is the rotational counterpart of the role played by mass in linear kinetics.

Mathematically, mass moment of inertia of a body with respect to its axis is a function of mass distribution in a body relative to the axis, measured in kilograms multiplied by square metres (kg.m^2). More about the methods of calculating mass moments of inertia of flywheels and similar objects can be found in Sections A3 and A4, Appendix A, at the end of this book.**

---

* Alternatively, J is sometimes used as a symbol for the mass moment of inertia. Both are acceptable symbols according to the International Standards Organisation.

** For the purposes of this chapter we shall note that the mass moment of inertia of a rigid disc or cylinder about its axis is given by $I = \frac{m.r^2}{2}$, where m is the mass and r is the radius of the disc (see Table A2 in Appendix A).

In rectilinear motion, force is the external agent capable of changing the conditions of rest or uniform linear motion of a body. In rotational kinetics, **torque** T is the analogue of force. Torque has previously been described as a pure turning effort measured in newton metres (N.m) or their multiples, such as kilonewton metres. (See Chap. 5.)

The relationship between mass moment of inertia of a rotating body, torque acting on it and the angular acceleration produced by the torque is identical to that between mass, force and acceleration in linear motion. If we substitute rotational terms, instead of linear terms, into Newton's second law equation, $F = ma$, we will have

$$T = I \times \alpha$$

where T is net unbalanced torque, N.m
I is mass moment of inertia, kg.m^2
α is angular acceleration, rad/s^2

Dimensional homogeneity of this equation is not immediately apparent, but can be demonstrated as follows

$$T = I \times \alpha$$

$$\text{N.m} = \text{kg.m}^2\ \text{rad/s}^2$$

After the metre cancels out and the radian is left out, as a dimensionless ratio, we have

$$\text{N} = \text{kg.m/s}^2$$

which is homogeneous by definition of the newton.

Example 9.7

Determine the net torque required to give a flywheel with a mass moment of inertia of 0.75 kg.m^2 an angular acceleration of 16 rad/s^2.

Solution

$$T = I \times \alpha = 0.75\ \text{kg.m}^2 \times 16\ \text{rad/s}^2 = 12\ \text{N.m}$$

Example 9.8

Determine the torque required to accelerate a turbine rotor undergoing a dynamic balancing test from rest to a speed of 15 000 r.p.m. in 80 seconds, if the mass moment of inertia of the rotor is 11.5 kg.m^2.

Solution

The angular acceleration required is found from $\omega = \omega_0 + \alpha t$, where $\omega = 15\,000 \times \frac{2\pi}{60} = 1571$ rad/s.

$$1571 = 0 + \alpha \times 80$$

$$\therefore \alpha = 19.63\ \text{rad/s}^2$$

Therefore, torque required:

$$T = I \times \alpha = 11.5\ \text{kg.m}^2 \times 19.63\ \text{rad/s}^2 = 225.8\ \text{N.m}$$

Example 9.9

Determine the angular acceleration of a flywheel in the form of a disc 400 mm diameter and having a mass of 60 kg, if the applied torque is 24 N.m.

Solution

The mass moment of inertia of the flywheel is:

$$I = \frac{mr^2}{2} = \frac{60\text{ kg} \times 0.2^2\text{ m}^2}{2} = 1.2\text{ kg.m}^2$$

Substituting into $T = I \times \alpha$ yields:

$$24\text{ N.m} = 1.2\text{ kg.m}^2 \times \alpha$$

Hence the angular acceleration is:

$$\alpha = \frac{24\text{ N.m}}{1.2\text{ kg.m}^2} = 20\text{ rad/s}^2$$

9.5 Acceleration against resistance

It is understood that in the equation $T = I \times \alpha$, the torque T is net accelerating torque, i.e. the resultant of all torques applied to the body. The resultant unbalanced torque, i.e. net accelerating torque, is the difference between the applied turning effort T_a and the resistance torque T_r (see Fig. 9.3).

$$T = T_a - T_r$$

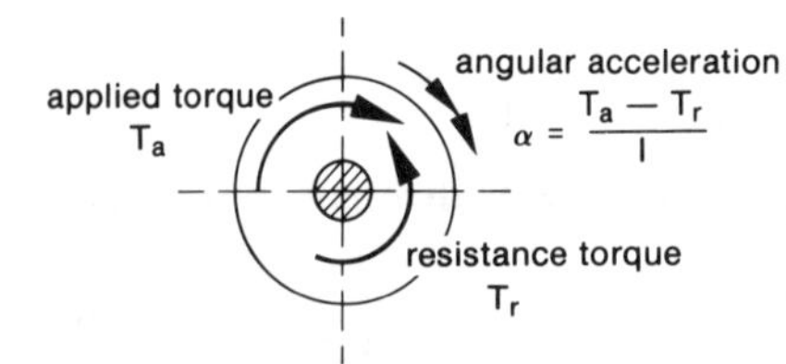

Fig. 9.3

The resistance is usually due to friction in bearings, axle friction, etc. Acceleration produced by the resultant torque is found from $T = I \times \alpha$.

$$\alpha = \frac{T}{I} = \frac{T_a - T_r}{I}$$

It follows from this equation that if the applied torque T_a is equal to the resistance torque T_r there will be no acceleration, i.e. the rotation, if any, will continue at constant angular velocity. Thus, a flywheel rotating at a constant speed requires a torque equal to the friction torque to maintain uniform rotation.

Any torque in excess of friction torque will accelerate the wheel. On the other hand, an applied torque which is less than friction torque will cause retardation.

Example 9.10

A flywheel of mass moment of inertia equal to 53 kg.m^2 is rotating at 300 r.p.m. The frictional resistance is 16 N.m. Calculate the torque that must be applied in order to accelerate the wheel to 500 r.p.m. in 15 seconds.

Solution

Angular velocity can be calculated using $\omega = \omega_0 + \alpha t$, where $\omega = 500 \times \dfrac{2\pi}{60} = 52.36$ rad/s and $\omega_0 = 300 \times \dfrac{2\pi}{60} = 31.42$ rad/s.

Substituting,

$$52.36 = 31.42 + \alpha \times 15$$

$$\alpha = 1.4 \text{ rad/s}^2$$

The net accelerating torque required to accelerate the wheel is

$$T = I \times \alpha = 53 \text{ kg.m}^2 \times 1.4 \text{ rad/s}^2 = 74 \text{ N.m}$$

Therefore, the total torque that must be applied is the sum of accelerating torque and resistance torque:

$$T_a = 74 \text{ N.m} + 16 \text{ N.m} = 90 \text{ N.m}$$

Problems

9.11 Calculate the angular acceleration of a cylindrical drum mounted on a shaft through its geometrical axis and subjected to a net accelerating torque of 60 N.m. The mass moment of inertia of the shaft and drum assembly is 12 kg.m^2.

9.12 A spindryer with a load of washing rotates at 200 r.p.m. when the power is turned off. It takes 2.5 minutes for it to come to rest. Given that the mass moment of inertia of the loaded spindryer is 1.72 kg.m^2, calculate the friction torque responsible for bringing the dryer to rest.

9.13 A wheel of a car has a moment of inertia of 1.6 kg.m^2. When tested for dynamic balance it is accelerated from rest to a speed of 1500 r.p.m. by a net torque of 20.6 N.m. Calculate the time taken and the number of turns made by the wheel to reach the final speed.

9.14 A flywheel in the form of a disc of uniform thickness, 500 mm in diameter and 150 kg mass, has frictional resistance of 5 N.m and an applied torque of 25 N.m acting on it. If it starts from rest, determine the angular velocity reached after one minute.

9.15 The flywheel in problem 9.14 rotates at a constant speed of 3000 r.p.m. It is to be brought to rest in 20 seconds by applying a force F to the brake lever as shown in Figure 9.4. If the coefficient of friction between the shoe and the wheel is 0.55, calculate the force F required. Assume that frictional resistance remains at 5 N.m.

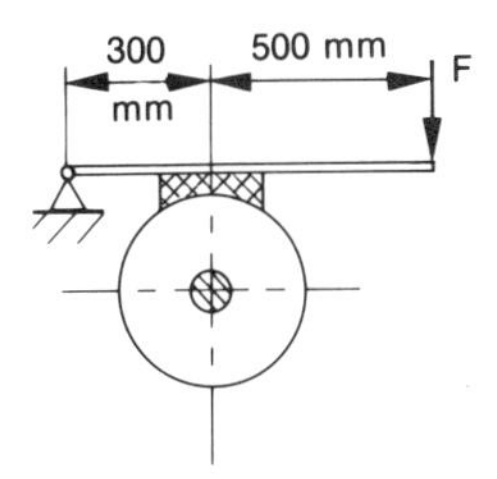

Fig. 9.4

9.16 A cast-iron pulley, 1.1 m in diameter, has a mass moment of inertia of 42.5 kg.m^2, and average resistance torque of 39 N.m. When accelerating from rest, the pulley is subjected to constant belt tensions of 800 N and 420 N on tight and slack sides of the belt respectively. Determine how long it takes to accelerate the pulley to 1375 r.p.m.

9.6 Systems of bodies in motion

Let us now consider systems of bodies in motion, which include rotating components. Our discussion will be limited to a pair of components constrained to move simultaneously and connected in a way which imposes a definite relationship between their motions.

The method of solving these problems is similar to the solution of systems of bodies in linear motion. It consists of first determining the relationship between aspects of motion of the two bodies, which depends on how they are interconnected, and then considering each component as a separate free body. This usually results in a system of simultaneous equations, which can readily be solved.

Example 9.11

The drum in Figure 9.5 has a mass moment of inertia of 25 kg.m^2. A mass of 2 kg is attached to the cord which is wrapped around the drum. Neglecting frictional resistance, determine the time taken for the mass to drop 2 m after being released from rest and the tension in the cord.

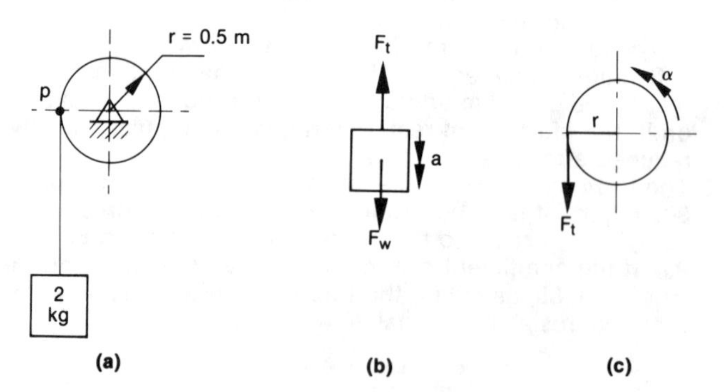

Fig. 9.5

Solution

In this system, when the 2 kg mass accelerates downwards, the drum will accelerate in a counter-clockwise direction at the rate related to the acceleration of the mass, because they are connected by the cord.

Point P on the cord has an instantaneous linear acceleration downwards, a, and an angular acceleration, α. These are related by

$$a = r.\alpha, \text{ or}$$

$$a = 0.5\alpha \qquad (1)$$

For the 2 kg mass as a free body we can write

$$F = ma$$

$$F_w - F_t = ma$$

where F_w is the weight, i.e. 2 kg $\times$ 9.81 $\frac{\text{N}}{\text{kg}}$ = 19.62 N

F_t is the tension in the cord

Therefore,

$$19.62 - F_t = 2a \qquad (2)$$

Likewise, for the drum, we write

$$T = I.\alpha$$

where T is the torque, i.e. $F_t \times r = F_t \times 0.5$
and $I = 25 \text{ kg.m}^2$

$$\text{or } F_t \times 0.5 = 25\alpha \qquad (3)$$

These three equations can easily be solved. For example, substitute $a = 0.5\alpha$ from Equation 1 into Equation 2

$$19.62 - F_t = 2 \times (0.5\alpha)$$

Now divide this new equation by Equation 3

$$\frac{19.62 - F_t}{0.5\ F_t} = \frac{2 \times 0.5\alpha}{25\alpha}$$

Simplifying and solving for F_t gives tension:

$$F_t = 19.24 \text{ N}$$

Substituting back into Equation 2 gives:

$$19.62 - 19.24 = 2a$$

Hence $\qquad a = 0.192 \text{ m/s}^2$

Now, $S = v_0t + \frac{at^2}{2}$ can be used to solve for time taken, since the distance dropped is S = 2 m and initial velocity is zero.

$$2 = 0 + \frac{0.192\ t^2}{2}$$

Hence, time taken is 4.56 seconds.

Problems

9.17 A mass of 12 kg is attached to a cord wrapped around a horizontal drum, of 0.8 m diameter, which has a mass moment of inertia of 18 kg.m². Neglecting friction, determine the tension in the cord and the downward acceleration of the mass after its release from rest.

9.18 If in the previous problem, bearing friction is equivalent to a resistance torque of 9 N.m, what will the acceleration of the mass be?

9.19 A hoist mechanism consists of a 450 mm diameter drum and a 950 mm diameter brake cylinder as shown in Figure 9.6. The mass moment of inertia of the rotating parts is 85 kg.m^2 and the coefficient of friction between the brake shoe and the cylinder is 0.6. Bearing friction is negligible. Determine the normal force F_N which must be applied to each shoe in order to lower a load of 400 kg at constant speed.

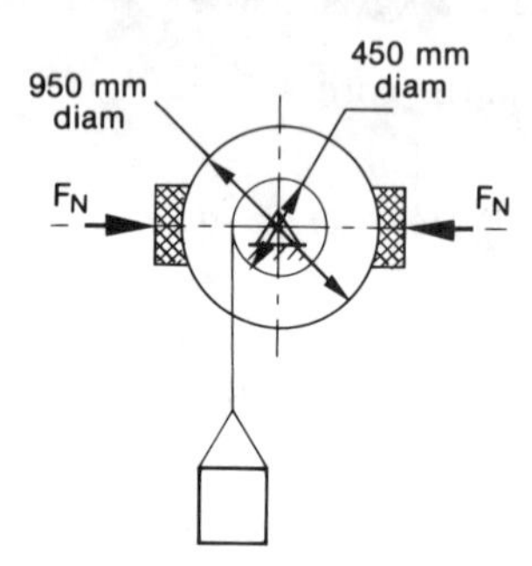

Fig. 9.6

9.20 If the brakes in the previous problem fail when the load is 25 m above ground and is moving down with a velocity of 2.4 m/s, determine the downward acceleration of the load and the velocity with which it will strike the ground.

9.21 If the hoist in problem 9.19 is used to lift a load of 600 kg, calculate the torque required to give the load an upward acceleration of 1 m/s^2.

9.22 A gear train consists of a pinion 150 mm in pitch circle diameter and 0.1 kg.m^2 mass moment of inertia, and a large gear 450 mm diameter and 1.2 kg.m^2 mass moment of inertia. Determine the torque which must be applied to the pinion in order to accelerate the large gear from rest to an angular velocity of 955 revolutions per minute in 20 seconds.

9.7 Centrifugal force

Imagine yourself as a passenger in a car which is negotiating a sharp right-hand curve on a level road. You have a feeling that a force of some kind tends to push you towards the left. The faster the speed the more distinct is your experience of the force, until you begin to worry about the possibility of the car's skidding on the road surface. This means that you are aware of the force acting not only on your body, but on the mass of the car in the direction at right angles to the curved path and away from the centre of curvature. (See Fig. 9.7.)

This force is called the **centrifugal force**.*

If we attach a ball of mass m to a light string and swing the ball in a horizontal circle with our hand as a centre, we have a primitive model of the car, or any other object, moving with circular motion, i.e. along a curved path of constant radius.

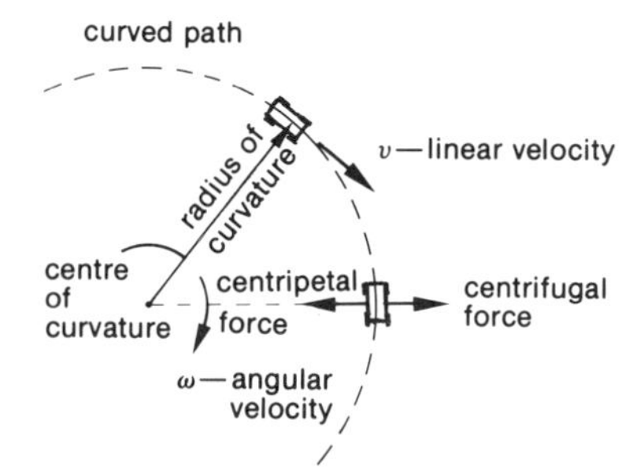

Fig. 9.7

It is possible to measure the magnitude of the centrifugal force acting on the mass by measuring the pull exerted by the string that is required to keep the ball moving in a circular path. The reaction provided by the string is a force, equal and opposite to the centrifugal force, which is acting on the mass and directed towards the centre of curvature. This force is called the **centripetal force**, meaning acting towards the centre.** In the case of a car, the centripetal force which holds it on the road is provided by the friction between its tyres and the road surface.

It can be shown by careful measurement, as well as analytical derivation, that centrifugal force is directly proportional to the mass and the square of the velocity of the body in circular motion and inversely proportional to the radius of curvature.

If m is the mass of the body, v is its velocity and r is the radius of its path, centrifugal force F_c is given by

$$F_c = \frac{mv^2}{r}$$

It is also helpful to recall that linear velocity v is related to angular velocity ω as $v = \omega r$. Substituting, we obtain two alternative expressions for centrifugal force.

$$\boxed{F_c = \frac{mv^2}{r} = mr\omega^2}$$

Example 9.12

Determine the centrifugal force acting on a passenger of mass 75 kg in a car travelling at 90 km/h around a curve of 100 m radius.

---

* From Latin *centrum* = centre and *fugere* = to fly from.

** From *centrum* and *petere* = to seek.

Solution

Convert velocity into base units:

$$v = 90 \times \frac{1000}{3600} = 25 \text{ m/s}$$

Centrifugal force is
$$F_c = \frac{mv^2}{r} = \frac{75 \times 25^2}{100} = 468.8 \text{ N}$$

Fig. 9.8 *Centrifugal force apparatus*

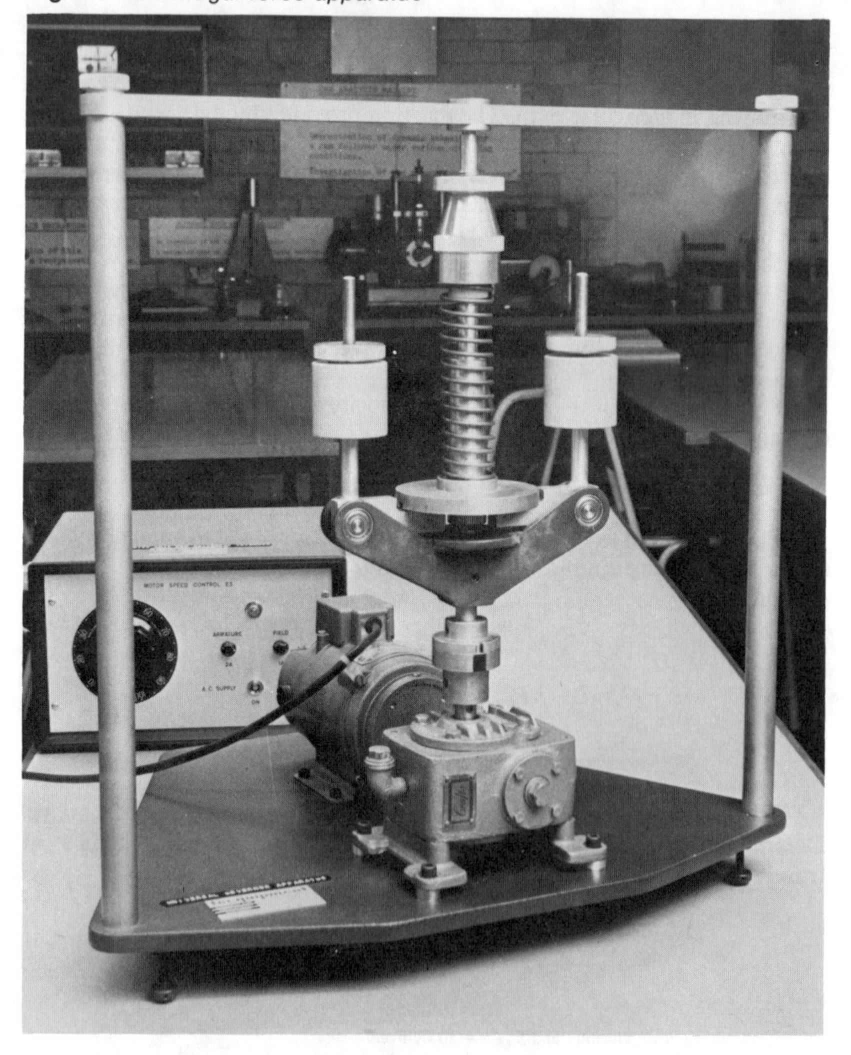

Often, the action of centrifugal force combines with that of gravity to exert a pull whose magnitude and direction is equal to the resultant of the two forces. The resultant force is responsible for the angle at which devices such as flyball engine governors, involving the principle of the conical pendulum, would operate, as the following example illustrates.

Example 9.13

A steam engine governor consists of two flyballs of 2 kg mass each as shown in Figure 9.9(a). The balls swing outwards under centrifugal force, so lifting a sleeve and progressively closing the engine throttle valve, which controls the admission of steam to the engine cylinders. Determine the speed, in revolutions per minute, at which the main arm, which is 250 mm long, makes an angle of 35° with the vertical shaft.

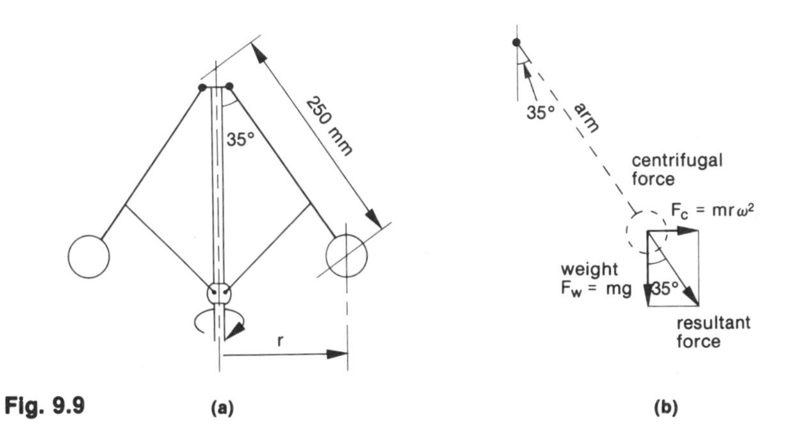

Fig. 9.9 **(a)** **(b)**

Solution

The relationship between the forces acting on the ball is shown in Figure 9.9(b).

$$\text{From the diagram, } \tan 35° = \frac{F_c}{F_w}$$

$$\text{or } \tan 35° = \frac{mr\omega^2}{mg}$$

where r = radius of rotation, i.e. 0.25 m × sin 35° = 0.1433 m and $g = 9.81$ N/m.

Hence:

$$0.7 = \frac{0.1433\omega^2}{9.81}$$

$$\therefore \omega = 6.92 \text{ rad/s}$$

$$\text{Converting to r.p.m. N} = \frac{60\omega}{2\pi} = \frac{60 \times 6.92}{2\pi} = 66.1 \text{ r.p.m.}$$

Problems

9.23 Calculate the centrifugal force acting on a 1.2 t vehicle when travelling at 60 km/h around a bend of 90 m radius.

9.24 If the coefficient of friction between the tyres of the vehicle in problem 9.23 and the road surface is 0.55, determine if skidding will occur.

9.25 Determine the speed at which the vehicle in the previous questions will start skidding on the curve.

9.26 A wheel out-of-balance is equivalent to a mass of 60 kg at an eccentricity of 0.5 mm. Calculate the centrifugal out-of-balance force at a speed of 600 revolutions per minute.

9.27 If the wheel in problem 9.26 is to be balanced by attaching a small balancing mass to the wheel at a radius of 300 mm, calculate the balancing mass required.

9.28 A car travels over a humpbacked bridge having a curvature of 25 m radius. At what speed will the wheels leave the road surface?

9.29 A turbine rotor is mounted on a horizontal shaft supported by two equally spaced bearings. If the 2500 kg rotor has an eccentricity of 0.12 mm, find the maximum and minimum load on each bearing at a speed of 1200 revolutions per minute.

9.30 A centrifuge for the separation of sugar crystals in molasses must exert a force of 6 N per gram mass. If the filter drum has a radius of 350 mm, determine the speed in revolutions per minute required.

9.31 A conical pendulum consists of a 2 kg ball attached to a cord and rotated around in a horizontal circle with a radius of 1.2 m at a speed of 30 revolutions per minute. Determine the tension in the cord and the angle it makes with the vertical.

9.32 A conical pendulum consists of a 3 kg mass attached to a 0.95 m length of a cord and rotated in a horizontal circle. If the breaking strength of the cord is 55 N, determine the speed in revolutions per minute at which the cord will break.

Review questions

1. Define angular displacement, angular velocity and angular acceleration and state the symbols and units used.
2. What is implied by the term "uniformly accelerated rotation"?
3. List four equations of rotational motion.
4. State the relationships between rotation and circular motion.
5. State the relationship between torque and angular acceleration.
6. State the unit of mass moment of inertia.
7. Explain centrifugal force and state the formula for calculating its value.
8. Give an example in which the action of centrifugal force combines with gravity to produce a resultant acting on a body.

10

Work and power

This chapter is concerned with the concepts of work and power as they apply to linear and rotational motion of moving objects and mechanical components.

10.1 Mechanical work

Mechanical work is defined in terms of linear or rotational effort, when the result produced by the effort is movement or rotation of a body.

In linear motion, if a force F is applied to a body which moves in a straight line a distance S in the direction of the force, then the work done by the force on the body is said to be the product of the force and the distance.

$$W = F \times S$$

The unit of work follows from this definition. If the units of force and distance are newtons and metres respectively, then the unit of work is the newton metre. Work is a new physical quantity different from both force and distance. It is therefore convenient to give the unit of work a special name. Thus, the SI unit of work equal to 1 N.m is called the **joule**, denoted by J.

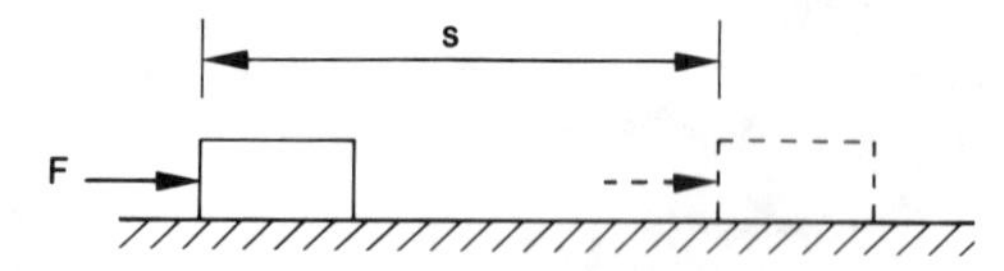

Fig. 10.1

Example 10.1

Determine the work done by a force of 50 N moving a distance of 3 m in the direction of the force.

Solution

Work done

$$\begin{aligned} W &= F \times S \\ &= 50 \text{ N} \times 3 \text{ m} \\ &= 150 \text{ J} \end{aligned}$$

Example 10.2

A hoist lifts a load of 1.5 tonnes through a vertical distance of 20 m. Determine the amount of work done against gravity.

Solution

The work is done against gravity, i.e. against the weight of the load, which is

$$F_w = m.g$$
$$= 1500 \text{ kg} \times 9.81\frac{\text{N}}{\text{kg}}$$
$$= 14\ 715 \text{ N}$$

The work done against this force is:

$$W = F \times S$$
$$= 14\ 715 \text{ N} \times 20 \text{ m}$$
$$= 294\ 300 \text{ J}$$
$$= 294.3 \text{ kJ}$$

In rotational terms, torque T is analogous to force, i.e. torque is a turning effort. Similarly, angular displacement, θ, is analogous to linear displacement. This analogy can be used to arrive at the expression for work done by torque on a rotating object, i.e.

$$\boxed{W = T \times \theta}$$

Since torque is measured in newton metres and angular displacement in radians, i.e. angular displacement is dimensionless, it can be seen that work done by torque can also be measured in joules.

Example 10.3

A flywheel makes 200 revolutions while the torque applied to it is 35 N.m. Determine the work done.

Solution

Angular displacement must be expressed in radians.

$$\theta = 200 \text{ rev} = 200 \times 2\pi = 1257 \text{ rad.}$$

Work done is:

$$W = T \times \theta = 35 \times 1257 = 43\ 980 \text{ J} = 44 \text{ kJ}$$

10.2 Power

When work is being done continuously over a period of time, the **time rate of doing work** is called **power,** P.

$$\boxed{P = \frac{W}{t}}$$

where P is average power
W is work done
t is time taken to do the work.

The unit of power follows from the definition and is equal to one joule of work per second of time, i.e. J/s. The name given to the unit of power is the **watt**,* denoted by W. (One has to be careful to avoid confusion between W as a symbol for work done and W as a unit symbol for power.)

Example 10.4

If it takes 27 seconds to lift the load in Example 10.2, what is the average power required?

Solution

$$\text{Power} = \frac{W}{t} = \frac{294.3 \text{ kJ}}{27 \text{ s}} = 10.9 \text{ kJ/s} = 10.9 \text{ kW}$$

When work is done by a force moving with a constant linear velocity v, we can substitute $S = v.t$ into the expression for power:

$$P = \frac{W}{t} = \frac{F \times S}{t} = \frac{Fvt}{t} = F.v$$

Hence, for power in linear motion of a force:

$$\boxed{P = F.v}$$

This relationship also applies if v is interpreted as average velocity.

Example 10.5

A train moving at 63 km/h requires 40 kN of tractive effort at this speed. Determine the driving power.

Solution

$$v = 63 \text{ km/h} = 17.5 \text{ m/s}$$

$$\begin{aligned} \text{Power } P &= F.v \\ &= 40 \text{ kN} \times 17.5 \text{ m/s} \\ &= 700 \text{ kN.m/s} \\ &= 700 \text{ kW} \end{aligned}$$

When work is done by a torque applied through a rotating member, such as a shaft, turning with a constant angular velocity $\omega = \frac{\theta}{t}$, substitution yields,

---

* The SI unit was named after the Scottish engineer James Watt who in the late eighteenth century, according to an historical anecdote, established another unit of power, called horsepower, after actual experiments with strong dray horses. The horsepower, equal to 746 watts, is about 50 per cent more than the rate that an average horse can sustain for a working day.

$$P = \frac{W}{t} = \frac{T \times \theta}{t} = \frac{T\omega t}{t} = T.\omega$$

Thus power, produced by a torque in rotational motion is

$$\boxed{P = T.\omega}$$

where ω is constant or average angular velocity in radians per second.

Example 10.6

An output shaft of an electric motor rotates at 1450 r.p.m. and produces a torque of 81 N.m. What is the shaft power?

Solution

$$\omega = 1450 \text{ r.p.m.} = \frac{2\pi \times 1450}{60} = 151.8 \text{ rad/s}$$

$$\text{Power} = T.w = 81 \text{ N.m} \times 151.8 \text{ rad/s} = 12\ 300 \text{ W} = 12.3 \text{ kW}$$

Problems

10.1 A locomotive applies a tractive effort of 120 kN over a distance of 2 km. Calculate the work done.

10.2 A crane lifts a crate of 2.5 t mass through a height of 12 m. Calculate the work done.

10.3 A box having a mass of 80 kg is hauled along a horizontal floor by a force parallel to the floor, for a distance of 25 m. If the coefficient of sliding friction between the floor and the box is 0.4, calculate the amount of work done.

10.4 A torque of 60 N.m is applied to a flywheel to turn it through 150 revolutions at constant speed. What is the work done?

10.5 A radar antenna of mass 3500 kg turns on rollers mounted on a horizontal circular track of 2 m diameter. If the frictional resistance of the rollers is 0.5 kN per tonne, calculate the work required to turn the antenna through one complete revolution.

10.6 Determine the power developed by an engine when it does 97.5 MJ of work in 25 minutes.

10.7 If a lift having a total mass of 1.6 t travels a vertical distance from the ground to the tenth floor in 17.5 seconds and the distance between floors is 3.5 m, calculate the average power required.

10.8 A train of total mass 650 t is hauled by a locomotive along a level track at constant speed of 60 km/h. If the tractive resistance is 85 N per tonne mass of the train, calculate the power developed at this speed.

10.9 The power output from an engine is measured by means of a belt brake arrangement as shown in Figure 10.2. The brake-drum diameter is 800 mm, over which a belt supports two masses of 55 kg each. One side of the belt is attached to a spring balance which reads 375 N. The engine speed is 450 r.p.m. Calculate the power dissipated in friction between the belt and the drum.

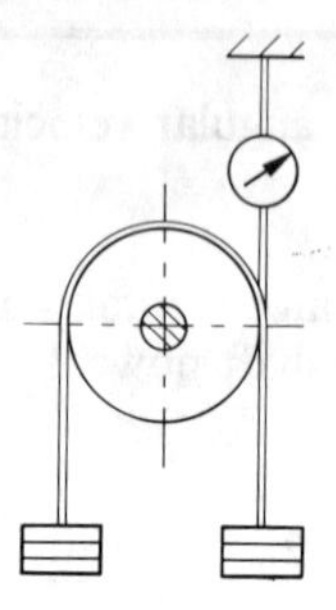

Fig. 10.2

10.10 A motor vehicle is travelling at 100 km/h against total air and frictional resistance of 950 N. Neglecting any transmission losses, estimate the torque which the engine must develop, if its rotational speed is 2200 r.p.m.

10.3 Work and acceleration

When a constant force or torque is available to accelerate a body in linear or rotational motion, the work done can be related to the acceleration produced.

Work required to accelerate a mass

If a body of mass m is acted upon by a net accelerating force F over a distance S, the work done is $W = F.S$.

At the same time,

$$S = v_{av} \times t = \left(\frac{v + v_0}{2}\right)t$$

and

$$F = ma = m\left(\frac{v - v_0}{t}\right)$$

Substituting into the work equation:

$$\begin{aligned} W &= F \times S \\ &= m\left(\frac{v - v_0}{t}\right) \times \left(\frac{v + v_0}{2}\right)t \\ &= \frac{m}{2}(v - v_0) \times (v + v_0) \\ &= \frac{m}{2}(v^2 - v_0^2) \end{aligned}$$

Thus work done in accelerating a mass m from an initial velocity v_0 to a final velocity v is given by:

$$W = \frac{m}{2}(v^2 - v_0^2)$$

It will be seen later, that this equation has a special significance with respect to the quantity called kinetic energy. At this stage, however, we will use it simply to calculate the work required to change the velocity of a body.

Example 10.7

Determine the force, work and average power required to accelerate a car of 1.2 t mass from rest to 72 km/h in 16 seconds.

Solution

Final velocity $v = 72 \text{ km/h} = 20 \text{ m/s}$

Acceleration
$$a = \frac{v - v_0}{t}$$
$$= \frac{20 - 0}{16} = 1.25 \text{ m/s}^2$$

Force required
$$F = ma$$
$$= 1200 \text{ kg} \times 1.25 \text{ m/s}^2 = 1500 \text{ N}$$
$$= 1.5 \text{ kN}$$

Work required
$$W = \frac{m}{2}(v^2 - v_0^2)$$
$$= \frac{1200}{2}(20^2 - 0) = 240\,000 \text{ J}$$
$$= 240 \text{ kJ}$$

Alternatively, distance travelled can be determined,
$$S = \left(\frac{v + v_0}{2}\right)t$$
$$= \left(\frac{20 + 0}{2}\right) \times 16 = 160 \text{ m}$$

Hence,
$$W = F \times S$$
$$= 1.5 \text{ kN} \times 160 \text{ m} = 240 \text{ kJ}$$

Average power can now be calculated:
$$P = \frac{W}{t} = \frac{240 \text{ kJ}}{16 \text{ s}} = 15 \text{ kW}$$

It should be understood that these results refer to the force, work and power associated with the acceleration of the mass only. Work and power due to other causes, such as frictional resistance, must be allowed for separately if required.

Work required for rotational acceleration

By analogy with linear motion, if a body of mass moment of inertia I is acted upon by a net accelerating torque T over an angular displacement θ, the work done is $W = T \times \theta$.

At the same time:

$$\theta = \omega_{av} \times t = \left(\frac{\omega + \omega_0}{2}\right)t$$

and

$$T = I \times \alpha = I\left(\frac{\omega - \omega_0}{t}\right)$$

Substituting into the work equation:

$$W = T \times \theta$$

$$= I\left(\frac{\omega - \omega_0}{t}\right)\left(\frac{\omega + \omega_0}{2}\right)t$$

$$= \frac{I}{2}(\omega - \omega_0)(\omega + \omega_0)$$

$$= \frac{I}{2}(\omega^2 - \omega_0^2)$$

Thus the work done in accelerating a body of mass moment of inertia I from initial angular velocity ω_0 to a final angular velocity ω is given by:

$$\boxed{W = \frac{I}{2}(\omega^2 - \omega_0^2)}$$

Example 10.8

Determine the torque, work and average power when a flywheel of mass moment of inertia of 53 kg.m^2 is accelerated from 700 r.p.m. to 1500 r.p.m. in 24 seconds.

Solution

Initial angular velocity $\omega_0 = 700 \text{ r.p.m.} = 73.3 \text{ rad/s}$

Final angular velocity $\omega = 1500 \text{ r.p.m.} = 157.1 \text{ rad/s}$

Angular acceleration $\alpha = \dfrac{\omega - \omega_0}{t} = \dfrac{157.1 - 73.3}{24}$

$= 3.49 \text{ rad/s}^2$

Torque required $T = I\alpha$

$= 53 \text{ kg.m}^2 \times 3.49 \text{ rad/s}^2$

$= 185 \text{ N.m}$

Work done
$$W = \frac{I}{2}(\omega^2 - \omega_0^2)$$
$$= \frac{53}{2}(157.1^2 - 73.3^2)$$
$$= 511\ 400 \text{ J}$$
$$= 511.4 \text{ kJ}$$

Alternatively, angular displacement can be determined

$$\theta = \left(\frac{\omega + \omega_0}{2}\right)t = \left(\frac{157.1 + 73.3}{2}\right) \times 24 = 2764 \text{ rad}$$

Hence,
$$W = T \times \theta$$
$$= 185 \text{ N.m} \times 2764 \text{ rad}$$
$$= 511\ 400 \text{ J}$$
$$= 511.4 \text{ kJ}$$

Average power can be calculated from

$$P = \frac{W}{t} = \frac{511.4 \text{ kJ}}{24 \text{ s}} = 21.3 \text{ kW}$$

Problems

10.11 A motor-car of mass 1 tonne is travelling at 40 km/h. Determine the average power required to accelerate the car to 80 km/h over a distance of 300 m. Neglect friction and air resistance.

10.12 If total resistance to motion of the car in the previous problem is 365 N, determine the average power required to accelerate the car, as specified above, taking into account the resistance force.

10.13 A flywheel having a mass moment of inertia of 65 kg.m^2 is to be accelerated from rest to a speed of 750 r.p.m. in 30 seconds. Calculate the work that must be done to accomplish this, if bearing friction is neglected.

10.14 If it is found that the average power actually required to accelerate the flywheel as described in the previous question is 9 kW, what is the magnitude of friction torque in the bearings?

10.15 A train of mass 200 t travelling at 60 km/h is brought to rest in a distance of 600 m by a constant braking force. If total tractive resistance is 18 kN, what is the value of the braking force and the average power dissipated by the brakes?

10.16 A flywheel revolving at 450 r.p.m. has its speed uniformly reduced to 150 r.p.m. in 3 minutes. The wheel, which is a disc of uniform thickness, has a mass of 880 kg and is 1200 mm in diameter. The bearing friction is 5 N.m. Calculate the braking torque applied and the average power dissipated by the brake.

Review questions

1. Define mechanical work.
2. State the relationships between
   (a) force and work for linear motion,
   (b) torque and work for rotational motion.
3. Define power.
4. State the relationships between
   (a) force and power for linear motion,
   (b) torque and power for rotational motion.
5. Explain how the work done in accelerating a body can be related to its change of velocity,
   (a) for linear motion,
   (b) for rotational motion.

11

Mechanical energy

The concept of mechanical energy is one of the most useful in engineering science. However, the colloquial meaning of the word "energy" is very different from its precise scientific meaning. The idea of energy, by itself, is very abstract and historically was a matter of great confusion among many of the ablest scientific minds until well into the nineteenth century.

The essential unity of the concept of energy in its different forms, e.g. mechanical, thermal, electrical, chemical, was not clear until the development of the steam engine prompted men such as James Watt, James Joule and Robert Mayer to explore the relationships between different forms of energy in action, such as work and heat, and to proclaim the law of conservation of energy.

In this chapter, we will not include non-mechanical forms of energy, such as heat, in our discussion, because they do not fit our description of mechanical systems of bodies at rest or in motion. Therefore our definition of energy, at this stage, is restricted to mechanical energy only.

11.1 Mechanical energy

Mechanical energy is usually defined as a physical quantity stored in a material body which is equivalent to its capacity for doing work.*

In the previous chapter, we saw how by doing work on a body it was possible to change the state or condition of the body in terms of (a) lifting the body to a higher level against the force of gravity, or (b) increasing its velocity against the inertia of the body.

It took many centuries of scientific thought to recognise that the work done in lifting a body or in accelerating it does not disappear, but is stored within the body by virtue of its increased level or velocity. It is true that, in many practical situations, friction appears to detract from the quantity of work done which can be recognised as stored energy in a body. However, if properly allowed for, frictional effects represent amounts of work converted into heat and dissipated to the surroundings.

---

* The word energy is derived from the Greek *energeia*, *en* = in and *ergon* = work.

Let us begin by ignoring friction and discussing two forms of mechanical energy called **potential energy** and **kinetic energy** and their relationship to the work done in the absence of frictional losses.

If energy is stored work, its units must correspond to the units of work. The SI unit of energy is, therefore, the **joule**.

11.2 Potential energy

The potential energy of a body is the energy which a body possesses due to its position in the gravitational field.

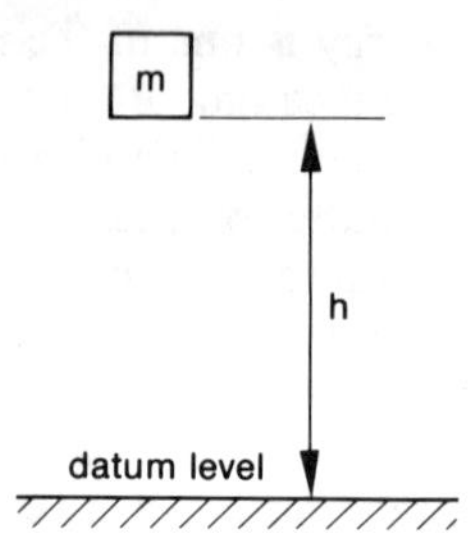

Fig. 11.1

As an example of this consider a block of mass m raised a distance h above the ground (Fig. 11.1). The force of gravity acting on the block is $F_w = mg$ and the work done in lifting the block is $W = F \times h = mgh$. This work is stored in the body as potential energy with respect to the ground as the datum. In symbols, potential energy can be stated as follows:

$$PE = mgh$$

Example 11.1

Calculate the potential energy of a drop hammer which has a mass of 1 tonne and is raised 1.5 m above the pile head before being allowed to drop freely in order to drive it into the ground.

Solution

Potential energy of the hammer relative to the pile is:

$$PE = mgh = 1000 \text{ kg} \times 9.81 \frac{\text{N}}{\text{kg}} \times 1.5 \text{ m} = 14\,715 \text{ J}$$

It is interesting to note that a rudimentary idea of potential energy goes back to Galileo, who recognised that, when a load is lifted with a pulley system, the force applied multiplied by the distance through which that force must be applied, i.e. the work done, remains constant even though the force and distance may vary.

11.3 Kinetic energy

In the previous chapter we found that, in order to accelerate a body in linear motion, work must be done on that body. As a result of doing W units of work, the velocity changes so that:

$$W = \frac{m}{2}(v^2 - v_0^2) = \frac{mv^2}{2} - \frac{mv_0^2}{2}$$

The quantity $\frac{mv^2}{2}$ was first recognised in the seventeenth century and was then called *vis viva* or living force. In the nineteenth century, it was finally accepted that *vis viva* was not a force, but a form of mechanical energy now called *kinetic energy*.

Kinetic energy of a body is the energy which it possesses due to its velocity. Kinetic energy is proportional to the mass and the square of the velocity. Thus:

$$KE = \frac{mv^2}{2}$$

Example 11.2

Calculate the kinetic energy of a vehicle of 1720 kg mass moving with a velocity of 80 km/h.

Solution

Velocity v = 80 km/h = 22.2 m/s

$$\text{Kinetic energy } KE = \frac{mv^2}{2}$$

$$= \frac{1720 \text{ kg } (22.2 \text{ m/s})^2}{2}$$

$$= 424\ 700 \text{ J}$$

$$= 424.7 \text{ kJ}$$

Kinetic energy of rotating bodies can also be calculated using rotational analogues of the linear terms mass and velocity, namely mass moment of inertia, I, and angular velocity ω. Thus for a rotating body:

$$KE = \frac{I\omega^2}{2}$$

Example 11.3

Calculate the kinetic energy of a flywheel of mass moment of inertia of 61 kg.m^2 rotating at 250 r.p.m.

Solution

Angular velocity ω = 250 r.p.m. = 26.18 rad/s.

Hence, kinetic energy stored in the flywheel at this speed is given by:

$$KE = \frac{I\omega^2}{2} = \frac{61 \times 26.18^2}{2}$$

$$= 20\,900 \text{ J} = 20.9 \text{ kJ}$$

Problems

11.1 Determine the potential energy of a lift cage which has a mass of 2.3 t when its position is 35 m above ground level.

11.2 Determine the amount of work to be done if the lift cage in problem 11.1 is to be lifted a further 10 m in elevation.

11.3 Calculate the height to which a drop hammer of mass 800 kg must be lifted in order to store 10.6 kJ of potential energy.

11.4 Calculate the kinetic energy of a 200 t train moving at 85 km/h.

11.5 Determine the change in the kinetic energy of the train in the previous question if its speed is reduced from 85 km/h to 60 km/h.

11.6 Calculate the kinetic energy of a car of mass 1.2 t at 60 km/h, 80 km/h and 100 km/h.

By comparing the energy stored at these speeds determine the work done in increasing the speed by 20 km/h,

(a) between 60 km/h and 80 km/h, and

(b) between 80 km/h and 100 km/h.

11.7 A car of mass 950 kg is initially travelling at 60 km/h. If the speed increases so that its kinetic energy increases by 165 kJ, determine the final speed of the car.

11.8 Calculate the kinetic energy of a flywheel of mass moment of inertia of 13.5 kg.m² rotating at 1200 r.p.m.

11.9 Estimate the mass moment of inertia of a flywheel, if a change in kinetic energy corresponding to a change in velocity from 500 r.p.m. to 900 r.p.m. is equal to 324 kJ.

11.10 A steel disc of 300 mm diameter and 50 mm thick is rotated at 400 r.p.m. Calculate the change in kinetic energy if the speed is doubled. Take density* of steel as 7800 kg/m³.

11.4 Conversion of potential and kinetic energy

One of the most fundamental laws of nature is that of conservation of energy, which states that energy can be converted from one form into another, but cannot be created or destroyed.**

* Density is mass per unit volume (see Ch. 15).

** In nuclear physics, scientists have discovered that matter and energy are related and that energy and matter are interchangeable. However, this new law does not concern ordinary concepts of mechanical engineering science.

The two forms of energy with which we are concerned now are potential and kinetic energy. If we let PE_1 and KE_1 be potential and kinetic energy amounts stored in a body in its initial state, and PE_2 and KE_2 be potential and kinetic energy stored in a body in its final state, the conservation of energy principle can be stated as follows:

$$PE_1 + KE_1 = PE_2 + KE_2$$

This equation is true only if no external work is done on the body and no loss of mechanical energy occurs due to friction.

Example 11.4

An object having a mass of 3 kg is dropped from a height of 12 m. Using the conservation of energy principle, calculate the velocity with which it strikes the ground (Fig. 11.2).

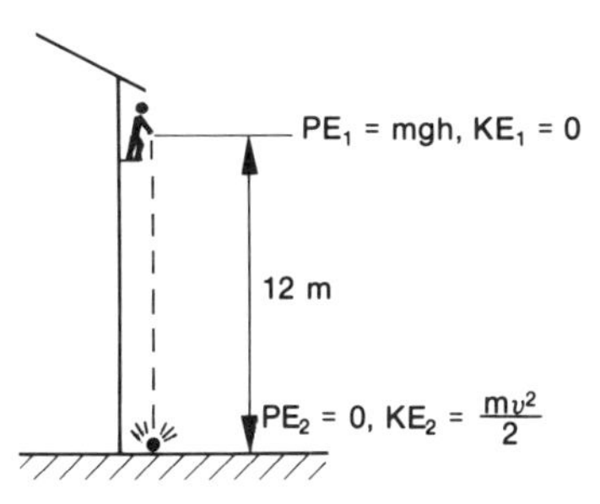

Fig. 11.2

Solution

Before the object is dropped, its kinetic energy KE_1 is zero and its potential energy relative to the ground is:

$$\begin{aligned} PE_1 &= mgh \\ &= 3 \times 9.81 \times 12 \\ &= 353.2 \text{ J} \end{aligned}$$

When it strikes the ground, its height above ground is zero, therefore $PE_2 = 0$. Its kinetic energy is:

$$KE_2 = \frac{mv^2}{2} = \frac{3v^2}{2} = 1.5v^2$$

Substitute into $PE_1 + KE_1 = PE_2 + KE_2$

$$353.2 + 0 = 0 + 1.5v^2$$

Hence, $v = 15.3$ m/s

Example 11.5

A car of a roller coaster has a mass of 500 kg and is released from a height of 18 m at the top of the first incline (Fig. 11.3). Calculate its velocity at the lowest

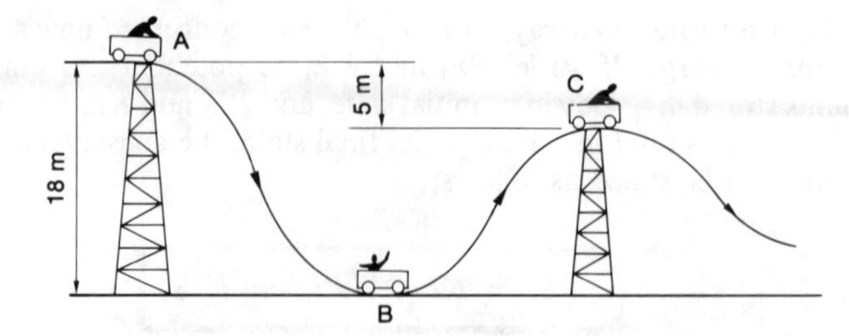

Fig. 11.3

point and also at the top of the second incline which is 5 m below the first incline.

Solution

Calculate potential and kinetic energy at points A, B and C.

$$PE_A = mgh = 500 \times 9.81 \times 18 = 88\,290 \text{ J}$$

$$KE_A = \frac{mv^2}{2} = 0 \text{ (since } v = 0)$$

$$PE_B = 0 \text{ (since } h = 0)$$

$$KE_B = \frac{mv^2}{2} = \frac{500}{2}v_B^2 = 250v_B^2$$

$$PE_C = mgh = 500 \times 9.81 \times 13 = 63765 \text{ J}$$

$$KE_C = \frac{mv^2}{2} = \frac{500}{2}v_C^2 = 250v_C^2$$

Equating total energy, $PE + KE$, for all points gives:

$$88\,290 + 0 = 0 + 250v_B^2 = 63\,765 + 250v_C^2$$

from which $v_B = 18.8$ m/s and $v_C = 9.9$ m/s.

Problems

(In the following problems assume that friction and other resistances are negligible.)

11.11 A drop hammer is raised to a height of 3 m above its striking point and released. Calculate its velocity at the instant of striking.

11.12 A ball is kicked vertically upwards with a velocity of 35 m/s. Determine the greatest height reached. (Compare with problem 8.14.)

11.13 A swinging pendulum of an Izod impact testing machine falls freely through a vertical height of 950 mm and then strikes a specimen at the lowest point of its travel. Determine the kinetic energy and the velocity of the 7 kg pendulum on impact.

11.14 A 40 kg boy has a velocity of 5.5 m/s as he passes the lowest position on a swing. If the effective length of the rope is 4.5 m, determine the angle it makes with the vertical when the boy reaches the maximum height.

11.15 If the hand brakes of a car parked on a 15° incline fail and the car rolls freely 100 m down the slope before hitting a tree, determine the velocity on impact.

11.16 A semi-trailer with disabled brakes is forced to use a safety ramp to avoid an accident. If the ramp is 100 m long at 20° to the horizontal, and the vehicle has a velocity of 25 km/h when it reaches the top of the ramp, what was its initial velocity?

11.17 A box slides down a smooth chute with an initial velocity of 2.9 m/s. If the chute is 10 m long and inclined at 35° to the horizontal, determine the velocity at the bottom of the chute.

11.18 Two masses of 5 kg and 3 kg are connected by a fine string passing over a smooth pulley. Determine the velocity with which the masses will be moving after each mass has moved one metre from rest. (Hint: The total energy of the system is the sum of the energies of its component parts.)

11.5 The work-energy method

The conservation of energy principle can be extended to situations where external work done on the body and work done against friction cannot be neglected and must form a part of the total energy account.

If we let:

$$(PE_1 + KE_1) = \text{initial total energy of the body}$$

$$(PE_2 + KE_2) = \text{final total energy of the body}$$

and

$$\pm W = \text{net work done on the body}$$

then, the final energy of the body is equal to the initial energy of the body increased by the amount of net work done on the body:

$$(PE_1 + KE_1) \pm W = (PE_2 + KE_2)$$

The net work is the difference between positive work done on the body by forces acting in the direction of motion and negative work done by forces, such as friction, which are opposing the motion.

Weight should not be included when determining the external work, since its effect has been allowed for in terms of potential energy.

Example 11.6

A 180 t train is climbing an incline of 1.5° for 2 km. Its initial velocity before the climb is 90 km/h. The tractive effort exerted by the engine is 53.2 kN and tractive resistance is 95 N per tonne. Determine the final speed after the climb (Fig. 11.4).

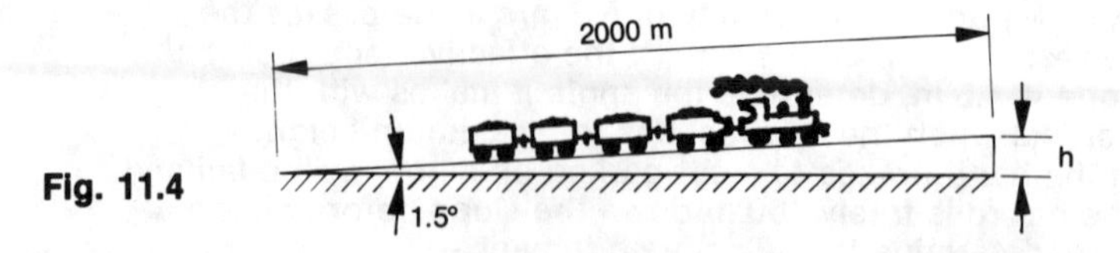

Fig. 11.4

Solution

Calculate all relevant terms for the work-energy equation:

Initial potential energy $PE_1 = 0$

The final elevation, $h = 2000 \text{ m} \times \sin 1.5° = 52.35 \text{ m}$

Final potential energy,
$$PE_2 = mgh$$
$$= 180\,000 \text{ kg} \times 9.81 \frac{\text{N}}{\text{kg}} \times 52.35 \text{ m}$$
$$= 92.45 \text{ MJ}$$

Positive work,
$$\text{effort} \times \text{distance} = 53\,200 \text{ N} \times 2000 \text{ m}$$
$$= 106.4 \text{ MJ}$$

Tractive resistance,
$$95 \frac{\text{N}}{\text{t}} \times 180 \text{ t} = 17100 \text{ N}$$

Negative work,
$$\text{resistance} \times \text{distance} = 17\,100 \text{ N} \times 2000 \text{ m}$$
$$= 34.2 \text{ MJ}$$

Net work,
$$W = \text{pos}W - \text{neg}W$$
$$= 106.4 - 34.2 = 72.2 \text{ MJ}$$

Initial velocity,
$$v_1 = 90 \text{ km/h} = 25 \text{ m/s}$$

Initial kinetic energy,
$$KE_1 = \frac{mv^2}{2} = \frac{180\,000 \text{ kg}(25 \text{ m/s})^2}{2}$$
$$= 56.25 \text{ MJ}$$

Substitute into:
$$(PE_1 + KE_1) \pm W = (PE_2 + KE_2)$$
$$0 + 56.25 + 72.20 = 92.45 + KE_2$$

Hence, final kinetic energy is:
$$KE_2 = 36 \text{ MJ}$$

Final velocity can now be calculated:
$$\frac{mv^2}{2} = 36 \text{ MJ} = 36 \times 10^6 \text{ J}$$

or
$$v = \sqrt{\frac{36 \times 10^6 \times 2}{180\,000}}$$
$$= 20 \text{ m/s or } 72 \text{ km/h}$$

The work-energy method can be applied to rotational as well as linear motion. Furthermore, it can be used successfully when more than one moving component is involved. The energy of a system of connected bodies, whose motions are related, is equal to the sum of the energies of the separate bodies.

Example 11.7

A block of mass 10 kg is attached to a cord which is wrapped around a drum of 800 mm diameter and mass moment of inertia 15 kg.m^2. After the block is released and has dropped a distance of 2.5 m, its velocity is 2 m/s. Determine the magnitude of the bearing-friction torque which resists rotation (Fig. 11.5).

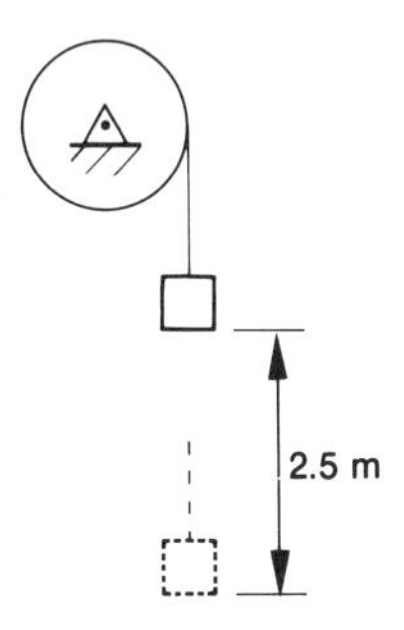

Fig. 11.5

Solution

Using the lowest position of the block as a datum:

$$PE_1 = mgh_1 = 10 \times 9.81 \times 2.5 = 245.3 \text{ J}$$

$$PE_2 = 0$$

$$KE_{1(block)} = 0$$

$$KE_{1(drum)} = 0$$

$$KE_{2(block)} = \frac{mv^2}{2} = \frac{10 \times 2^2}{2} = 20 \text{ J}$$

Final angular velocity of the drum:

$$\omega_2 = \frac{v_2}{r} = \frac{2}{0.4} = 5 \text{ rad/s}$$

$$KE_{2(drum)} = \frac{I\omega^2}{2} = \frac{15 \times 5^2}{2} = 187.5 \text{ J}$$

Substitute into:

$$(PE_1 + KE_{1B} + KE_{1D}) + W = (PE_2 + KE_{2B} + KE_{2D})$$

$$245.3 + 0 + 0 + W = 0 + 20 + 187.5$$

$$W = -37.8 \text{ J}$$

This is negative work done by resistance torque equal to $W = T \times \theta$, where:

$$\theta = \frac{S}{r} = \frac{2.5 \text{ m}}{0.4 \text{ m}} = 6.25 \text{ rad}$$

Substitute: $37.8 = T \times 6.25$

Hence, bearing friction torque $T = 6.05$ N.m.

Problems

11.19 A 25 kg box is pushed 10 m along a horizontal surface by a horizontal force of 80.3 N. If the box acquires a velocity of 1.2 m/s at the end of its travel, determine the coefficient of friction between the box and the surface.

11.20 A car of 1.4 t mass is travelling along a level road when the brakes are suddenly applied, causing the car to skid and come to rest in 51.6 m. Assuming the coefficient of friction between the road and the tyres is 0.5, calculate the initial speed of the vehicle.

11.21 A 50 kg block is pushed up an inclined plane, at 25° to the horizontal, by a force of 450 N parallel to the incline. The coefficient of friction is 0.35. Determine the distance measured along the plane in which the velocity will change from 1 m/s to 3 m/s.

11.22 Calculate the torque that must be applied to a flywheel of mass moment of inertia of 475 kg.m^2 in order to accelerate it to 650 r.p.m. from rest in 220 turns. The bearing friction is 63 N.m.

11.23 A flywheel of mass moment of inertia 30 kg.m^2 has constant torque of 170 N.m applied to it in order to accelerate it from 200 r.p.m. to 1000 r.p.m. while it makes a total of 180 revolutions. Calculate the bearing friction torque.

11.24 A mass of 16 kg resting on a smooth horizontal table is connected by a fine string passing over a smooth pulley on the edge of the table, with a mass of 3 kg hanging freely. With what velocity will the 3 kg block strike the floor if it is released from a height of 1 m above floor level?

11.25 What additional mass must be added to the 3 kg mass in the previous question, in order to increase the velocity with which the mass strikes the ground to 2 m/s?

11.26 A system of a 120 kg empty trolley and a 100 kg counterweight is as shown in Figure 11.6.

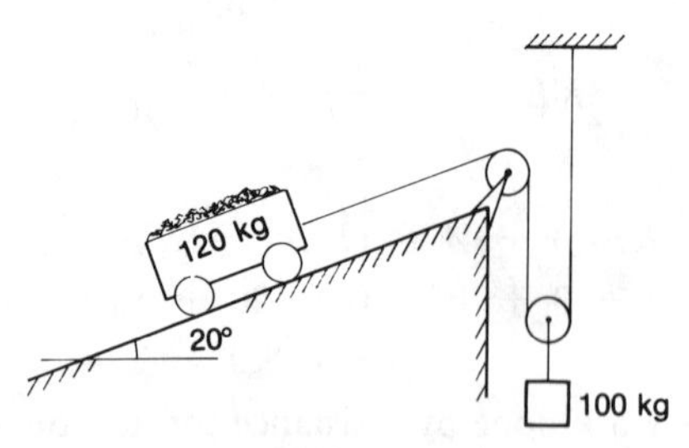

Fig. 11.6

After the counterweight has dropped 5 m from rest, the velocity of the trolley is 3 m/s. If all frictional resistances can be regarded as a single force F acting on the trolley parallel to the plane, calculate the resistance force.

11.27 For the system in problem 11.26, use the work-energy principle to calculate the correct load which can be:
(a) raised in the trolley, and
(b) lowered in the trolley,
at constant velocity.

Assume that the resistance to motion is proportional to the mass of the loaded trolley.

11.28 A load of 150 kg is being lowered with a velocity of 5 m/s, as shown in Figure 11.7.

The mass moment of inertia of the rotating assembly is 37.5 kg.m^2.

Determine the force F required to bring the load to rest after travelling an additional 8 m, if the coefficient of friction is 0.45.

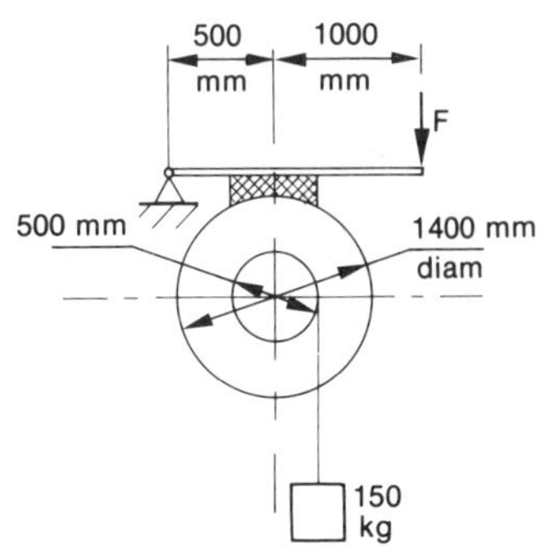

Fig. 11.7

Review questions

1. Define mechanical energy and state the SI unit of energy.
2. Define and state the formula for potential energy.
3. Define and state the formula for kinetic energy for
   (a) linear motion, and
   (b) rotation.
4. Explain the principle of conservation of energy in terms of potential and kinetic energy.
5. Explain the work-energy method of problem solving.
6. What is the effect of friction on energy conversion?

12

Momentum

Momentum is a concept used in mechanics for the solution of certain types of problems which are difficult to solve by the force-acceleration or by the work-energy methods introduced earlier.

These problems usually involve direct relationships between force, mass, velocity and time.

12.1 Momentum

Momentum,* sometimes described as the quantity of motion, is the product of the mass of a body, m, and its velocity at any given instant, v.

No special symbol is used here for momentum, hence

$$\text{Momentum} = m.v$$

Unlike energy, momentum is a vector quantity, i.e. it has a direction which corresponds to the direction of the velocity.

The unit of momentum is the kilogram metre per second, kg.m/s, when mass and velocity are expressed in SI base units.

Example 12.1

A rocket of 2.5 t mass is fired vertically upwards with a velocity of 250 m/s. What is its momentum?

Solution

$$\text{Momentum} = m \times v = 2500 \text{ kg} \times 250 \text{ m/s}$$
$$= 625\,000 \text{ kg.m/s upwards.}$$

According to Newton's first law of motion, the velocity of a body does not change unless an external force is applied to change the velocity. This law, therefore, implies conservation of momentum in the absence of an external force.

---

* In this book, only the linear momentum will be considered. A similar concept, called **angular momentum**, may also be used in rotational dynamics where it is expressed in rotational terms.

This principle enables us to solve certain types of problems directly, where the use of other methods is not convenient.

Example 12.2

A block of wood of mass 2 kg is freely suspended on a string. A bullet of mass 75 g is fired horizontally into the block. If the velocity of the bullet before impact is 415 m/s, calculate the velocity of the block, with the bullet imbedded in it, immediately after the impact.

Solution

Momentum before impact:

Bullet $0.075 \text{ kg} \times 415 \text{ m/s} = 31.13 \text{ kg.m/s}$

Block $= 0$

Total, before impact $= 31.13 \text{ kg.m/s}$

Momentum after impact:

Bullet and block $(2 + 0.075)\text{kg} \times v = 2.075v$

Conservation of momentum requires that the momentum after impact be equal to the momentum before impact.

$$\therefore 2.075v = 31.13$$

Hence velocity immediately after impact:

$$v = 15 \text{ m/s.}$$

12.2 Impulse

According to Newton's second law of motion, force and acceleration are related, i.e. $F = ma$.

If a constant force F is applied during a time interval t, the acceleration produced by the force is:

$$a = \frac{v - v_0}{t}$$

and substitution yields:

$$F = m\left(\frac{v - v_0}{t}\right)$$

This can be rearranged as follows:

$$\boxed{F.t = mv - mv_0}$$

The right-hand side of this equation can be recognised as the change of momentum from mv_0 to mv, where v_0 and v are the initial and final velocities of the body on which force F is acting.

The product of the force F and the time t during which it acts is called **impulse**. Because it contains force, impulse is a vector quantity. The unit of impulse is the newton second, N.s.

Example 12.3

When a golf ball having a mass of 50 g is struck by a club, the ball and club are in contact for 0.001 s. Immediately after impact, the ball travels at 45 m/s. Determine the average force of the collision.

Solution

Momentum before impact $= 0$

Momentum after impact $mv = 0.05 \text{ kg} \times 45 \text{ m/s} = 2.25 \text{ kg.m/s}$

Substitute into $F.t = mv - mv_0$

$$F \times 0.001 = 2.25 - 0$$

Hence $F = 2250 \text{ N} = 2.25 \text{ kN}$

The concept of impulse is also helpful in calculating thrust developed by a continuous flow of fluids, such as exhaust gases from a rocket, steam from a nozzle, or water from a garden hose.

Example 12.4

The exhaust gases from a rocket have a velocity of 1200 m/s and flow at the rate of 5 kg/s. Determine the thrust produced by the gases.

Solution

The exhaust jets accelerate the gases from rest to 1200 m/s. Therefore the initial momentum is zero and final momentum is $mv = 5 \text{ kg} \times 1200 \text{ m/s} = 6000 \text{ kg.m/s}$ every second.

Substitute into $F.t = mv - mv_0$,

$$F \times 1 \text{ s} = 6000 \text{ kg.m/s} - 0$$

Hence, the thrust $F = 6000 \text{ N} = 6 \text{ kN}$

Problems

12.1 A car of mass 1.4 t is travelling at 60 km/h. Calculate its momentum.

12.2 Determine the change in the momentum of the car in the previous problem, if its velocity (a) increases by 20 km/h and (b) decreases by 20 km/h.

12.3 A vehicle of mass 3.5 t is moving with a velocity of 90 km/h. Determine how long it will take to bring it to rest with a braking effort of 7 kN.

12.4 A gun fires a shell of mass 8 kg in a horizontal direction with a velocity of 375 m/s. The mass of the gun is 2 t and the shell takes 0.012 s to leave the barrel. Calculate the velocity of recoil and the average force on the gun.

12.5 Two men of masses 80 kg and 60 kg sit facing each other in two light boats holding the ends of a rope between them. If they pull with a force of 50 N for 4 seconds, calculate the velocity acquired by each boat. Neglect water resistance and the mass of the boats.

12.6 An inflated balloon contains 12.8 g of air which is allowed to escape from a nozzle with a velocity of 17.2 m/s. If the balloon deflates at a steady rate in 6.2 seconds, determine the force exerted on the balloon.

12.7 A rocket of mass 6 t is to be launched vertically. If the flow rate of the gas is 100 kg/s, determine the minimum velocity of the gas to just lift off the launching pad.

If the velocity is 900 m/s, what is the net upward accelerating force on the rocket?

12.8 A railway wagon of mass 14 t travelling at 18 km/h collides with a second wagon of mass 12 t which is at rest. If immediately after collision both wagons travel on coupled together, calculate their common velocity.

12.9 A ballistic pendulum consisting of a large wooden block of mass 3 kg is suspended by cords. When a bullet of mass 7.5 g and unknown velocity v is fired horizontally into it, the block with the bullet embedded in it swings, rising a maximum vertical distance of 200 mm.

Calculate the increase in the potential energy of the system and the kinetic energy immediately after impact. Hence, determine the velocity of the bullet using conservation of momentum principle.

12.10 A drop hammer of mass 120 kg falls 2.5 m on to a pile of mass 250 kg and drives it 70 mm into the ground. Calculate:

(a) the velocity with which the hammer strikes the pile (use conservation of energy principle),

(b) the velocity immediately after the impact (use conservation of momentum principle assuming the hammer does not rebound on impact),

(c) the average ground resistance (use work energy method).

12.3 Impact

An **impact** is a collision between two bodies which occurs in a very short interval of time and involves relatively large forces which the two bodies exert on each other.

We will only consider direct central impact, i.e. a kind of collision in which the two bodies move along the same straight line before and after collision.

The conservation of momentum principle applies during impact, which enables us to write an equation relating total momentum before impact to total momentum after impact.

If we let m_A and m_B be the masses of the bodies A and B and v_{0A} and v_{0B} their initial velocities, the total momentum before impact is given by:

$$m_A v_{0A} + m_B v_{0B}$$

Similarly if v_A and v_B are the final velocities, total momentum after impact is:

$$m_A v_A + m_B v_B$$

For the system, initial momentum equals final momentum:

$$m_A v_{0A} + m_B v_{0B} = m_A v_A + m_B v_B$$

This equation, by itself, is not sufficient for the solution of problems involving the impact between two bodies, as it contains two final velocities which are not usually known.

We have to examine the effect of deformation and subsequent restitution of the colliding bodies during impact. The extent of restoration of the original shape immediately after collision depends on the elastic* properties of the material. If the bodies are completely elastic, they will rebound and return to their original shape, like billiard balls. If, on the other hand, the bodies are completely plastic, they will stay permanently deformed and will travel together with the same velocity, like two lumps of putty, after collision.

The measure of the ability of the bodies to regain their original shape is called the **coefficient of restitution**. Mathematically it is defined in terms of relative velocities before and after impact. The equation defining the coefficient of restitution ϵ is:**

$$\epsilon(v_{0A} - v_{0B}) = (v_B - v_A)$$

The value of the coefficient varies from $\epsilon = 0$ for completely plastic impact to $\epsilon = 1$ for completely elastic collisions. For example, for steel on lead ϵ is about 0.12, for lead on lead, 0.2, for glass on glass, 0.93.

Before an illustrative example is attempted, one very important point must be emphasised. We know that velocities are vectors. It is, therefore, necessary to choose a sign convention, e.g. positive to the right and negative to the left, and to be absolutely consistent when applying the momentum and restitution equations.

Example 12.5

A railway car of mass 18 t moving at a speed of 10 m/s to the right collides with another car of 12 t mass which is moving to the left at 3 m/s. The coefficient of restitution is 0.6. Determine the final velocities of the two cars (Fig. 12.1).

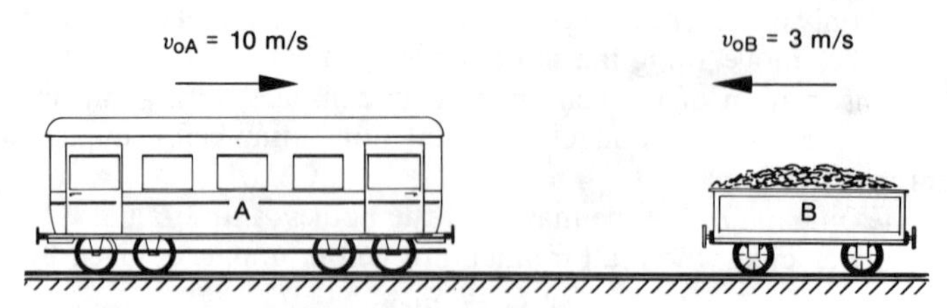

Fig. 12.1

* See more about elasticity of materials in Chapter 15.
** The symbol is another Greek letter called *epsilon*.

Solution

The following information is given:

$m_A = 18$ t $\qquad v_{0A} = 10$ m/s

$m_B = 12$ t $\qquad v_{0B} = -3$ m/s

$\epsilon = 0.6$

Substitute:

(a)
$$m_A v_{0A} + m_B v_{0B} = m_A v_A + m_B v_B$$
$$18 \times 10 - 12 \times 3 = 18 v_A + 12 v_B$$
$$144 = 18 v_A + 12 v_B$$

(b)
$$\epsilon(v_{0A} - v_{0B}) = v_B - v_A$$
$$0.6(10 + 3) = v_B - v_A$$
$$7.8 = v_B - v_A$$

Solving the two equations yields:

$$v_A = 1.68 \text{ m/s} \quad \text{and} \quad v_B = 9.48 \text{ m/s}$$

Both answers are positive, meaning that the two cars will move to the right after collision, but with new velocities.

Example 12.6

A tennis ball of mass 150 g is dropped from a height of 1 m and rebounds to a height of 0.8 m (Fig. 12.2). What was the coefficient of restitution between the ball and the ground?

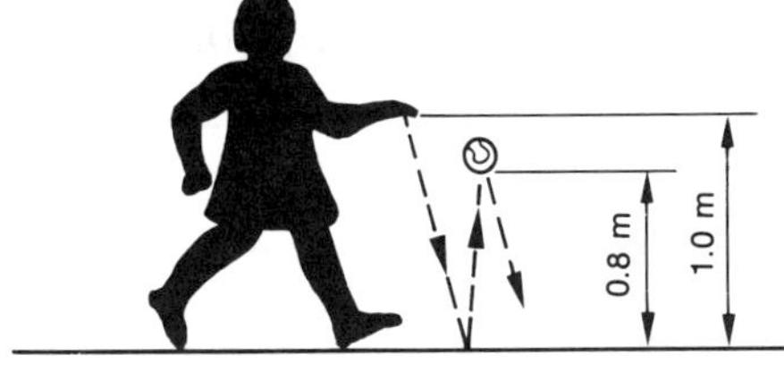

Fig. 12.2

Solution

Kinetic energy before impact:

$$KE_1 = PE_1 = mgh_1 = 0.15 \times 9.81 \times 1 = 1.47 \text{ J}$$

Velocity before impact:

$$v_{0A} = 4.43 \text{ m/s (negative)}$$

Kinetic energy after impact:

$$KE_2 = PE_2 = mgh_2 = 0.15 \times 9.81 \times 0.8 = 1.18 \text{ J}$$

Velocity after impact:

$$v_A = 3.96 \text{ m/s (positive)}$$

Velocity of the ground:

$$v_B = v_{0B} = 0$$

Therefore
$$\epsilon(v_{0A} - v_{0B}) = (v_B - v_A)$$
$$\epsilon(-4.43 - 0) = (0 - 3.96)$$
$$\epsilon = 0.89$$

Problems

12.11 A car of 1 t mass is moving at 15 m/s when a 2 t truck runs into its back at 25 m/s. Immediately after collision, the velocity of the truck is 20 m/s in the same direction. What is the velocity of the car after collision?

12.12 What is the coefficient of restitution between the two vehicles in the previous problem during collision?

12.13 Two bodies *A* and *B* of mass 5 kg and 4 kg collide with initial velocities 6 m/s to the right and 2 m/s to the left respectively. After the collision, the velocity of body *A* is 0.75 m/s to the left. What is the velocity of body *B* after collision?

12.14 Calculate the coefficient of restitution for the bodies in the previous problem.

12.15 Two bodies *A* and *B* of mass 8 kg and 14 kg move in the same direction along a straight line with 10 m/s and 1.2 m/s velocities respectively. If the coefficient of restitution is 0.6, calculate
(a) the velocities of the bodies after collision,
(b) the amount of kinetic energy lost during collision.

12.16 Repeat problem 12.15, if the collision is fully elastic, i.e. $\epsilon = 1$.

12.17 Repeat problem 12.15, if the collision is fully plastic, i.e. $\epsilon = 0$.

12.18 In order to determine the coefficient of restitution of a material, a ball of mass 0.5 kg is dropped on to a hard surface from a height of 2.5 m and is observed to bounce to a height of 1.8 m. What is the value of the coefficient?

Review questions

1. Define linear momentum.
2. Define impulse.
3. How is impulse related to the change of linear momentum?
4. Explain the principle of conservation of linear momentum.
5. State the relationship expressing the conservation of momentum in the case of impact between two bodies.
6. State the mathematical definition of the coefficient of restitution.
7. What is the numerical value of the coefficient of restitution in the case of
   (a) completely plastic collision, and
   (b) completely elastic collision?

PART FOUR

Mechanics of machines

Give me a lever long enough, a fulcrum and a place to stand and I can move the Earth .

Archimedes
on the law of the lever.

13
Machines

This chapter is devoted to a discussion of simple machines.

A **machine** can be defined as a mechanical device, consisting of one or more rigid components, designed and used for transmitting force and motion and for doing work.

The general purpose of machines is to augment or replace human effort for the accomplishment of physical tasks. In particular, the fundamental feature of most machines is that a large resistance is overcome by means of a relatively small force or effort.

Simple machines include the lever, inclined plane, wheel and axle, pulley and screw.

Fig. 13.1 *A class in machine tools laboratory*

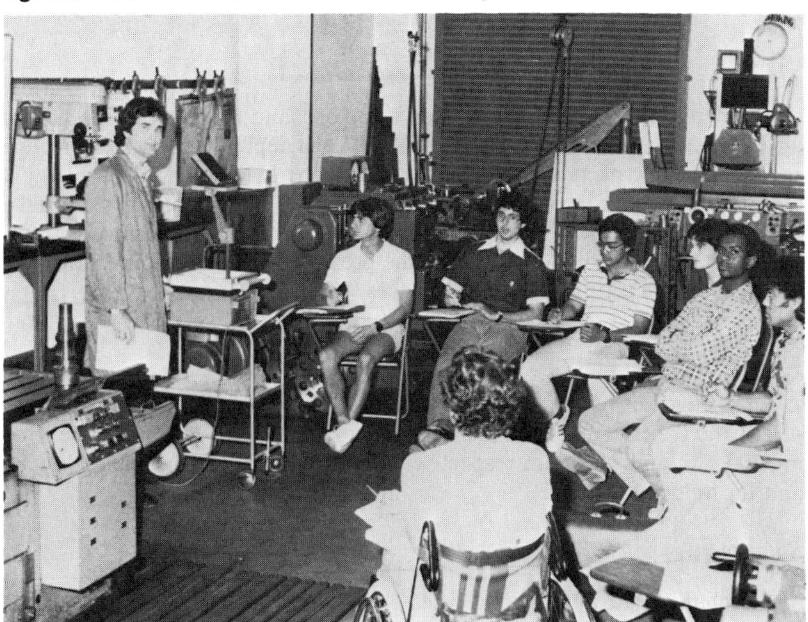

13.1 Mechanical advantage and velocity ratio

All machines have an input side and an output side. The force exerted on the machine on the input side is known as the **effort**, F_E. The resistance to be overcome, or the force on the output side of the machine, is called the **load**, F_L.

Usually machines are of such design that by application of a small effort a large load can be moved. The relationship between the load and the effort, which gives an indication of the advantage which can be obtained by using the machine, is called the **mechanical advantage** of the machine.

$$\text{Mechanical advantage} = \frac{\text{load}}{\text{effort}}$$

$$MA = \frac{F_L}{F_E}$$

Mechanical advantage is usually greater than one and depends upon the type of machine which is being used and varies with the load.

In order to do work, both the load and the effort must move. In some machines the motion is linear, while in others it is rotational.

The ratio between the distance moved through by the effort on the input side (S_E) and the distance moved through by the load on the output side (S_L) is called the **velocity ratio** of the machine.

$$\text{Velocity ratio} = \frac{\text{distance moved by effort}}{\text{distance moved by load}}$$

$$VR = \frac{S_E}{S_L}$$

Velocity ratio is usually greater than one and, unlike the mechanical advantage, is constant for a given machine, i.e. it depends only on the arrangement of moving parts and is independent of the load.

Ideally, the mechanical advantage of any given machine should be equal to its velocity ratio. However, for a real machine the actual mechanical advantage is always less than the ideal, due to the presence of friction between moving parts, such as bearing or sliding surfaces.

Example 13.1

A simple machine is represented diagrammatically in Figure 13.2. The load is 450 N and the effort is 50 N. The distances moved by the load and by the effort are 100 mm and 1200 mm respectively. Calculate the mechanical advantage and the velocity ratio.

Solution

Mechanical advantage, $$MA = \frac{F_L}{F_E} = \frac{450\ \text{N}}{50\ \text{N}} = 9$$

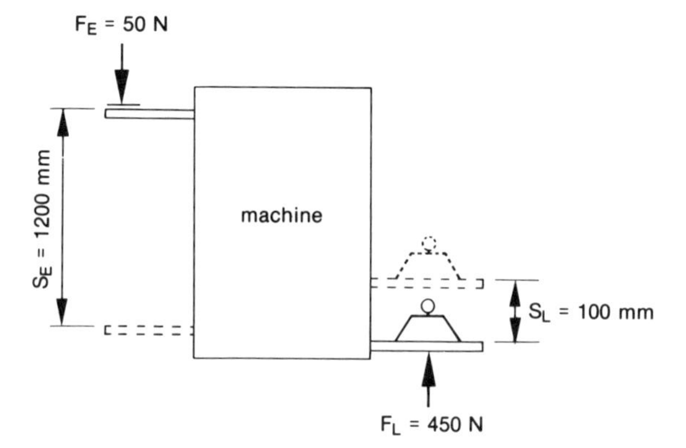

Fig. 13.2

Velocity ratio, $$VR = \frac{S_E}{S_L} = \frac{1200 \text{ mm}}{100 \text{ mm}} = 12$$

Note that both mechanical advantage and velocity ratio have no units.

13.2 Work and efficiency

Whenever a force moves through a distance, the product of force and distance is the work done.

On the input side of a machine, the work done by the effort is equal to:

$$W_E = F_E \times S_E$$

Similarly, on the output side, the work done in moving the load is given by:

$$W_L = F_L \times S_L$$

The ratio of the useful work done by the machine in moving the load on the output side to the work put into the machine by the effort on the input side is called the **efficiency*** of the machine.

$$\therefore \text{Efficiency} = \frac{\text{work done in moving load}}{\text{work done by the effort}}$$

$$\eta = \frac{W_L}{W_E} = \frac{F_L \times S_L}{F_E \times S_E}$$

Since $MA = \dfrac{F_L}{F_E}$ and $VE = \dfrac{S_E}{S_L}$,

$$\eta = \frac{MA}{VR}$$

Usually, efficiency is expressed as a percentage and in the absence of friction should ideally be equal to 100 per cent. In actual machines efficiency is always less than 100 per cent.

* The symbol for efficiency is a Greek letter called *eta*.

Example 13.2

For the machine in the previous example, calculate the input and output work and the efficiency.

Solution

Input work, $W_E = F_E \times S_E = 50\text{ N} \times 1.2\text{ m} = 60\text{ J}$

Output work, $W_L = F_L \times S_L = 450\text{ N} \times 0.1\text{ m} = 45\text{ J}$

Efficiency $\eta = \dfrac{W_L}{W_E} = \dfrac{45}{60} = 0.75 = 75\%$

Alternatively, $\eta = \dfrac{MA}{VR} = \dfrac{9}{12} = 0.75 = 75\%$

13.3 Friction effort

If the machine was perfect, no work would have to be done against friction and the efficiency would be 100 per cent. This would mean that for a perfect machine:

$$\frac{MA}{VR} = 100\%$$

or that the ideal mechanical advantage is equal to the velocity ratio, $MA = VR$.

It also means, that if there is no friction to be overcome, it takes a smaller effort to move the same load. The effort required to move a given load F_L if the machine is 100 per cent efficient is called the **theoretical effort**, F_{Th}.

Substituting into $MA = VR$, we have:

$$\frac{F_L}{F_{Th}} = VR \qquad \text{or} \qquad F_{Th} = \frac{F_L}{VR}$$

The difference between the actual effort F_E and the theoretical effort F_{Th} is the effort wasted in overcoming friction and is known as the **frictional effort**, F_F.

$$\boxed{F_F = F_E - F_{Th}}$$

Example 13.3

For the machine in the previous examples, calculate the theoretical and frictional efforts.

Solution

Theoretical effort, $F_{Th} = \dfrac{F_L}{VR} = \dfrac{450\ N}{12} = 37.5\text{ N}$

Frictional effort, $F_F = F_E - F_{Th} = 50\text{ N} - 37.5\text{ N} = 12.5\text{ N}$

13.4 The law of a machine

The law of a machine is an equation which expresses the relationship between load F_L and effort F_E. In many cases this relationship, when plotted as a graph of effort against load, is a straight line. Its mathematical equation is of the linear form:

$$F_E = aF_L + b$$

where F_E is the effort
F_L is the load
a is the slope of the graph
b is the value of F_E where the graph cuts the F_E axis.

After the constants have been determined for a particular machine, the law of the machine can be used to predict the effort required to move any load by the machine.

Example 13.4

The machine in the previous questions has been tested under different loads and the following efforts were recorded for each of the loading conditions:

Load F_L (newtons)	0	200	400	600	800	1000
Effort F_E (newtons)	5	25	45	65	85	105

Plot the load-effort graph (Fig. 13.3) and determine the law of the machine. Use the law to estimate the effort required to move a load of 700 N.

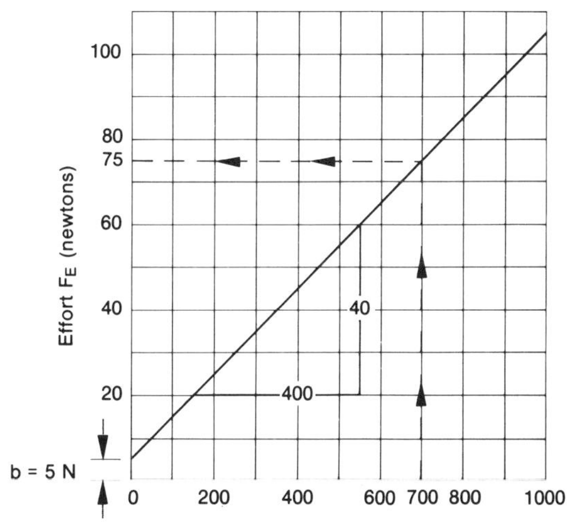

Fig. 13.3

Solution

In Figure 13.3, representing the load-effort graph, the line cuts the effort axis at $F_E = 5$. This is the value of b.

The slope is $a = \frac{40}{400} = 0.1$

The law of the machine is $F_E = 0.1F_L + 5$

For a load of 700 N, the effort required is:

$$F_E = 0.1 \times F_L + 5 = 0.1 \times 700 + 5 = 75 \text{ N}$$

13.5 Limiting efficiency

If we now calculate and plot efficiency for the experimental results under different load conditions, it will be found that the efficiency increases with the load. However, the increase is not proportional to the load.

There is a limiting value to the efficiency of a particular machine, which is always less than 100 per cent.

The value of the limiting efficiency can be found by combining the law of a machine with the definition of efficiency, as follows:

$$\eta = \frac{MA}{VR}$$

$$= \frac{F_L}{F_E \times VR}$$

also

$$F_E = aF_L + b$$

Substitute:

$$\eta = \frac{F_L}{(aF_L + b)VR}$$

$$= \frac{1}{\left(a + \frac{b}{F_L}\right).VR}$$

$$= \frac{1}{a.VR + \left(\frac{b.VR}{F_L}\right)}$$

As the load F_L increases, the term $\frac{b.VR}{F_L}$ becomes smaller, tending towards zero at very large loads, when the limiting efficiency becomes:

$$\boxed{\eta = \frac{1}{a.VR}}$$

Example 13.5

For each of the test results in the previous example, calculate efficiency and show that it tends towards a limiting value at large loads.

Solution

Efficiency
$$\eta_1 = \frac{F_L}{F_E . VR} = \frac{0}{5 \times 12} = 0\%$$

$$\eta_2 = \frac{200}{25 \times 12} = 66.7\%$$

$$\eta_3 = \frac{400}{45 \times 12} = 74.1\%$$

$$\eta_4 = \frac{600}{65 \times 12} = 76.9\%$$

$$\eta_5 = \frac{800}{85 \times 12} = 78.4\%$$

$$\eta_6 = \frac{1000}{105 \times 12} = 79.4\%$$

Limiting efficiency
$$\eta = \frac{1}{a . VR} = \frac{1}{0.1 \times 12} = 83.3\%$$

This relationship can best be illustrated by a graph as shown in Figure 13.4.

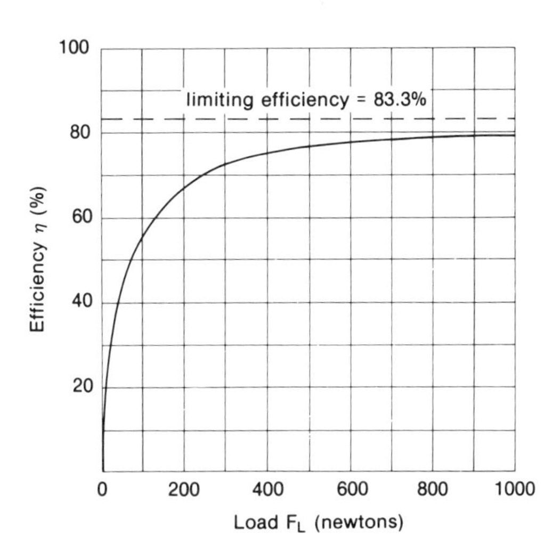

Fig. 13.4

13.6 Velocity ratio of simple machines

Problems dealing with machines usually involve the calculation of the velocity ratio.

The velocity ratio of a particular machine is independent of load and friction and depends only on the dimensions and arrangement of the moving components. The velocity ratio of a machine can be determined from the dimensions alone, and methods for calculating *VR* for various simple machines are summarised below.

The lever

The lever is the simplest machine, which consists of a rigid bar pivoted at a point called the **fulcrum**, a load being applied at one point on the bar and an effort sufficient to move or balance the load at another.

A lever which is bent is called a **bell-crank lever**. It is used to provide a change in the direction of the applied forces.

All early people used the lever in some form for moving or lifting heavy stones and other objects. A beam balance, which originated in Egypt about 5000 BC, is still widely used in its accurate modern forms. Many mechanisms used in modern machinery consist of a combination of straight and bent lever arrangements connected together.

The velocity ratio of a single lever arrangement is equal to the ratio of the perpendicular distances from the fulcrum to the line of action of the effort (d_E) and load (d_L) (see Fig. 13.5).

$$VR = \frac{d_E}{d_L}$$

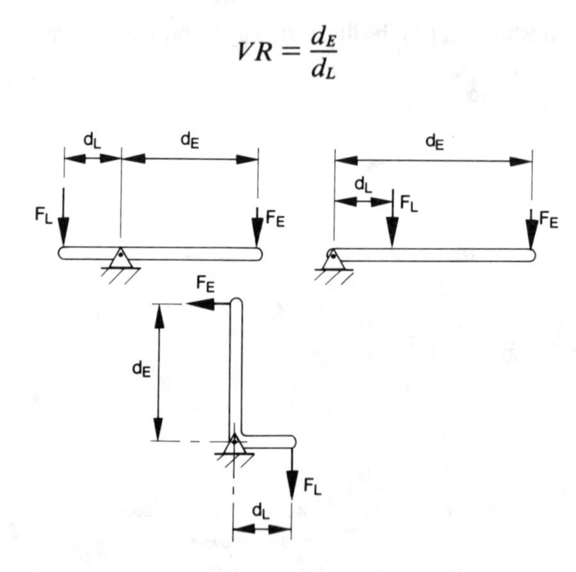

Fig. 13.5

The inclined plane

The inclined plane when used as a simple machine is a surface inclined at an angle to the horizontal and used to lift a load by an effort acting parallel to the plane.

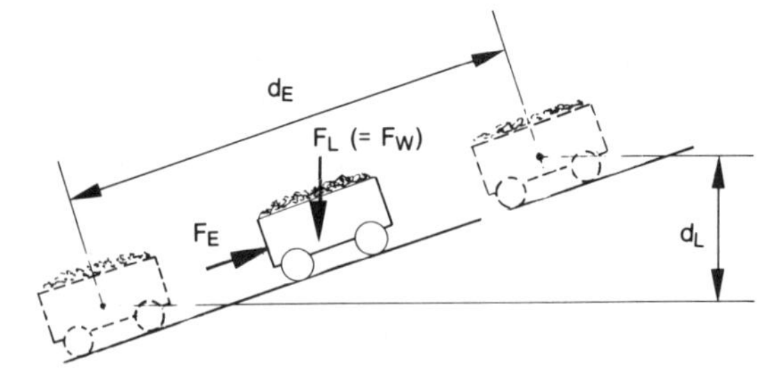

Fig. 13.6

The inclined plane is said to have been used in Egypt for moving heavy stones in the construction of pyramids. Modern machinery often employs the principle of the inclined plane for various sliding surfaces, cams and metal-cutting tools.

The velocity ratio is the ratio of the distance moved by the effort to the distance moved by the load in the same time. When the effort moves along the plane a distance d_E, the load moves through a vertical distance d_L, i.e. in the direction in which the weight of the load acts.

Hence:

$$VR = \frac{d_E}{d_L}$$

If the angle of inclination is θ, the ratio d_E/d_L is equal to cosec θ. The inclined plane may therefore be regarded as a machine having a velocity ratio of cosec θ.

Wheel and axle

In a mechanism known as the wheel and axle, the effort is applied at the circumference of the wheel while the load is raised by a rope wound around the axle whose diameter is smaller than that of the wheel.

The wheel and axle operate basically on a leverage principle. The velocity ratio is equal to the ratio of the radius of the wheel and the radius of the axle. It is usually more convenient to express this ratio as the ratio of the two diameters, D_w and D_A.

$$VR = \frac{D_W}{D_A}$$

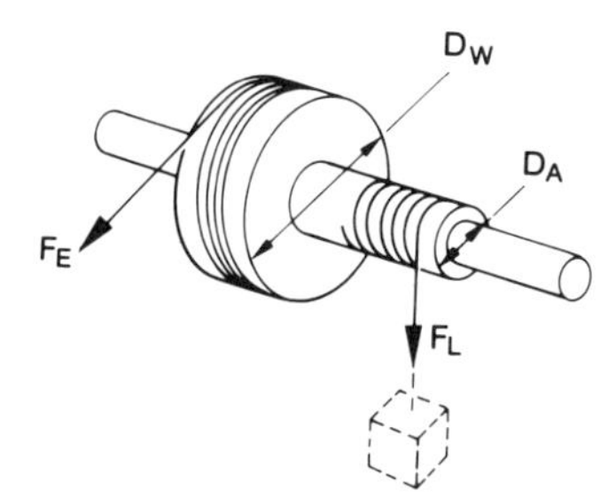

Fig. 13.7

The situation is similar to that for the lever. A lever, however, can move a load for only short distances, while the wheel and axle can move the load for a distance limited only by the available length of the rope (see Fig. 13.7).

Pulley and pulley block

The pulley is one of the most useful of the basic simple machines. It consists essentially of a wheel with a grooved rim carrying a rope or chain and supported in either a fixed or a movable bearing block.

A system of fixed and movable pulleys with a continuous rope can provide a useful mechanical advantage.

From the geometry of block and tackle arrangements, such as the one shown in Figure 13.8, it can be shown that with pulley blocks using one continuous rope, the **velocity ratio is equal to the number of falls of rope supporting the load.**

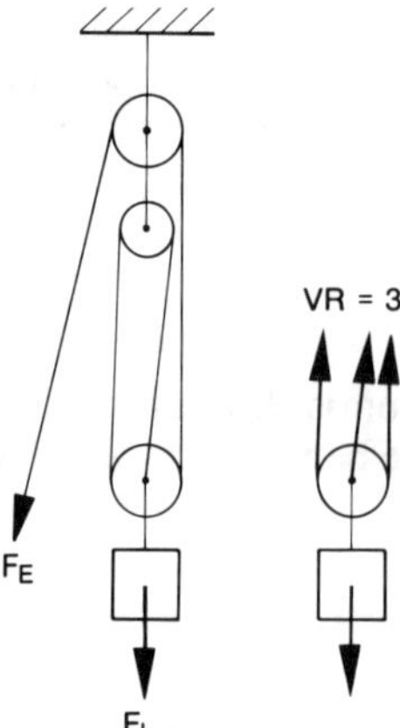

Fig. 13.8

Screw jack

A simple screw jack is a portable lifting machine for raising heavy loads through a short distance, consisting of a screw raised by a nut rotated by the effort applied at the end of a long arm (Fig. 13.9). In some designs, the screw rotates within the nut which is part of the main body of the jack.

In either case, for each revolution of the arm, i.e. the distance moved by the effort, the load is raised by an amount equal to the lead of the screw. Thus when the distance moved by the effort is $d_E = 2\pi r$, where r is the length of the arm, the load moves through a vertical distance l, equal to the lead of the screw.* Thus velocity ratio is given by

$$VR = \frac{d_E}{d_L} = \frac{2\pi r}{l}$$

* In the case of a single start thread, which is common in simple screw jacks, the lead is equal to the pitch of the screw, i.e. the distance between corresponding points on the successive convolutions of the thread.

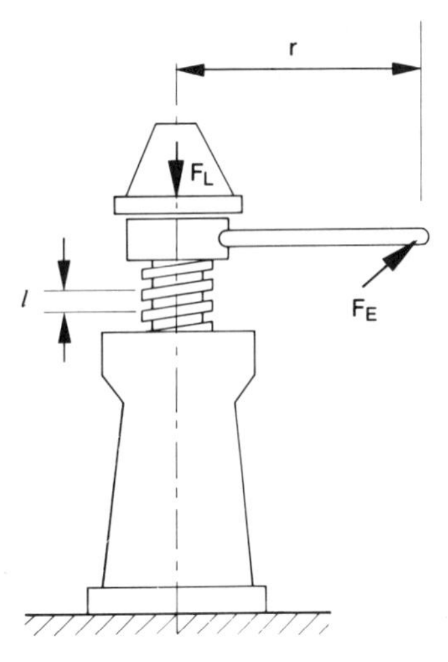

Fig. 13.9

The invention of the screw is attributed to the Pythagorean philosopher Archytas of Tarentum (fifth century BC). However, the date of its first use as a mechanical device is not known. The screw press, similar in principle to the operation of the screw jack, was probably invented in Greece in the second century BC and was used as a wine and olive-oil press in the time of the Roman Empire. In addition to its use in screw jacks and presses, the screw is used in modern machinery for a variety of purposes, including continuous transmission of motion.

Problems

13.1 The following questions refer to the same machine.
(a) Given that the effort required to lift a load of 5 t is 343 N, calculate the mechanical advantage.
(b) If the effort moves 200 mm for every millimetre moved by the load, calculate the velocity ratio.
(c) If the load is lifted a total distance of 1.37 m, calculate the output work and the input work.
(d) Calculate the efficiency at this load.
(e) Calculate the frictional effort at this load.

13.2 If the law of the machine in the previous question is $F_E = \frac{F_L}{150} + 16$, in newtons, calculate:
(a) the mechanical advantage and efficiency when the load is (i) 2 t and (ii) 8 t;
(b) the limiting efficiency.

13.3 A test on a machine gave the following results:

Load (kN)	0	40	80	120	160	200
Effort (kN)	2	4.5	7.0	9.5	12.0	14.5

Draw a graph and determine the law of the machine, in kilonewtons. If the velocity ratio is 18, calculate the efficiency when the load is 100 kN.

13.4 A simple wheel-and-axle mechanism has a wheel of 700 mm diameter and an axle of 100 mm diameter. If an effort of 50 N is required to lift a load of 29 kg, determine the mechanical advantage, efficiency and the frictional effort at this load.

13.5 A mechanism for operating a valve consists of an arrangement of two levers connected together as shown in Figure 13.10.

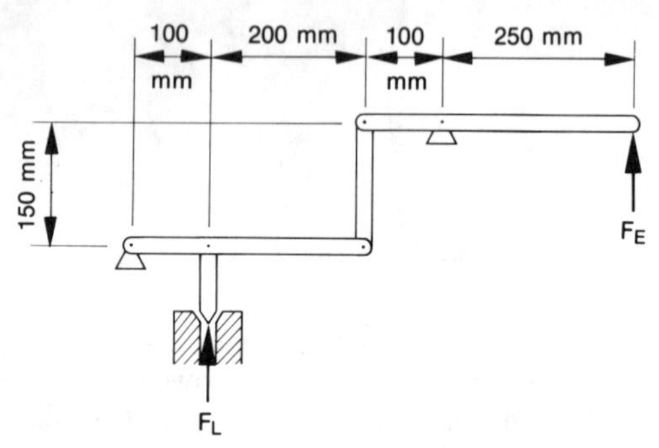

Fig. 13.10

If the load is 30 N and the frictional effort is 20 per cent of the actual effort, determine the velocity ratio, the mechanical advantage and the actual effort required.

13.6 Determine the effort parallel to the plane, required to move a body of mass 225 kg up a plane inclined at 20° to the horizontal if the coefficient of friction is 0.25, and hence calculate the efficiency.

13.7 A lifting tackle consists of two pulley blocks containing two and three pulleys respectively. How must this tackle be used to obtain the greatest velocity ratio? What is the value of this ratio?

13.8 With the best arrangement of pulleys in the previous problem and efficiency of 85 per cent, determine the effort required to lift a load of 250 kg.

13.9 A screw jack is used to raise a load of 1.2 t with an effort of 100 N at the end of an arm 250 mm long. The lead of the screw is 10 mm. Calculate the efficiency of the device.

13.10 Determine the load in tonnes that can be lifted with a screw jack, if an effort of 75 N is applied to the arm of 300 mm length and the lead of the screw is 5 mm. The efficiency is 65 per cent.

What is the number of revolutions required to lift the load through a vertical distance of 200 mm?

Review questions

1. State the definition of a machine.
2. Define:
   (a) mechanical advantage,
   (b) velocity ratio.
3. How can mechanical advantage and velocity ratio be related to the efficiency of a machine?
4. Explain the terms theoretical effort and frictional effort and state the relationship between them.
5. Explain what is meant by the law of a machine.
6. Explain limiting efficiency of a machine and show how it can be calculated.
7. For each of the following simple machines draw a diagram and explain how the velocity ratio can be determined:
   (a) the lever,
   (b) the inclined plane,
   (c) the wheel and axle,
   (d) the pulley block,
   (e) the screw jack.

14

Mechanical drives

Mechanical drives are devices, such as gear boxes and belt or chain drives, used for the purpose of mechanical power transmission. Unlike simple machines discussed in the previous chapter, whose action is usually intermittent, mechanical drives are designed for continuous operation involving rotating components.

Transmission of power by mechanical drives is often associated with changes in the speed or direction of rotational motion of mechanical components.

Some of the fundamental principles of simple machines, such as efficiency and velocity ratio, are also applicable to the study of mechanical drives. However, in the case of mechanical drives, instead of work done it is usually more convenient to refer to power transmitted, i.e. to the time rate of doing work, as will be seen from the following discussion.

14.1 Mechanical power and drive efficiency

All mechanical drives have an input side and an output side, usually involving a rotating shaft, sprocket or pulley on each side.

The torque exerted on the drive on the input, or driver, side is called the **input torque**, T_{in}. The torque transmitted by the drive to its output side is called the **output torque**, T_{out}.

Mechanical power associated with continuous rotation of a component is given by:

$$P = T.\omega = \frac{2\pi NT}{60}$$

where P is power in watts
T is torque in newton metres
ω is rotational speed in radians per second
N is speed in revolutions per minute.

If the input and output power are calculated using the corresponding values of torque and speed, drive efficiency can be defined as the ratio of output power (P_{out}) to input power (P_{in}).

$$\eta = \frac{P_{out}}{P_{in}}$$

Example 14.1

The input shaft of a gear box rotates at 1450 r.p.m. and transmits a torque of 65.9 N.m. The output shaft rotates at 500 r.p.m. and transmits a torque of 143.3 N.m.

Determine the input and output power and the efficiency of the device.

Solution

Input power $P_{in} = \frac{2\pi NT}{60} = \frac{2 \times \pi \times 1450 \times 65.9}{60} = 10 \text{ kW}$

Output power $P_{out} = \frac{2\pi NT}{60} = \frac{2 \times \pi \times 500 \times 143.3}{60} = 7.5 \text{ kW}$

Efficiency $\eta = \frac{P_{out}}{P_{in}} = \frac{7.5 \text{ kW}}{10 \text{ kW}} = 75\%.$

14.2 Gear drives

Gear wheels operate in pairs to transmit torque and motion from one shaft to another, by means of specially shaped projections or teeth. The teeth on one gear mesh with corresponding teeth on the second gear so that motion is transferred from one to the other without any slip taking place.

There are four main types of gears: spur, helical, worm and bevel. Gear types are determined largely by the relative position of the input and output shafts.

The most common type is the **spur gear**, which has tooth elements that are straight and parallel to its axis. A spur gear pair can be used to connect parallel shafts only. Parallel shafts, however, can also be connected by gears of another type. **Helical gears**, for example, have a higher load carrying capacity than spur gears, when connecting parallel shafts.

Bevel gears are commonly used for transmitting rotary motion and torque around corners. The connected shafts, whose axes would intersect if extended, are usually, but not necessarily, at right angles to each other.

Worm gear is a gear of high reduction ratio connecting shafts whose axes are at right angles but do not intersect. It consists of a screw-like component carrying a helical thread of special form, the worm, meshing in sliding contact with a concave face gear wheel. Because of their similarity, the operation and efficiency of a worm and gear depend on the same factors as that of a screw.

The velocity ratio of a gear drive is equal to the ratio of the revolutions of the driver wheel (the input) to the revolutions of the driven wheel (the output) in the same time. In any interval of time, the same number of teeth from both

Fig. 14.1 *Gear train on a lathe*

gears come in contact with each other. It can be seen that the velocity ratio is the ratio of the number of teeth in the driven wheel to the number of teeth in the driver wheel.

Example 14.2

Figure 14.2 shows a gear drive in a certain machine. If the input shaft rotates at 660 r.p.m. and transmits a torque of 12 N.m and the efficiency is 80 per cent, determine output speed, torque and power.

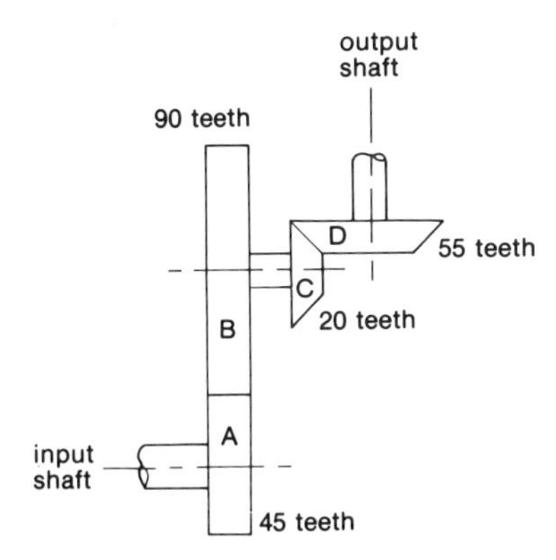

Fig. 14.2

Solution

Gear A rotates at 660 r.p.m.

$$\text{Gear B rotates at 660 r.p.m.} \times \frac{45}{90} = 330 \text{ r.p.m.}$$

Gear C rotates at 330 r.p.m.

$$\text{Gear D rotates at 330 r.p.m.} \times \frac{20}{55} = 120 \text{ r.p.m.}$$

Therefore, the output speed is 120 r.p.m.

$$\text{Input power is} \qquad P_{in} = \frac{2\pi NT}{60} = \frac{2\pi 660 \times 12}{60} = 829.4 \text{ W}$$

With an efficiency of 80 per cent, output power is:

$$P_{out} = \eta \times P_{in} = 0.8 \times 829.4 = 663.5 \text{ W}$$

Hence output torque can be calculated from:

$$P_{out} = \frac{2\pi N T_{out}}{60}$$

$$663.5 = \frac{2\pi \times 120 \times T_{out}}{60}$$

$$T_{out} = \frac{663.5 \times 60}{2\pi \times 120} = 52.8 \text{ N.m}$$

The following points should be noted very carefully:

1. Velocity ratio is a function of the dimensions and arrangement of the moving parts, e.g. number of teeth in meshing gears. It is independent of torque and efficiency.
2. Efficiency is the ratio of power output and power input which depends on power losses which occur within the gear drive due to friction.
3. Output torque depends both on velocity ratio and on efficiency and, if not given, should be calculated from output power and speed.

14.3 Chain drives

A chain drive consists of an endless chain of links meshing with the driving and driven sprockets. A very familiar example is a bicycle drive (Fig. 14.3).

The chain fits into specially shaped teeth cut in the sprockets, which prevent the chain from slipping. The tension force in the tight side of the chain is responsible for the transmission of power between the two sprockets.

The velocity ratio is inversely proportional to the number of teeth in the sprockets.

The torque on each sprocket is equal to the force in the chain multiplied by the radius of the sprocket.

In many respects the operation of a chain drive is similar to that of a pair of spur gears, with the exception that the centre distance between two parallel shafts is limited only by the length of the chain.

Efficiency of chain drives is usually high and for our purposes can be assumed to be 100 per cent.

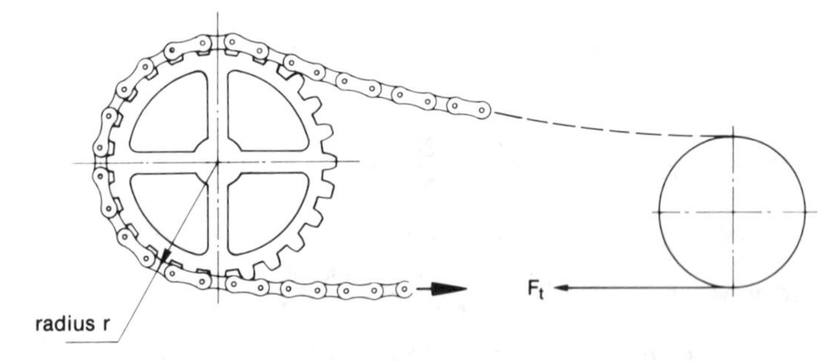

Fig. 14.3

Example 14.3

A chain drive transmits 20 kW of power from a driver sprocket 80 mm diameter having 18 teeth to a driven sprocket of 200 mm diameter with 45 teeth.

If the speed of the driver is 500 r.p.m., calculate the input and output torque and the force in the chain.

Solution

From $P = \frac{2\pi NT}{60}$, input torque is equal to:

$$T_{in} = \frac{60P}{2\pi N} = \frac{60 \times 20\,000}{2\pi \times 500} = 382 \text{ N.m}$$

The velocity ratio is $VR = \frac{45}{18} = 2.5$

Therefore the velocity of the driven sprocket is:

$$500 \div 2.5 = 200 \text{ r.p.m.}$$

If the efficiency is 100 per cent, the output power is undiminished, i.e. it is equal to 20 kW.

Hence, output torque:

$$T_{out} = \frac{60P}{2\pi N} = \frac{60 \times 20\,000}{2\pi \times 200} = 955 \text{ N.m}$$

The force of tension in the chain is found from $T = F_t \times r$, where T is torque and r is the corresponding radius.

$$F_t = \frac{T}{r} = \frac{382}{0.04} = 9550 \text{ N} = 9.55 \text{ kN}$$

14.4 Flat belt drives

Power transmission by belts is possible only with sufficient friction between the belt and its pulleys. In order to provide the necessary grip on the pulley, both sides of the belt must be in tension, i.e. there must be a tension force in both sides of the belt (see Fig. 14.4).

As a result of applied torque, and friction between the belt and the pulley, there is a difference between the tight-side tension F_t and the slack-side tension F_s.

It can be shown that friction limits the ratio between these two belt tensions according to the equation:

$$\frac{F_t}{F_s} = e^{\mu\theta}$$

where $e = 2.718$*

μ is coefficient of friction between the belt and the pulley

θ is angle of contact in radians

At the same time, torque is equal to the algebraic sum of the moments of the two forces about the centre line of the pulley,

* e is a mathematical constant and is usually available as a function key on most calculators.

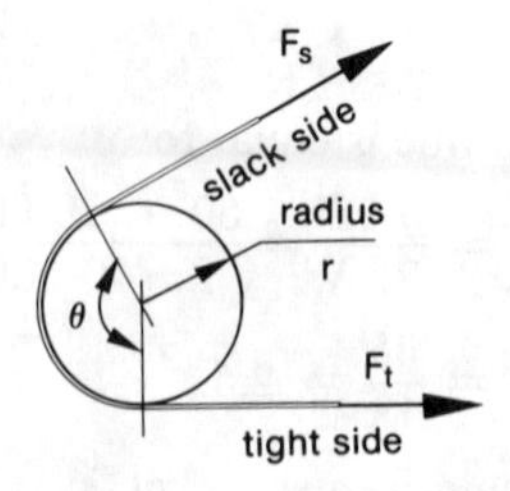

Fig. 14.4

$$T = F_t r - F_s r = r(F_t - F_s)$$

where r is radius.

Example 14.4

Determine the maximum torque that can be transmitted by a flat belt if the maximum tension is 500 N, the coefficient of friction is 0.25, the angle of contact is 150° and the diameter of the pulley is 300 mm.

Solution

Given $\mu = 0.25$

$$\theta = 150° = 150 \times \frac{\pi}{180} = 2.618 \text{ rad}$$

$$F_t = 500 \text{ N}$$

we can find the slack-side tension from:

$$\frac{F_t}{F_s} = e^{\mu\theta}$$

$$\frac{500}{F_s} = e^{0.25 \times 2.618}$$

$$F_s = 260 \text{ N}$$

Therefore, torque $T = r(F_t - F_s) = 0.15(500 - 260) = 36$ N.m

14.5 V-belt drives

The V-belt drive benefits from the wedging effect produced by the rubber belt in the specially shaped groove of the pulley. The wedging action of the belt tension increases the normal force on the belt and hence increases friction which provides the grip.

The flat belt drive equation given above can be modified by the inclusion of the sine of one half of the wedge angle as shown.

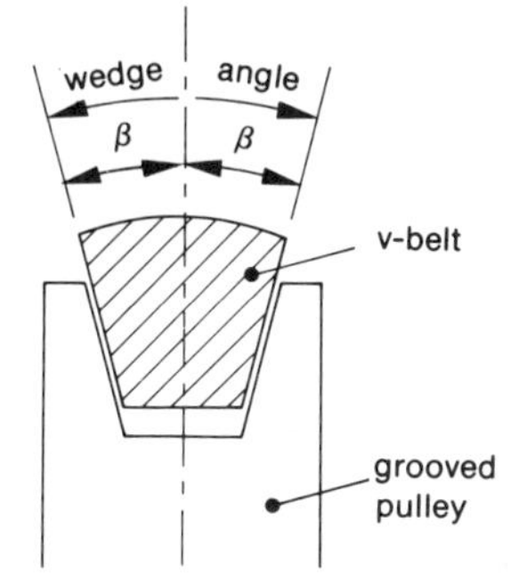

Fig. 14.5 *V-belt drive*

$$\frac{F_t}{F_s} = e^{(\mu\theta/\sin\beta)}$$

where β = one half of the wedge angle.*

Example 14.5

If in the previous example, a V-belt drive with a wedge angle of 40° was used, other conditions being the same, calculate the maximum torque.

* β is another letter from the Greek alphabet, *beta*.

Solution

$$\beta = 20°$$

Therefore,
$$\frac{F_t}{F_s} = e^{\left(\frac{\mu\theta}{\sin\beta}\right)}$$

$$\frac{500}{F_s} = e^{\left(\frac{0.25 \times 2.618}{\sin 20°}\right)}$$

$$\frac{500}{F_s} = e^{1.914}$$

$$F_s = 73.8 \text{ N}$$

Hence, torque $T = 0.15(500 - 73.8) = 63.9$ N.m

Compared with the flat belt drive of the previous example, the torque is almost doubled. Therefore, power transmitted* at a given speed would also be nearly doubled.

Problems

14.1 If the input shaft of a gearbox rotates at 1450 r.p.m. and the input torque is 50 N.m, while the output shaft rotates at 500 r.p.m. transmitting 116 N.m of output torque, calculate the input and output power and the efficiency.

14.2 A gearbox reduction unit contains four pairs of gears having the following particulars:
Pair 1—20 teeth, 131 teeth
Pair 2—20 teeth, 106 teeth
Pair 3—20 teeth, 72 teeth
Pair 4—20 teeth, 40 teeth
The input speed is 1000 r.p.m. Calculate the output speed.

14.3 The hoist of a crane consists of a 10 kW electric motor running at 1440 r.p.m. driving a 300 mm diameter drum through a 60:1 gear reduction unit. If the efficiency is 90 per cent, calculate the load in tonnes that can be lifted at the rated motor capacity and the lifting speed.

14.4 A bicycle drive has 36 teeth on the crank sprocket and 12 on the wheel sprocket. The cranks are 180 mm long and the road wheel is 600 mm in diameter.

What is the velocity ratio of the road-wheel rim to the pedal in terms of (a) rotational and (b) linear speed?

---

* The efficiency of a V-belt drive ranges from 90 to 98 per cent with a generally accepted average of 95 per cent. However, for the sake of simplicity in presentation, we have assumed 100 per cent efficiency for all calculations involving flat belt and V-belt drives in this book.

Likewise, we have ignored the effect of centrifugal force, which at high velocities tends to lift the belt off the pulley, thus reducing the frictional grip.

14.5 A small pump is driven through a chain drive by an electric motor running at 950 r.p.m. and developing 1.91 kW of power at this speed.

Details of the sprockets are as follows:
Motor, 60 mm diameter, 16 teeth
Pump, 180 mm diameter, 48 teeth
Calculate the input and output torque and the force in the chain.

14.6 A 400 mm diameter pulley is driven at 750 r.p.m. by a flat belt with a tight-side tension of 300 N and slack-side tension of 45.3 N. Determine the power transmitted by the belt.

14.7 Determine the number of V-belts required to transmit 45 kW to a 500 mm diameter pulley at 850 r.p.m., if the tight-side tension in each belt is not to exceed 775 N and the tight to slack-side tension ratio is estimated to be 8:1.

14.8 A flat-belt makes contact with a 350 mm diameter pulley over an angle of 160°. The coefficient of friction is 0.3 and the speed of the pulley is 1200 r.p.m. If the maximum allowable tension in the belt is 550 N, calculate the maximum torque and maximum power that can be transmitted by the belt.

14.9 If instead of a flat belt in the previous problem, a V-belt with a wedge angle of 28° was used, other conditions being equal, what will the maximum torque and power be?

14.10 A V-belt drive consists of a driven pulley of 250 mm diameter and a driver pulley of 100 mm diameter, at a centre-to-centre distance of 270 mm. The driver pulley rotates at 1440 r.p.m. transmitting 30 kW through 4 belts with a wedge angle of 40° and $\mu = 0.35$.

Determine total torque on each pulley and the maximum tension in each belt.

Review questions

1. State the formula used for calculating mechanical power associated with continuous rotation of a component.
2. Define drive efficiency.
3. Briefly describe different types of gear drives.
4. Explain how velocity ratio of a gear train is related to the number of teeth.
5. What is meant by gearbox efficiency?
6. Show how tension in a bicycle chain is related to the power transmitted to the wheel.
7. What is the formula relating belt tensions in
   (a) a flat belt drive?
   (b) a V-belt drive?
8. What are the advantages of V-belt over flat belt drives?

PART FIVE

Strength of materials

Have you heard of the wonderful one-hoss shay,
That was built in such a logical way,
It ran for a hundred years to a day,
And then, of a sudden, it . . .
. . . went to pieces all at once,—
All at once, and nothing first,—
Just as bubbles do when they burst.

The Deacon's Masterpiece
by Dr Oliver Holmes

15

Properties of solid materials

A major concern of the engineer lies in estimating the ability of engineering materials to withstand such factors as tension, compression, torsion and bending without failure or excessive distortion. That branch of engineering science which is concerned with the mechanical properties of materials, as they relate to problems of strength and stability of structures and mechanical components, is known as **strength of materials**.

We are not concerned here with the methods of materials testing to obtain information about the physical properties of engineering materials or with the theory and practice of materials processing. However, it is undeniable that a good understanding of the wide range of materials now available in industry is essential to the engineer. It is, therefore, assumed that the student of engineering science will also direct attention to the study of **properties and testing of materials**.

The purpose of this chapter is to provide a general introduction to physical and mechanical properties of solid materials, which include density, strength under different loading conditions, hardness and the ability to undergo or resist deformation.

15.1 Density

Density ρ (rho)* of a material is defined as its *mass per unit volume*. The SI unit of density derived from the units of mass (kg) and volume (m^3), is kilogram per cubic metre (kg/m^3).

$$\rho = \frac{m}{V}$$

Alternatively tonnes per cubic metre may be used as a convenient unit of density.

Typical values of density for some solid materials are given in Table 15.1. It should be understood that these are approximate or typical values only, the actual density of each material will depend on its exact composition.

---

* *Rho* is another letter of the Greek alphabet.

Table 15.1 Densities of various solids

Material	*Density (kg/m³)*
Aluminium	2780
Asbestos	3070
Balsa wood	160
Brass	8250
Brick	2080
Bronze	8670
Cast iron	7200
Concrete	2240
Copper	8870
Ice	920
Oregon pine timber	530
Rubber	920
Sand	1470
Steel	7800
Zinc	7020

Example 15.1

Determine the weight of a tubular steel column 120 mm outside diameter, 100 mm inside diameter and 3 m high.

Solution

Cross-sectional area, $A = \frac{\pi \times 0.12^2}{4} - \frac{\pi 0.1^2}{4}$

$= 3.456 \times 10^{-3}\ \text{m}^2$

Volume, $A \times L = 3.456 \times 10^{-3}\ \text{m}^2 \times 3\ \text{m} = 0.0104\ \text{m}^3$

Mass, $V \times \rho = 0.0104\ \text{m}^3 \times 7800\ \frac{\text{kg}}{\text{m}^3} = 80.86\ \text{kg}$

Weight, $m.g = 80.86\ \text{kg} \times 9.81 \frac{\text{N}}{\text{kg}} = 793.3\ \text{N}$

15.2 Strength

Strength is the ability of the material to withstand applied force without failure. As such, strength is one of the most important mechanical properties of engineering materials.

Strength varies, depending on the nature of the load applied to the material and we often have to distinguish between tensile, compressive and shear strengths of a particular material.

Tensile strength

Tensile strength of a material is determined by the tension test which consists of the gradual application of an axial tensile force, i.e. pulling apart of the specimen, until fracture occurs.

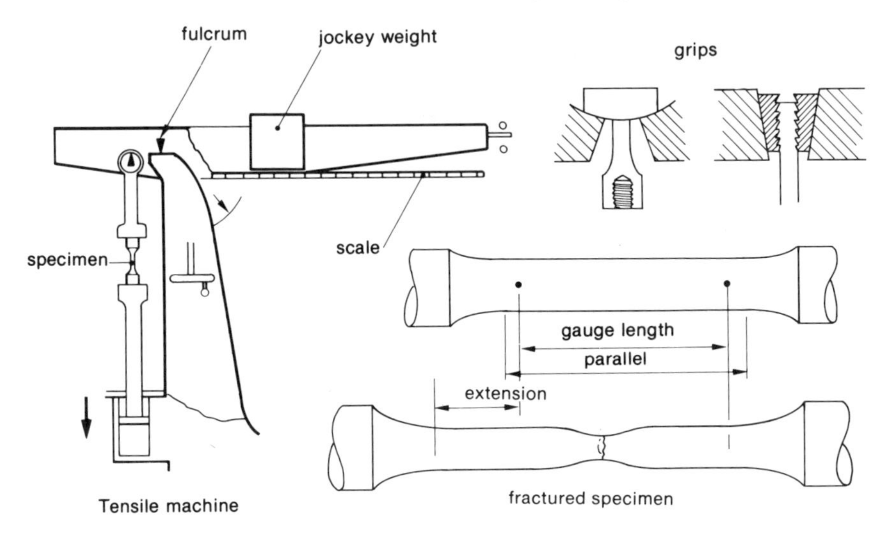

Fig. 15.1

The equipment for tensile testing of materials usually consists of a test specimen, generally cylindrical, with a middle section of smaller diameter than the ends as shown in Figure 15.1, an appropriate set of grips to grasp the test piece, and a machine that applies, measures and records the load.

It is obvious that for the result to be meaningful it should depend on the strength of the material and be independent of the actual size of the test piece. It is therefore expressed not as a force but as force per unit of cross-sectional area of the test specimen.

Much engineering design is based on this measure of tensile strength, known as **ultimate tensile strength*** (UTS) and defined as the ratio of the maximum tensile force applied before fracture occurs to the initial cross-sectional area of the test specimen.

$$\text{UTS} = \frac{\text{maximum tensile force}}{\text{initial cross-sectional area}}$$

Since diameters of test pieces are usually measured in millimetres, it is often convenient to calculate cross-sectional areas in square millimetres and to express ultimate tensile strength in units of force, i.e. newtons, per square millimetre of area.

* Also known as ultimate tensile **stress**. The term "stress" will be introduced in the next chapter.

Table 15.2 Ultimate tensile strength

Material	*UTS* *(N/mm²)*
Tool steel	1000
High tensile steel	590
Mild steel	470
Copper wire	415
Copper sheet	210
Brass	190
Cast iron	180
Timber (pine)	105*
Aluminium	90
Nylon	70

* Wood strength varies considerably with the direction of application of the load. The strength shown above is for parallel (i.e. along the grain) direction of the force. Transverse (i.e. perpendicular to grain) strength is only 3 N/mm².

Example 15.2

A steel test specimen, 10 mm diameter, ruptured under a tensile load of 37 kN. What was the ultimate tensile strength of the steel?

Solution

$$\text{Cross-sectional area} = \frac{\pi \times 10^2}{4} = 78.54 \text{ mm}^2$$

$$\text{Tensile force} = 37\ 000 \text{ N}$$

$$\text{UTS} = \frac{37\ 000 \text{ N}}{78.54 \text{ mm}} = 471 \text{ N/mm}^2$$

Typical average values of ultimate tensile strengths of some materials are given in Table 15.2.

Compressive strength

Compression may be regarded as opposite to tension in so far as the axial force is applied in the opposite direction, tending to crush, rather than pull apart, the material.

Most materials of a brittle nature, such as concrete, brick and cast iron, have relatively low tensile strength and are not used in structures or components subjected to large tensile forces. However, these materials can withstand high compressive loads and are usually used in building construction (brick, concrete,* masonry) and for machine-tool framework (cast iron). For these materials the **ultimate compressive strength** (UCS) is more important than tensile strength.

* Reinforced concrete is an application in which high compressive strength of concrete is combined with the high tensile strength of steel reinforcing rods embedded in the concrete.

Compressive strength of cast iron is generally about four times its tensile strength, depending on the composition of alloy elements, or an average of 700 N/mm^2.

Compressive strength of concrete and especially that of bricks varies so greatly that no single figure can be given. Typical values lie somewhere between 10 N/mm^2 and 80 N/mm^2, with the average somewhere around 20 N/mm^2.

Example 15.3

A portable testing machine for carrying out crushing tests on concrete, used for quality control on a large construction site, applies an axial force of 433 kN which causes compression failure in a concrete specimen, 150 mm diameter and 300 mm high. What is the ultimate compressive strength of concrete? (See Fig. 15.2.)

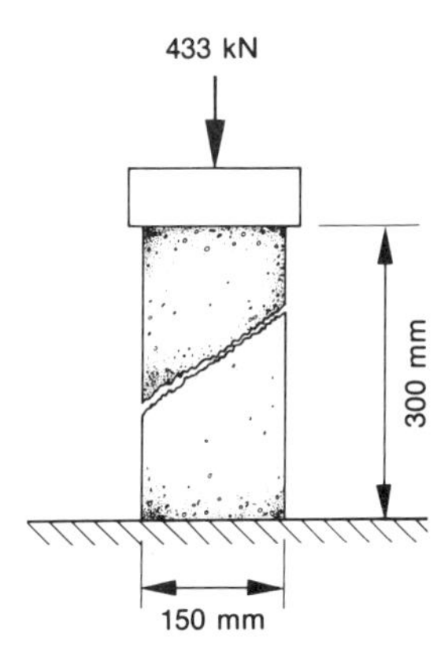

Fig. 15.2

Solution

$$\text{UCS} = \frac{\text{force causing crushing failure}}{\text{cross-sectional area}}$$

$$= \frac{433\,000 \text{ N}}{\left(\dfrac{\pi 150^2}{4}\right)\text{mm}^2} = 24.5 \text{ N/mm}^2$$

Shear strength

When a steel plate is cut by a guillotine or a hole is punched in it, the failure of the material occurs not in a plane normal to the force, as is the case in tension or compression, but as a sliding failure parallel to the load applied. The material is said to shear under the action of a shear force.

Apart from direct applications of shear forces for cutting or punching of materials, consideration of shear strength has relevance to the design of bolted and welded connections, bending of beams and torsion in shafts.

Typical values of ultimate shear strength (USS) of some materials are: mild steel—360 N/mm^2, brass—150 N/mm^2, aluminium—125 N/mm^2.

Example 15.4

Determine the force required to punch a 12 mm diameter hole in a mild steel plate 6 mm thick.

Solution

The area resisting shear is measured by the product of circumference (πD) and plate thickness (t). (See Fig. 15.3.)

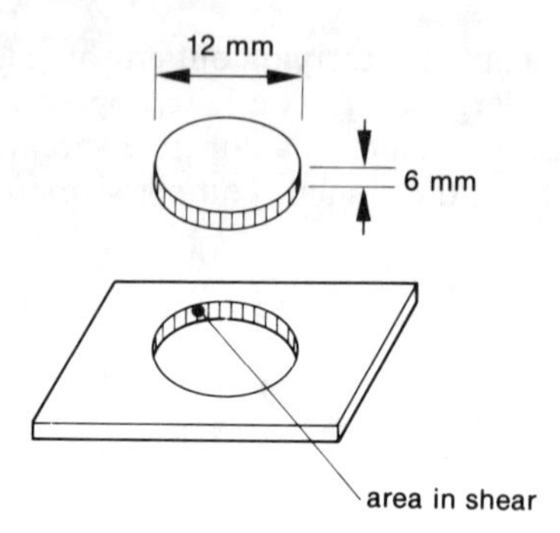

Fig. 15.3

$$A = \pi \times D \times t = \pi \times 12 \text{ mm} \times 6 \text{ mm} = 226.2 \text{ mm}^2$$

The ultimate strength of mild steel is 360 N/mm², i.e. a force of 360 N is required to shear through every square millimetre of area subjected to shear.

Therefore, the total force required to punch the hole is:

$$F = 360 \text{ N/mm}^2 \times 226.2 \text{ mm}^2 = 81\ 432 \text{ N} = 81.4 \text{ kN}$$

Example 15.5

Determine the greatest thickness of brass sheet, 850 mm wide, that can be sheared by a straight cutting* guillotine capable of applying a force of 510 kN.

Solution

The ultimate shear strength of brass is 150 N/mm².

The area in shear is given by the product of the width and thickness of the sheet, i.e.

$$850 \text{ mm} \times t = (850t) \text{ mm}^2$$

The force available is 510 000 N, therefore:

$$510\ 000 \text{ N} = 150 \text{ N/mm}^2 \times (850t) \text{ mm}^2$$

$$\text{Hence } t = 4 \text{ mm}$$

---

* If the guillotine knife edge cuts obliquely, the cutting action progressing gradually along the line of the cut requires a somewhat different force.

15.3 Other properties

There are many other mechanical properties of solid materials which are important for engineering design and manufacturing technology. Some of these are briefly outlined below. The student should refer to one of the many texts on the science of engineering materials for further study.

Impact strength

Many materials, sensitive to the presence of cracks or notches, fail suddenly under impact. In the most common impact testing machines, a hammer-like pendulum strikes a notched bar specimen of standard size. The energy expended to fracture the specimen is calculated from the heights of the pendulum before and after impact. The amount of energy used is a measure of the impact strength of the material tested.

Impact strength is important in the design of machines and machine tools in which sudden application of a load is likely to occur.

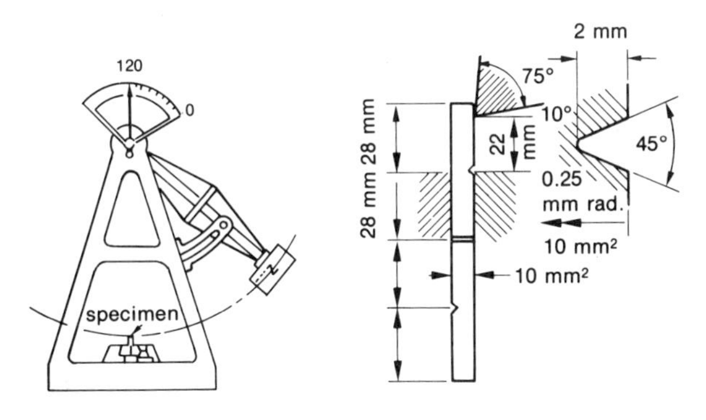

Fig. 15.4 izod

Fatigue

It is found that materials that can survive a single application of a large force, frequently fail under a smaller load if such a load is applied repeatedly. The phenomenon of the failure of materials under the repeated application of a load cycle is called fatigue.

Three factors are involved: the range of load variation, the mean value of the load and the frequency of the cycle. Fatigue is usually measured by means of subjecting a rotating test specimen to some form of eccentric force producing a cyclically varying loading condition. A material is generally considered to suffer from fatigue if it fails in less than 10 000 cycles.

Fatigue is an important factor in aircraft design and wherever repeated load cycles occur, e.g. in turbine blades, railway wheels, gear teeth, engine parts and bearings. It is said that about four in every five machine component failures occur due to metal fatigue.

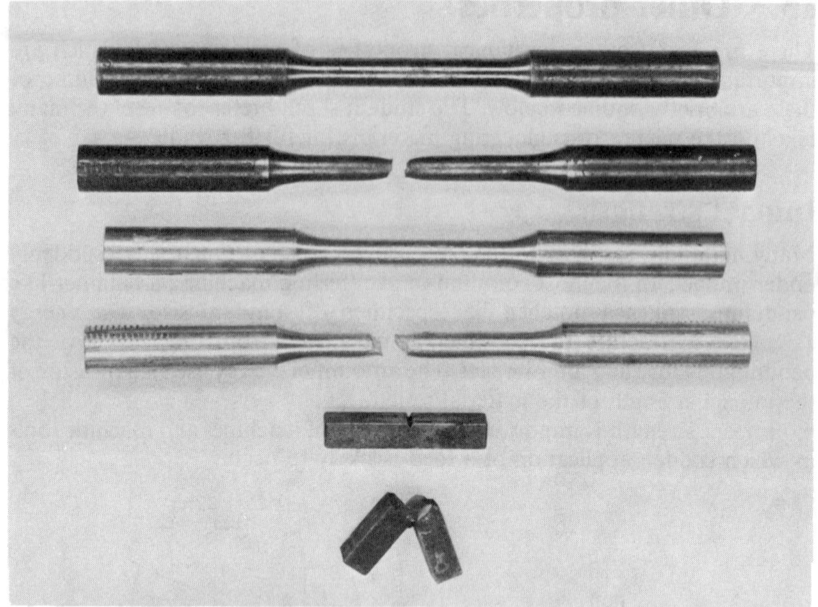

Fig. 15.5 *Tensile and impact test specimens*

Hardness

Hardness of a material is defined as its ability to resist surface scratching, abrasion, indentation or local penetration. Diamond is very hard, while gypsum is soft. The hardness of most engineering materials lies between these extremes and is somewhat related to their strength. In engineering applications, material hardness has an important bearing on the problem of wear.

The most common hardness testing methods, known as Vicker's, Rockwell or Brinell tests, are all based on pressing a standard indenter, of pyramid, cone or spherical shape, into the surface of the material, and relating the force and the size of indentation to a standardised scale of hardness.

Deformation

No material is perfectly rigid. When subjected to a force, no matter how small, every material will in fact suffer some degree of deformation in size or shape, or both. Different materials respond differently to applied forces and below is a brief summary of terminology used to describe the ability of a material to undergo, or to resist, deformation.

Stiffness is the ability of a material to resist deformation under load. The exact measure of stiffness, known as Young's modulus, will be discussed in the following chapter.

Toughness is the ability of a material to absorb energy when being deformed and thus resist deformation and failure.

Elasticity is the ability of a material to return to its original dimensions after having been deformed, upon removal of a deforming force.

Plasticity is the ability of a material to undergo permanent deformation without failure.

Ductility is the ability of a material to be permanently deformed by predominantly tensile forces.

Malleability is the ability of a material to be plastically deformed by predominantly compressive forces.

Problems

15.1 An aluminium column is 160 mm outside diameter, 140 mm inside diameter and 2.5 m high. Determine its mass.

15.2 Determine the mass of the component shown in Figure 15.6 if the material is cast iron.

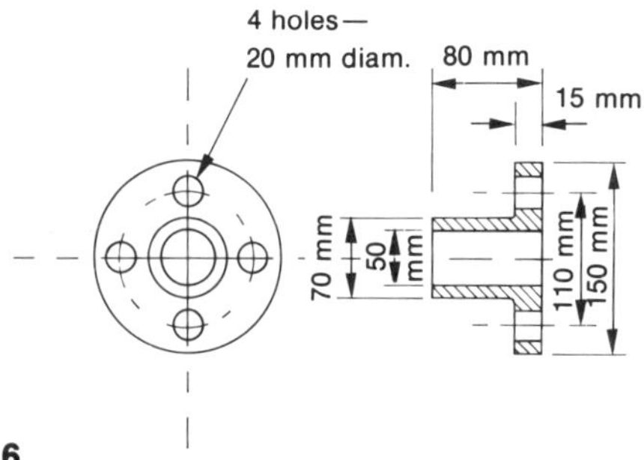

Fig. 15.6

15.3 A tensile test specimen, 15 mm diameter, requires a maximum force of 70 kN to cause failure. What is the ultimate tensile strength of the material?

15.4 A tensile testing machine is capable of a maximum pull of 125 kN. What can be the maximum diameter of test specimens, to ensure that materials with ultimate tensile strength of up to 1100 N/mm^2 could be tested to the point of failure?

15.5 A tensile test on a 10 mm diameter specimen registered failure under a load of 32.6 kN. What force will be sufficient to break a 1.5 mm diameter wire made from this material?

15.6 What is the ultimate compressive strength of brick if a test cube, 50 mm × 50 mm × 50 mm, was crushed by a force of 77.5 kN during a compression test?

15.7 Determine the force required to punch a 15 mm diameter hole in a 6 mm thick aluminium plate.

15.8 Determine the force required to punch 20 mm × 15 mm rectangular holes in a mild steel plate 8 mm thick.

15.9 A press designed to punch 25 mm diameter holes in 10 mm thick brass plate is to be used to punch 15 mm × 12 mm rectangular holes in mild steel using the same force. What is the maximum thickness of mild steel plate that can be used?

15.10 Determine the force required to shear an aluminium strip, 250 mm wide by 3 mm thick, by a straight cutting guillotine knife?

Review questions

1. Define density.
2. Explain what is meant by ultimate strength.
3. Distinguish between:
   (a) ultimate tensile strength,
   (b) ultimate compressive strength,
   (c) ultimate shear strength.
4. Briefly explain the significance of
   (a) impact strength,
   (b) fatigue,
   (c) hardness.
5. List and explain terms used to describe material properties related to deformation.

16

Stress and strain

So far we have discussed ultimate strength of materials, i.e. strength up to the instant of failure. We must now consider behaviour of materials under loads which do not cause rupture, crushing or any form of permanent damage or deformation. We must also learn how to calculate the allowable load on a structure or a component which will ensure a sufficient degree of safety.

The first part of this chapter is devoted to simple applications of tension and compression, and the second to materials subjected to shear. The following two chapters will examine some of the more specialised applications of tension, compression and shear.

16.1 Direct axial stress

We will start by considering a bar of solid material, forming a component of a machine or structure, subjected to an axial push or pull, i.e. tension or compression. Let us call the force acting on the bar direct axial load, F.

If the bar is of uniform cross-sectional area, A, the force F may be assumed to be distributed uniformly over the cross-section, requiring a certain degree of adhesion between the particles of the material in order to keep the material intact.

The intensity of force distribution over the cross-sectional area of the material subjected to direct load is called **direct stress**. Thus, direct stress can be defined as the share of the total axial load carried by each unit of cross-sectional area:

$$f = \frac{F}{A}$$

where f stands for the stress and can be identified as tensile or compressive stress by means of appropriate subscripts, f_t or f_c, if necessary.

The SI unit of stress is the **pascal**, with the symbol, Pa. The definition of the pascal follows from the definition of stress, i.e. force per unit area, or newton per square metre.

The pascal is a very small unit, not very suitable for measuring stresses normally encountered in engineering applications. For the majority of such applications a larger prefixed unit called megapascal (MPa) is used (1 MPa = 1 000 000 Pa).

It is also very useful to remember that if force is expressed in newtons and area in square millimetres, the stress will automatically be found in megapascals. This is convenient, first because linear dimensions of mechanical components are normally measured in millimetres, hence area can easily be expressed in square millimetres, and secondly the order of magnitude of stress will conveniently be between zero and 1000 MPa. This helps to avoid the use of small decimal fractions or very large numbers.*

Example 16.1

If a bar of mild steel, 20 mm × 10 mm in cross-section, is subjected to a tensile force of 18.8 kN, determine the stress in the material.

Solution

$$\text{Stress } f_t = \frac{F}{A} = \frac{18\ 800 \text{ N}}{20 \text{ mm} \times 10 \text{ mm}} = 94 \text{ MPa}$$

16.2 Factor of safety

It is interesting to compare the result of the above example with the ultimate tensile strength of the material (UTS for mild steel is 470 N/mm^2, i.e. 470 MPa). Obviously, the stress is not sufficient to rupture the material of the bar. The bar is safe under the applied load. But how safe? If we compare the UTS and the actual stress in the material of the bar as a ratio, it is:

$$\frac{470 \text{ MPa}}{94 \text{ MPa}} = 5$$

This ratio is an indication of the degree of safety built into the situation by virtue of the fact that the actual working stress in the material is considerably below the ultimate tensile strength. Thus we say that the factor of safety in the previous example is equal to 5.

The **factor of safety** can be defined as the ratio of ultimate strength to the actual working stress in the material.**

$$FS = \frac{\text{ultimate strength}}{\text{working stress}}$$

* For all calculations in this and the following chapters, wherever stresses are involved we shall use newtons (N) for force, millimetres (mm) for linear dimensions, square millimetres for area and megapascals (MPa) for stress. The student may prefer to work in base units; the results should be the same.

** This definition of the factor of safety is equally valid for tension, compression or shear, provided that the appropriate ultimate strength is used.

Table 16.1 Factors of safety

Material	*Static load*	*Cyclic load*
Steel, ductile materials	3–4	8
Cast iron, brittle materials	5–6	10–12
Timber	7	15

The factor of safety is a dimensionless number, always greater than one, and depends on a number of considerations, including: possible defects in materials or workmanship, the exactness with which probable loads are known, possibility of shock or impact loads, safety to human life or property; and an allowance for decay, wear, corrosion, etc.

Typical factors of safety used in design practice are given in Table 16.1.

In design problems the aim may be to determine a suitable cross-section of a material for a given load using a recommended safety factor. The working stress under these conditions is usually referred to as design or allowable stress.

Example 16.2

Determine the minimum required diameter of a high tensile steel rod to carry a tensile load of 26 kN with a safety factor of 3.5.

Solution

From $\text{FS} = \dfrac{\text{UTS}}{f}$, allowable stress is:

$$f = \frac{\text{UTS}}{\text{FS}} = \frac{590\ \text{N/mm}^2}{3.5} = 168.6\ \text{MPa}$$

Now, from $f = \dfrac{F}{A}$, the area required is:

$$A = \frac{F}{f} = \frac{26\,000\ \text{N}}{168.6\ \text{MPa}} = 154.2\ \text{mm}^2$$

But area $A = \dfrac{\pi D^2}{4} = 154.2\ \text{mm}^2$, hence

$$D = \sqrt{\frac{154.2 \times 4}{\pi}} = 14.0\ \text{mm}$$

16.3 Axial strain

When a material is subjected to direct axial load, its size is changed in the direction of the applied force. A member subjected to a tensile force tends to stretch, and a member in compression tends to contract. It is found that, for a given magnitude of tensile load, the elongation produced is proportional to the length of the member.

The elongation per unit of original length, called the **axial strain**, is a convenient relative measure of the change in the longitudinal dimension. If the total elongation produced in a member is designated by x, and the original length by l, then the axial strain, expressed by e, is given by:

$$e = \frac{x}{l}$$

It can easily be shown that since strain is the ratio between two lengths, it is dimensionless, provided that both x and l are expressed in the same units, usually millimetres.

Example 16.3

If the mild steel bar in Example 16.1 is 2.7 m long and extends 1.27 mm under the load, what is the axial strain?

Solution

$$\text{Strain } e = \frac{x}{l} = \frac{1.27 \text{ mm}}{2700 \text{ mm}} = 0.47 \times 10^{-3} \text{ (or 0.47 per 1000).}$$

16.4 Hooke's law

All solid materials exhibit some degree of stiffness, or the ability to resist deformation under load. Load-extension diagrams recorded during tensile testing of metals and other materials all indicate that in order to produce elongation a force must be applied. Furthermore, the greater the force the greater the amount of elongation produced by that force.

The property of stiffness is closely related to the elastic and plastic behaviour of materials. It is found that up to a certain point, known as the elastic limit, the material can be stretched without taking up any permanent deformation. It can be said that, within the elastic limit, the material possesses the property of elasticity, or the ability to return to its original size after the force has been removed.

If after the material is stretched to its elastic limit the force continues to increase, the material ceases to be elastic and displays plasticity, i.e. it will suffer permanent deformation, until eventually rupture occurs at what we already know to be the ultimate stress.

A further observation reveals that there is another limit, known as the limit of proportionality, within which the elongation produced by a tensile force is directly proportional to the force.*

The concept of proportionality between load and elongation of elastic materials was originally expounded in 1676 by the famous English physicist,

---

* For most materials, particularly metals, the elastic limit and the limit of proportionality almost coincide and are regarded as one, for most practical purposes, often known as the "proportional elastic limit".

Robert Hooke (1635–1703). Hooke, who was the curator of experiments to the Royal Society of London and a distinguished member of the Society, conducted scientific research in a remarkable variety of fields, ranging from astronomy and mathematics, of which he was a professor, to the study of microscopic fossils, which made him one of the first proponents of a theory of evolution. Hooke's experimental study of elastic materials led to his discovery of the law,* bearing his name, which laid the foundation for our understanding of the relationship between stress and strain.

In its original form, Hooke's Law simply stated that for a bar of elastic material of uniform cross-section, subjected to a progressively increasing tensile load, the elongation is directly proportional to the deforming force. This formulation is limited to a particular bar of known dimensions. Furthermore, it is only true provided that the elongation produced does not exceed the limit of proportionality. A more general statement of Hooke's Law, expressed in terms of stress ($f = F/A$) and strain ($e = x/l$), is that **within the limit of proportionality, the strain is directly proportional to the stress producing it.**

The mathematical equation expressing Hooke's Law is

$$\frac{\text{stress}}{\text{strain}} = E$$

or

$$\frac{f}{e} = E$$

The constant, *E*, relating stress and strain, is a measure of the material's ability to resist stretching, within the limit of proportionality and is sometimes appropriately described as the modulus of stiffness. Historically, however, this modulus was introduced as a result of experimental study of elastic materials by another English scientist, Thomas Young (1773–1829), and was named in his honour, as **Young's modulus of elasticity**.

From definitions of stress and strain it follows that:

$$E = \frac{f}{e} = \frac{F/A}{x/l} = \frac{F.l}{A.x}$$

where E is Young's modulus of elasticity, MPa
F is axial force, N
A is cross-sectional area, mm^2
x is elongation, mm
l is original length, mm
f is stress, MPa
e is strain, mm/mm, i.e. dimensionless.

The modulus of elasticity can be determined experimentally. Table 16.2 gives typical values for common engineering materials.

* In 1676, not yet sure of all the facts, but wishing to establish priority of dates, Hooke tucked the anagram "CEIIINOSSTTUV" into a scientific publication on an entirely different subject. Unscrambled this reads "Ut tensio, sic vis" and translated from Latin "As the stretch, so is the force", implying proportionality between elongation and force.

Table 16.2 Young's modulus of elasticity

Material	*E (MPa)*
Aluminium	70 000
Brass	90 000
Bronze	105 000
Cast iron	120 000
Copper	112 000
Steel	200 000
Timber	12 000

Example 16.4

The stress in the mild steel bar of Example 16.1 is 94 MPa and the corresponding strain is 0.47×10^{-3}. What is Young's modulus?

Solution

$$E = \frac{f}{e} = \frac{94 \text{ MPa}}{0.47 \times 10^{-3}} = 200\,000 \text{ MPa}$$

Example 16.5

A 40 mm diameter aluminium tie-rod 500 mm long is turned down to 30 mm diameter over 200 mm of its length. The rod is then subjected to 20 kN axial pull. Determine the total amount of elongation and the safety factor.

Solution

From Tables 15.2 and 16.2, ultimate tensile strength of aluminium is 90 N/mm^2 and Young's modulus is 70 000 MPa.

For the 40 mm diameter section, which is 300 mm long,

$$\text{stress,} \quad f_1 = \frac{F}{A_1} = \frac{20\,000 \text{ N}}{\dfrac{\pi \times 40^2}{4} \text{ mm}^2} = 15.92 \text{ MPa}$$

$$\text{strain,} \quad e_1 = \frac{f_1}{E} = \frac{15.92}{70\,000} = 0.227 \times 10^{-3}$$

$$\text{elongation,} \quad x_1 = e_1 . l_1 = 0.227 \times 10^{-3} \times 300 \text{ mm} = 0.0682 \text{ mm}$$

Similarly, for the 30 mm diameter, 200 mm length,

$$\text{stress,} \quad f_2 = \frac{F}{A_2} = \frac{20\,000 \text{ N}}{\dfrac{\pi \times 30^2}{4} \text{ mm}^2} = 28.29 \text{ MPa}$$

$$\text{strain,} \quad e_2 = \frac{f_2}{E} = \frac{28.29}{70\,000} = 0.404 \times 10^{-3}$$

$$\text{elongation,} \quad x_2 = e_2 . l_2 = 0.404 \times 10^{-3} \times 200 \text{ mm} = 0.0808 \text{ mm}$$

The total elongation of the rod is made up of the extensions of the two parts.

$$\text{Total elongation, } x = x_1 + x_2 = 0.0682 \text{ mm} + 0.0808 \text{ mm} = 0.149 \text{ mm}$$

The safety factor must be based on the maximum stress in the material, which occurs in the reduced section and is equal to 28.29 MPa.

$$\text{Safety factor FS} = \frac{\text{UTS}}{f} = \frac{90}{28.29} = 3.18.$$

Problems

(Refer to Tables 15.2, 16.1 and 16.2 for information.)

16.1 Determine the stress in a concrete column, 150 mm x 150 mm, if it supports an axial load of 112.5 kN.

16.2 What is the strain in a structural member which is 3.5 m long, if its elongation is 1.05 mm?

16.3 What is Young's modulus of a material in which a stress of 20 MPa produces a strain of 0.167 per 1000?

16.4 What is the maximum allowable stress in a component made from high tensile steel, designed for a cyclic load?

16.5 A steel measuring tape, 5 mm wide x 0.3 mm thick, is stretched with a pull of 50 N when used to measure a length of 40 m. Determine the stress in the material and the elongation produced.

16.6 A vertical timber column, 200 mm × 200 mm, 3 m high is subjected to an axial compression load of 120 kN. Calculate the stress in the timber and the amount of axial deformation.

16.7 A short cast-iron column is 100 mm outside diameter, 75 mm inside diameter. Calculate the maximum load it may carry, if the factor of safety is 6 and the ultimate compressive strength of cast iron is 700 N/mm^2.

16.8 A brass bar, 25 mm × 25 mm × 350 mm long, is to carry an axial tensile load with a safety factor of 4. Calculate the maximum load it may carry and the elongation produced under this load.

16.9 During a test on a new synthetic material, a tensile force of 4170 N acting on a 10 mm diameter specimen produced an elongation of 0.17 mm on a gauge length* of 100 mm.

What should be the square section of a component made from this material if it is to carry a tensile load of 12.5 kN with a strain not exceeding 0.001?

16.10 A brass rod is 500 mm long and 20 mm diameter. If it is turned down to 15 mm diameter for a length of 100 mm at each end, calculate the total elongation when subjected to an axial pull of 7 kN. Calculate also the factor of safety at this load.

* Gauge length is the exact length between two marks on the specimen measured along its axis before loading is begun.

16.5 Shear stress

In the previous chapter some examples of shear force were considered, including punching of holes and cutting of metal plate by a guillotine.

There are also numerous applications where shear force is not sufficient to overcome the ultimate shear strength of a material and no failure occurs. However, shear stress may exist. Shear stress is produced by equal and opposite forces whose lines of action do not coincide.

Figure 16.1(a) shows a simple pin coupling in which the pin is placed in double shear by the force F_S, along the two shear planes as indicated. Shear stress acts along the planes, parallel to the lines of action of the applied forces. The tendency is to shear through the pin as shown in Figure 16.1(b). Naturally, if the size and strength of the pin are sufficient no such failure would occur.

Shear stress (f_s) is assumed to be distributed uniformly over the total area subjected to shear force and may be defined as shear force per unit of the total area:

$$f_s = \frac{F_s}{A_s}$$

Care should be exercised to distinguish between areas in single shear, such as those of a bolt or rivet holding two plates in a simple lap-joint, and areas in double shear, as illustrated by the following examples.

Example 16.6

If the diameter of the mild steel pin in Figure 16.1 is 10 mm and the maximum force applied to the coupling is 11.3 kN, what is the shear stress in the material of the pin and the factor of safety?

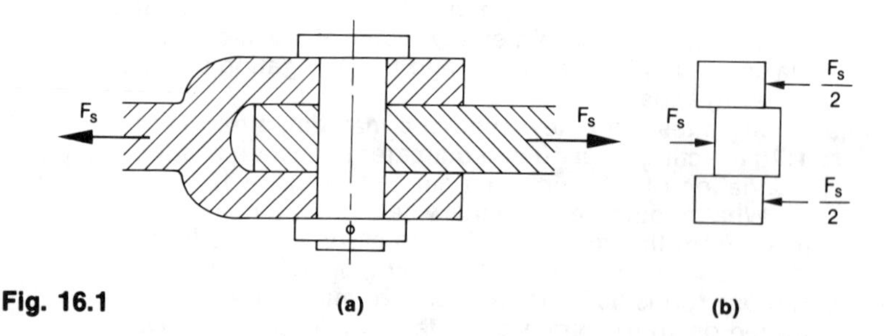

Fig. 16.1

Solution

This is a case of double shear. Therefore, the total area is:

$$A_s = 2 \times \left(\frac{\pi \times 10^2}{4}\right) = 157 \text{ mm}^2$$

Shear stress,
$$f_s = \frac{F_s}{A_s} = \frac{11\,300 \text{ N}}{157 \text{ mm}^2} = 72 \text{ MPa}$$

From Chapter 15, ultimate shear strength of mild steel is 360 N/mm^2. Hence, the factor of safety

$$FS = \frac{\text{USS}}{f_s} = \frac{360}{72} = 5$$

Example 16.7

Determine the required diameter of a single mild steel bolt, holding two overlapping strips of metal, against a shear force of 4.5 kN, if the allowable stress in shear is 90 MPa.

Solution

From $f_s = F_s/A_s$, the required area is:

$$A_s = \frac{F_s}{f_s} = \frac{4500 \text{ N}}{90 \text{ MPa}} = 50 \text{ mm}^2$$

This is a case of single shear, only one cross-sectional area need be considered, hence:

$$D = \sqrt{\frac{4A_s}{\pi}} = \sqrt{\frac{4 \times 50}{\pi}} = 7.98 \text{ mm}$$

Therefore an 8 mm diameter bolt will be satisfactory.

16.6 Shear modulus of rigidity

Elasticity of solid materials is not limited to tension and compression. When an elastic material is subjected to moderate shear force, some degree of deformation occurs. However, unlike tension or compression, shear stress is accompanied by deformation of shape, rather than size.

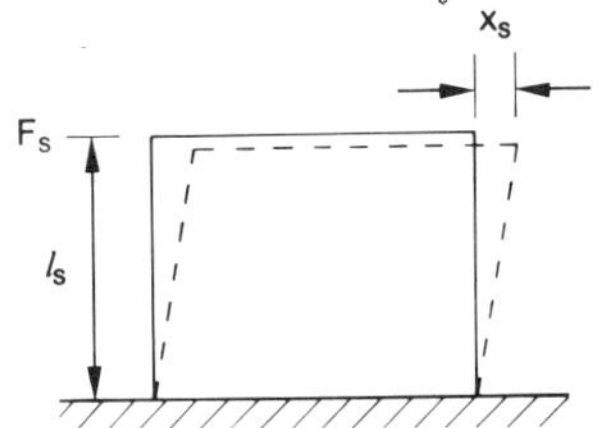

Fig. 16.2

Let us consider, for example, a solid block, as shown in Figure 16.2, subjected to shear force F_s. The distortion of its shape, indicated by dotted lines, can be described in terms of deformation x_s relative to a dimension, l_s, at right angles to x_s. The ratio of x_s to l_s is called **shear strain** e_s.

$$e_s = \frac{x_s}{l_s}$$

It should be clearly understood that although the form of the definition, and the equation arising from it are almost identical, shear stress is distinctly different from axial stress, in that deformation is that of shape and not size.

Having defined shear stress (f_s) and shear strain (e_s), we can relate the two as a ratio called **modulus of rigidity**, usually denoted by G.

$$G = \frac{f_s}{e_s}$$

Modulus of rigidity for any given material is found to be constant, within its limit of proportionality. Furthermore, it is found that the numerical value of the modulus of rigidity of a particular material is equal to a fixed percentage of its Young's modulus (E). For isotropic materials, i.e. materials which have the same elastic properties in all directions,* this percentage, determined theoretically by using the molecular theory of structure of the material, is 40 per cent.**

Therefore,

$$G = 0.4E$$

Example 16.8

What is the value of the modulus of rigidity of steel?

Solution

Young's modulus for steel is $E = 200\,000$ MPa. Therefore, $G = 40\%$ of $E = 0.4 \times 200\,000 \text{ MPa} = 80\,000$ MPa.

Example 16.9

An anti-vibration mounting for a machine is in the form of rubber pads, 200 mm × 200 mm × 10 mm thick, each subjected to a periodic horizontal force, reaching a maximum of 600 N, distributed over its area.

Determine the amount of shear deformation, x_s, corresponding to the maximum force, if the modulus of rigidity of rubber is 1.5 MPa.

Solution

Shear stress in the material of the pads is:

$$f_s = \frac{F_s}{A_s} = \frac{600\text{N}}{200 \text{ mm} \times 200 \text{ mm}} = 0.015 \text{ MPa}$$

From $G = f_s/e_s$, shear strain is:

$$e_s = \frac{f_s}{G} = \frac{0.015}{1.5} = 0.01$$

---

* This limitation obviously excludes timber.

** Experimental investigations show that for engineering materials, particularly metals, the actual values of the modulus of rigidity generally vary between 37 per cent and 42 per cent of the elastic modulus. This variation is not very significant and 40 per cent will be used as a sufficiently accurate value for our purposes.

Therefore, from $e_s = x_s/l_s$, shear deformation, x_s, is

$$x_s = e_s \times l_s = 0.01 \times 10 \text{ mm} = 0.1 \text{ mm}$$

Shear modulus of rigidity is an important factor in the design and analysis of power transmitting shafts and other cylindrical components subjected to torque, or torsional effort, which tends to produce some degree of twisting in the material of the component. The angle of twist is inversely proportional to the shear modulus of rigidity. However, the relationship also involves other parameters and is best left to more advanced texts in strength of materials.

Problems

(Where appropriate, use ultimate shear strength of mild steel = 360 MPa, brass = 150 MPa and aluminium = 125 MPa.)

16.11 A mild steel bolt, 14 mm diameter, is subjected to double shear by a force of 27.7 kN. Determine the shear stress and the factor of safety.

16.12 Determine the required number of 10 mm diameter mild steel bolts to hold two overlapping strips of metal, against a total shear force of 75.4 kN if the allowable stress in shear is 120 MPa.

16.13 Calculate the required diameter of a brass pin which is to be subjected to double shear under a load of 1.7 kN, if the factor of safety is to be 5.

16.14 A 10 mm diameter specimen of hardened tool steel was tested in double shear and failed under a load of 106.8 kN. If a machine-tool component of square cross-section, 12 mm × 12 mm, is made from the same material, determine the maximum shear load in single shear allowed if the factor of safety is to be 4.

16.15 A flanged coupling is used to transmit a torque of 1500 N.m between two shafts. The coupling has four 8 mm diameter bolts equally spaced on a pitch circle diameter of 175 mm. Determine the shear stress in the bolts.

16.16 Determine the maximum torque that can be transmitted by a flanged coupling using six 10 mm diameter mild steel bolts equally spaced on a pitch circle diameter of 250 mm, if the allowable stress is 90 MPa.

16.17 The following specification is used for the design of a flanged coupling between two coaxial shafts:

Speed—650 r.p.m.
Power transmitted—550 kW
Bolt diameter—12 mm
Pitch circle diameter—200 mm
Material—mild steel
Factor of safety—4

Determine the number of bolts required, assuming the bolts are equally loaded.

16.18 A rectangular hole, 15 mm × 10 mm, is to be punched in a 4 mm thick brass sheet. Calculate the force required and the compressive stress in the punch.

If the ultimate compressive strength of the material of the punch is 650 MPa, what is the factor of safety during punching?

16.19 Determine the maximum thickness of an aluminium sheet in which a 50 mm diameter hole is to be punched, if the allowable compressive stress in the punch is not to exceed 80 MPa.

16.20 A square block of rubber, 350 mm × 350 mm × 25 mm thick is fastened by adhesive between two parallel horizontal plates 25 mm apart.

If a horizontal force of 29.4 kN is applied to one of the plates, while the other is held firmly in its original position, what will be the magnitude of relative motion between the plates?

Under these conditions state the values of shear stress and shear strain in the rubber. (Assume G = 1.5 MPa.)

Review questions

1. Define direct axial stress.
2. What is the SI unit of stress?
3. Explain the meaning and significance of factor of safety.
4. What are some of the factors influencing the choice of factors of safety?
5. Define axial strain.
6. State Hooke's Law.
7. What is Young's modulus of elasticity?
8. Define shear stress.
9. Explain the difference between single shear and double shear.
10. Define and explain shear strain.
11. What is the modulus of rigidity?
12. What is the relationship between Young's modulus and the modulus of rigidity?

17

Bolted and welded connections

In this chapter we apply the concept of stress to the analysis of connections used to join together components of building frames, such as beams and columns, and in the construction of boilers and pressure vessels.

The methods of joining two structural members, or two plates of metal, together can be divided into two groups:

1. bolted (and riveted) connections,
2. welded connections.

Bolted connections have the advantage of being semi-permanent, i.e. capable of being disassembled. Bolts are the basic screw fasteners generally used with through holes and the appropriate nuts.

A rivet, on the other hand, provides a permanent joint. A rivet is inserted in a hole drilled or punched through the plates or members to be joined, and closed by forming a head usually with a pneumatic hammer.

There are some differences in analysis and design criteria between bolted and riveted joints. However, for our purposes here these differences will be ignored. The discussion of bolted connections is generally applicable to similarly constructed riveted joints.

When a permanent structure or a tight boiler joint is required, construction methods presently used in industry tend to favour welding in preference to bolted or riveted connections. Welding is often the least expensive process and has the advantage of not requiring holes to be drilled in the parts being joined, thus not reducing their original strength.

However, the choice of the most suitable construction method depends on many design variables, which the engineer must understand. This chapter provides an introduction to the analysis of bolted or riveted joints and welded connections.

17.1 Bolted and riveted joints

We shall consider two types of joints here. A **lap joint** is formed when two plates are lapped over one another and the bolt or rivet goes through both plates as in Figure 17.1(a). In a **butt joint**, the plates are placed edge to edge, and the joint is made with the use of straps (Fig. 17.1(b)).

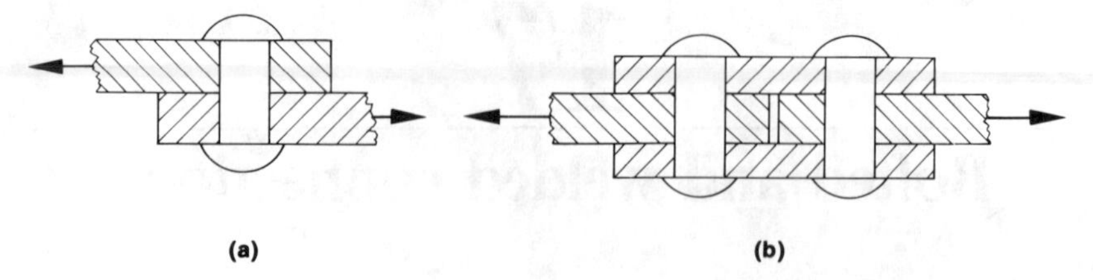

Fig. 17.1

There are single, double and multiple joints, according to whether there are one or more rows of bolts or rivets. Our discussion will be limited to single-row joints only.*

When a bolted or riveted joint is subjected to a force which is in excess of its strength, the joint will usually fail in one of three ways. These are **shearing of the bolts or rivets** in single or double shear, **tearing apart of the plate** weakened by the presence of holes or by **compression or crushing failure in bearing** between the bolts or rivets and the plate, as illustrated in Figure 17.2. Other types of failure, e.g. bending of the bolts or rivets, are possible but less common and are not discussed in this book.

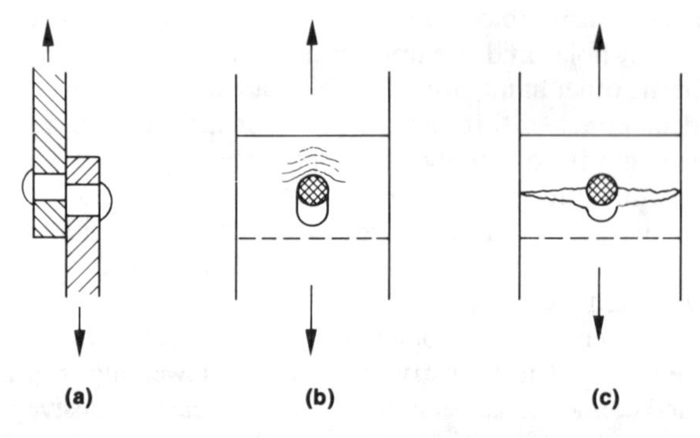

Fig. 17.2(a) *Shearing failure* **(b)** *Crushing failure* **(c)** *Tearing failure*

The analysis of a bolted or riveted joint involves calculation of each type of stress, i.e. in shear, tension (tearing) and compression (crushing), and comparing these with allowable stresses for the materials used. For the purposes of illustration in this chapter we shall adopt the following typical values** of allowable stress:

Allowable stress in shear = 90 MPa
Allowable stress in tension = 110 MPa
Allowable stress in compression = 220 MPa

---

* The student should refer to books on mechanical or structural design for more detailed treatment of bolted and riveted joints.

** These are only average values based on permissible stress recommended for structural joints in certain types of steel construction. For any other material, or for detailed design purposes, appropriate codes and handbooks should be consulted.

In calculations, it is convenient to consider the allowable load per bolt or rivet calculated as the product of the area under stress and the corresponding value of the allowable stress. Each joint must be checked separately with respect to shear, tearing and crushing and the least allowable load resulting from these calculations can then be taken to be the allowable strength of the joint.

Example 17.1

A simple lap joint is composed of two steel straps, 40 mm wide × 8 mm thick, held together by a single bolt 16 mm in diameter.

Calculate the allowable strength of the joint.

Solution

(a) Strength of bolt in shear.

The area in single shear is $A_s = \frac{\pi D^2}{4} = \frac{\pi \times 16^2}{4} = 201.1 \text{ mm}^2$

Allowable shear stress, $f_s = 90$ MPa

Allowable load in shear, $A_s \times f_s = 201.1 \times 90$

$= 18\,100 \text{ N} = 18.1 \text{ kN}$

(b) Strength of straps in tension (tearing).

The net area of the plate subjected to tearing is:

$$A_t = \text{net width} \times \text{thickness}$$

$$= (40 - 16)\text{mm} \times 8 \text{ mm} = 192 \text{ mm}^2$$

Allowable stress $f_t = 110$ MPa

Allowable load in tension $= A_t \times f_t = 192 \times 110$

$= 21\,120 \text{ N} = 21.1 \text{ kN}$

(c) Bearing strength.

The projected area of the bolt subjected to crushing is:

$$A_c = \text{plate thickness} \times \text{bolt diameter}$$

$$= 8 \text{ mm} \times 16 \text{ mm} = 128 \text{ mm}^2$$

Allowable compressive stress, $f_c = 220$ MPa

Allowable bearing load, $A_c \times f_c = 128 \times 220$

$= 28\,160 \text{ N} = 28.2 \text{ kN}$

The joint can only be as strong as its weakest element, in this case the bolt in shear. Thus the allowable load on the bolt represents the strength of the joint.

Strength of joint = 18.1 kN.

17.2 Efficiency of bolted and riveted joints

The strength of a bolted or riveted joint is always somewhat less than that of the unpunched* parent metal.

If the strength of the joint is compared with the original strength of the unpunched plate, the efficiency of the joint can be calculated from the following definition:

$$\text{Joint efficiency} = \frac{\text{strength of joint}}{\text{strength of unpunched plate}}$$

Example 17.2

Determine the efficiency of the joint in Example 17.1.

Solution

Cross-sectional area of the unpunched plate in tension is $40 \text{ mm} \times 8 \text{ mm} = 320 \text{ mm}^2$.

Therefore, the strength of unpunched plate, using the same allowable stress in tension as before, is

$$f \times A = 110 \text{ MPa} \times 320 \text{ mm}^2 = 35\,200 \text{ N} = 35.2 \text{ kN}$$

Strength of the joint, as calculated previously is 18.1 kN.

Therefore, the efficiency of the joint is:

$$\text{Efficiency} = \frac{18.1 \text{ kN}}{35.2 \text{ kN}} = 0.514 = 51.4\%.$$

It should be understood that in a continuous joint bolts or rivets are usually equally spaced, or form a pattern which repeats itself along the length of the joint. In such case a section equal in length to a repeating section is used as a unit for calculations.

Problems

(Use allowable stress in shear, tension and compression of 90 MPa, 110 MPa and 220 MPa respectively.)

17.1 Determine the shear stress in the bolts of a lap joint held by eight 10 mm diameter bolts if the total force on the joint is 31.4 kN.

17.2 Determine the tearing stress in a 200 mm width of a 10 mm thick plate with a single row of four 14 mm diameter bolt holes drilled in it, if the total force is 115.2 kN.

17.3 Determine the crushing stress in the four bolts in the previous question. Does it exceed the allowable compressive stress?

---

* Depending on the plate thickness and other design and manufacturing criteria, the holes may be punched or drilled. For our purposes here this difference is not significant.

17.4 Determine the strength per section and the efficiency of a lap joint in which each repeating section 100 mm wide is held by a 30 mm diameter rivet holding two plates 10 mm thick each.

17.5 A butt joint between two 8 mm thick plates is made by using two 6 mm thick straps with a single row of 10 mm diameter rivets spaced at a repeating interval of 50 mm along the joint.

If a force on the joint is 10 kN for every repeating section of 50 mm, determine the magnitude of existing stress in rivet shear, plate tearing and crushing of the rivets.

17.6 Determine the strength and efficiency of the joint described in the previous question.

17.7 Design a connection composed of two steel straps held together by a single bolt. The straps are to be of the same width and thickness and should transmit a force of 2500 N.

Use the following sequence in your calculations:

(a) Determine the required bolt diameter for the allowable shear stress of 90 MPa. Round off to the next millimetre up.

(b) Using the result of step (a), calculate the required plate thickness for the allowable compression (crushing) stress of 220 MPa. Round off to the next millimetre up.

(c) Using diameter and thickness selected in steps (a) and (b), calculate the minimum width of the straps required for plate tension (tearing) stress of 110 MPa. Round off to the next millimetre up.

17.8 Calculate the efficiency of the joint designed in problem 17.7.

17.9 A single riveted butt joint is to be designed for a force of 81 kN per 100 mm pitch. If the plates which are placed edge to edge are to be twice as thick as each of the two straps, determine:

(a) the minimum diameter, to the nearest mm, of the rivets required for allowable shear.

(b) the minimum thickness, to the nearest mm, of the plate and straps (check both crushing and tearing).

17.10 Determine the efficiency of the joint constructed in accordance with the results of problem 17.9.

17.3 Welded connections

Welding is a manufacturing process for permanent joining of metal parts by fusion. The most common method of joining is achieved by striking an electric arc between a rod of similar metal and the pieces to be joined, metal being melted from the electrode into the joint. As a method of construction, welding is widely used in structural work and for repair and fabrication of boilers, pressure vessels and heavy machinery.

The two types of welds most frequently used are fillet welds and butt welds as illustrated in Figure 17.3.

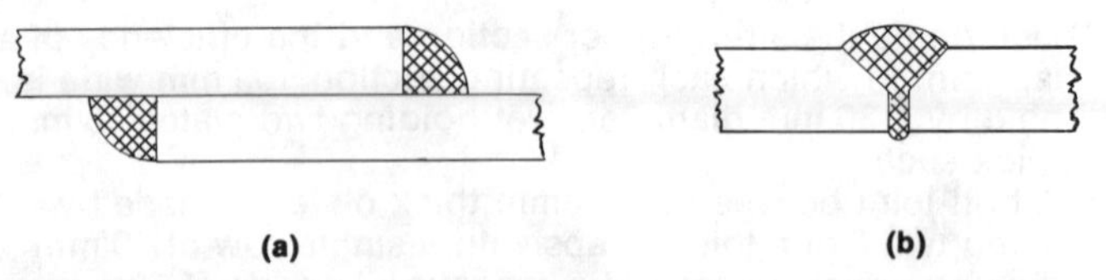

Fig. 17.3(a) *Double-fillet lap joint* **(b)** *Butt weld joint*

Butt welds

The plates for butt welds may be unbevelled for thin plates or bevelled on one or both sides. They are most frequently used for manufacturing of boiler shells, air receivers, etc., and are usually subjected to tension and compression, not to shear. The thickness of the weld is at least equal to the thickness of the plates joined, and its strength is thought of in relation to the strength of the plate.

Tests show that good butt welds have about the same strength as the plates being joined. In practice it is safe to assume an efficiency of the joint of approximately 90 per cent. However, for our purposes it is possible to regard the strength of a butt weld as equal to that of the plates joined.

Fillet welds

A standard full fillet weld has a section of an isosceles right triangle as shown in Figure 17.4 with the legs of the triangle equal to the thickness of the plate. The size of the weld is its leg length and the throat is 0.707 times that length.* In a fillet weld the throat is the critical dimension.

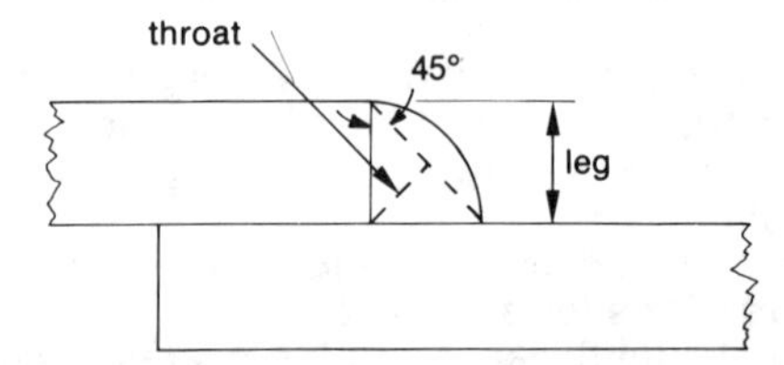

Fig. 17.4

It is common practice to take one-third of the nominal tensile strength of electrode used as the permissible working stress, i.e. allowable stress, in fillet welds, and to refer it to the area based on the length of weld and throat thickness. The strength of fillet weld is therefore given by

$$F = f.l.t = 0.707.f.l.s$$

where F is maximum allowable load on weld
f is allowable stress
l is length of weld
s is nominal size of weld
t is throat thickness

* $\sin 45° = 0.707$.

It should be understood that stress distribution within the weld may be a complex combination of shear, tension and sometimes bending. However, for many practical purposes, it can be assumed that stress in the weld is uniformly distributed shearing stress.

Example 17.3

For a fillet weld of 8 mm nominal size and electrode strength of 410 MPa, determine:

(a) the throat thickness,
(b) the allowable stress,
(c) the allowable load per mm length,
(d) the length of weld required to carry a load of 52.6 kN.

Solution

(a) throat $= 0.707s = 0.707 \times 8$ mm $= 5.66$ mm

(b) allowable stress, $f = \dfrac{410}{3} = 136.7$ MPa

(c) allowable load per mm $= \dfrac{F}{l} = f.t$

$= 136.7 \text{ MPa} \times 5.66 = 773 \text{ N/mm}$

(Note that here, as before, if stress is expressed in megapascals and linear dimensions in millimetres, force will be in newtons.)

(d) $l = \dfrac{F}{f.t} = \dfrac{52\,600}{136.7 \times 5.66} = 68$ mm

It is customary to add an allowance for end-craters, i.e. for starting and stopping, equal to about twice the nominal weld size. Thus, the required length of weld specified in the previous example would be

$$68 \text{ mm} + 2 \times 8 \text{ mm} = 84 \text{ mm}.$$

In this book, we will omit this allowance and will simply regard the computed answers, e.g. 68 mm, as representing the effective length of weld required in each particular case.

Eccentricity of welded joints is a common problem in structural design, e.g. when non-symmetrical members, such as structural angles, are welded to gusset plates. The total load is presumed to act along the centroidal axis, while the required resistances offered by the welds are inversely proportional to the distances d_1 and d_2 from the axis, determined by summation of moments as illustrated by the following example.

Example 17.4

A structural steel angle is to be welded to a gusset plate as shown in Figure 17.5.

Determine the length of 8 mm weld required to withstand a load of 100 kN, if the allowable stress is 136.7 MPa.

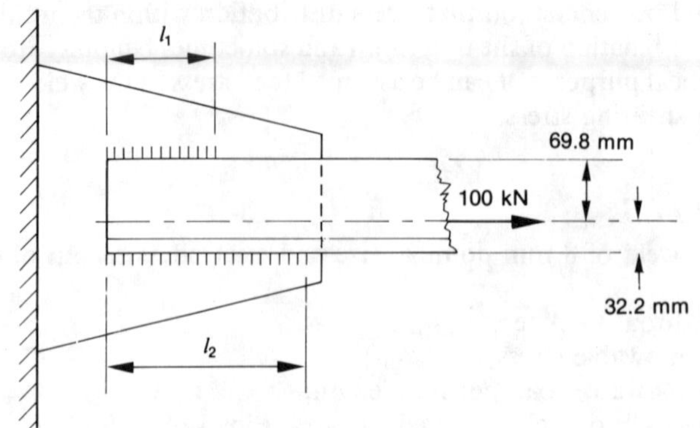

Fig. 17.5

Solution

Strength of weld per millimetre is:

$$\frac{F}{l} = 0.707 \times f \times s = 0.707 \times 136.7 \times 8 = 773.2 \text{ N/mm}$$

The required resistance at the toe of the angle is found by taking moments about the heel of the angle:

$$F_1 = \frac{100\,000 \text{ N} \times 32.2 \text{ mm}}{102 \text{ mm}} = 31\,570 \text{ N}$$

The resistance of the weld at the heel is:

$$F_2 = 100\,000 \text{ N} - 31\,570 \text{ N} = 68\,430 \text{ N}$$

The corresponding lengths of weld required are:

$$l_1 = \frac{31\,570 \text{ N}}{773.2 \text{ N/mm}} = 40.8 \text{ mm}$$

$$l_2 = \frac{68\,430 \text{ N}}{773.2 \text{ N/mm}} = 88.5 \text{ mm}$$

Problems

17.11 For the following nominal sizes of fillet welds determine:
(a) the throat thicknesses (mm),
(b) the safe load (N/mm) based on weld material with ultimate tensile strength = 410 MPa.

Nominal size (mm)	4	6	8	10	12	16	20	24
Throat thickness (mm)								
Safe load (N/mm)								

Note: Use these results to solve the following problems.

17.12 A 100 mm × 10 mm steel strap is to be welded to a heavy steel plate as shown in Figure 17.6. Determine the length l of 8 mm fillet weld to withstand a tensile force of 150 kN.

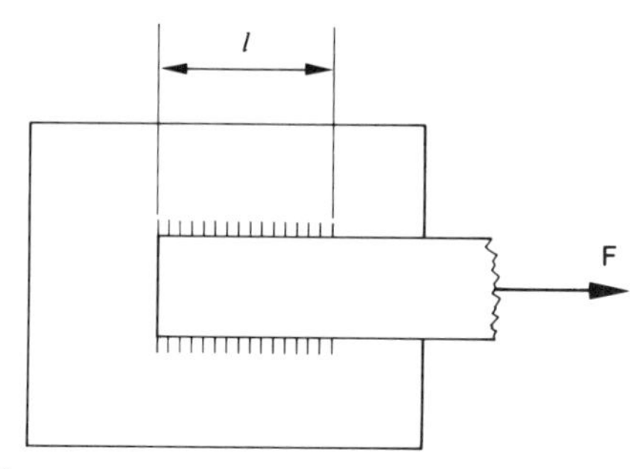

Fig. 17.6

17.13 A double-fillet lap joint is made using 12 mm welds (Fig. 17.7). Determine the load allowed on 500 mm width of plate.

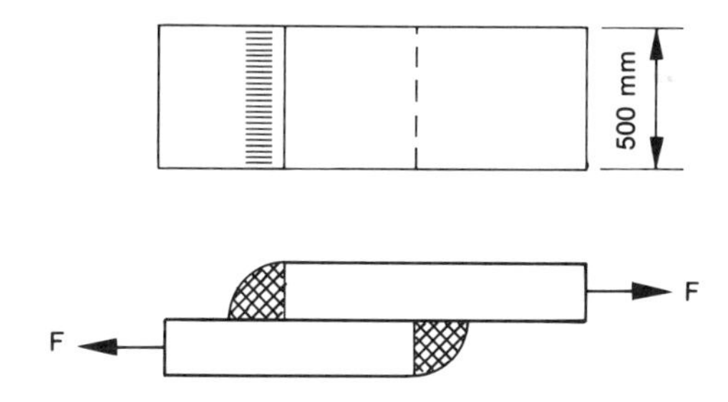

Fig. 17.7

17.14 Determine the minimum size weld that can be used for the connection in problem 17.12 if the length of weld is not to exceed 100 mm on each side and the load is 115 kN.

17.15 A steel strap 12 mm thick by 60 mm wide is welded to a steel plate by means of 10 mm fillet welds, as shown in Figure 17.6. Determine the length l required to match the full tensile strength of the strap, if the allowable stress for the strap in tension is 120 MPa.

17.16 A steel bracket is to be welded as shown in Figure 17.8. Determine the maximum mass that can be supported from this bracket.

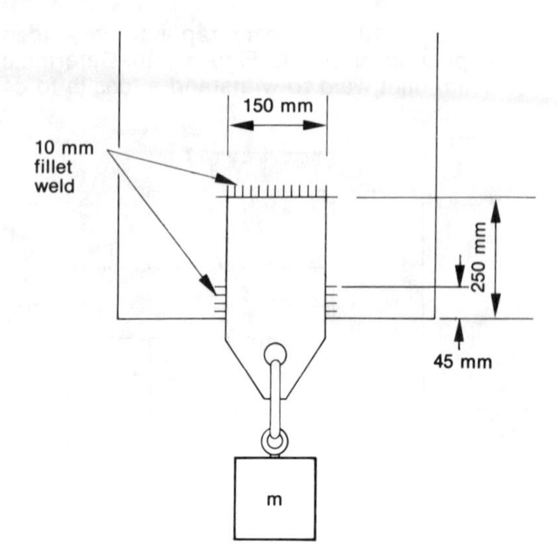

Fig. 17.8

17.17 A structural steel angle is to be welded to a gusset plate as shown in Figure 17.9. Determine the lengths of weld l_1 and l_2, required to withstand a load of 250 kN.

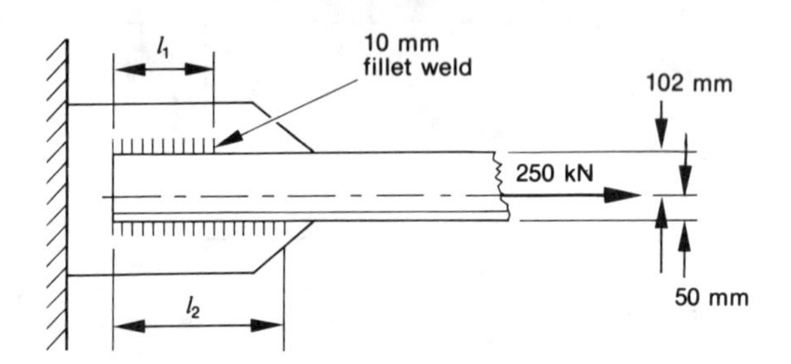

Fig. 17.9

Review questions

1. Briefly describe bolted, riveted and welded connections.
2. What is the difference between a lap joint and a butt joint?
3. List and explain the three most common types of failure in bolted or riveted joints.
4. Define joint efficiency.
5. Distinguish between fillet welds and butt welds.
6. State the formula for calculating the strength of a fillet weld.
7. Explain the effect of load eccentricity on weld design.

18

Pressure vessels

Pressure vessels are cylindrical or, less frequently, spherical containers used to hold fluids, i.e. gases or liquids, under pressure.

The most common form of pressure vessels, as used in steam boilers and compressed air receivers, is the cylinder of welded construction with curved, but not quite hemispherical, ends. Spherical vessels are sometimes used, particularly for the containment of liquefied gases at high pressure and low temperature.

A pressure vessel containing a gas is subject to uniform internal pressure* normal to its walls. The material of the wall, as well as the welded seams, must be sufficiently strong to resist stresses set up within the material due to the gas pressure.

The pressure responsible for the stress in the walls of a vessel is the difference between internal pressure and atmospheric pressure outside the vessel. It is in fact the pressure as measured directly by a pressure gauge fitted to a pressure vessel and is known as gauge pressure.

For the purposes of this chapter, pressure in a fluid can be regarded as similar to stress in a solid material, defined as force per unit area and measured in pascals, or its derivatives, kilopascals and megapascals. It is assumed here that everyday experiences, e.g. pumping an automobile tyre, have taught the student at least some basic understanding of pressure and that its analogy with stress is sufficient to make the material of this chapter comprehensible. However, if a better understanding of pressure seems necessary or if revision of the concept of pressure and its units is desired, it is suggested that Chapter 21 be referred to before proceeding any further.

18.1 Stresses in cylindrical shells

If we consider a cylindrical pressure vessel of welded construction, as shown in Figure 18.1, there are two types of seams that need to be considered for possible stress. These are the **longitudinal seam** along the length of the cylinder and the **circumferential seam** (see Fig. 18.2). It should be noted at the outset that stresses induced in the material of the shell are found throughout the

---

* If the fluid in a container is a liquid, such as water in a tank, the pressure varies with depth as discussed in Chapter 22.

material of the shell, whether there is an actual welded joint at a particular point or not. However, it is often useful to imagine two halves of the shell as if they were connected by some form of joint, as will be seen from the following discussion.

Fig. 18.1 *Cylindrical pressure vessels*

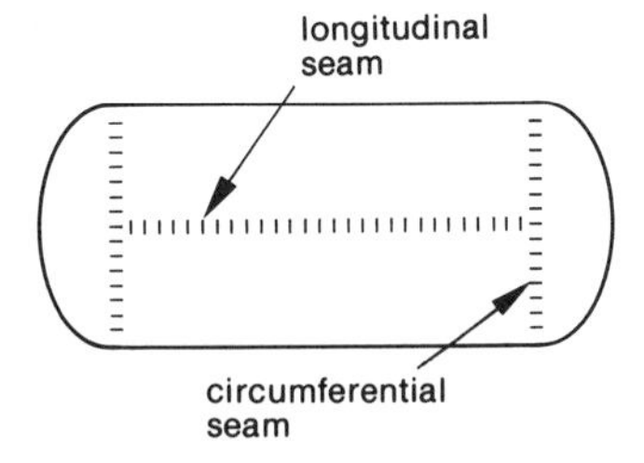

Fig. 18.2

Stress on longitudinal seam: Hoop stress

Let us consider one half of a cylindrical shell separated from the other along its longitudinal seams,* as shown in Figure 18.3. For a given length, l, and wall thickness, t, the area subjected to stress is twice $l \times t$:

$$A = 2l.t$$

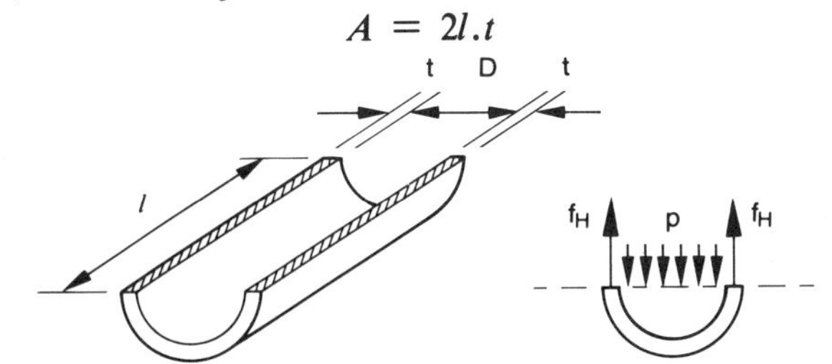

Fig. 18.3

The pressure, p, acts in the radial direction upon all elements of the exposed internal surface. However, it can be shown that after summation of all components of pressure perpendicular to the plane of the section, the total force due to pressure is equal to the product of the pressure and the projected area $l \times D$, where D is the shell internal diameter.

$$F = p.l.D$$

The stress in the material of the shell can, therefore, be found as force per unit area:

$$f = \frac{F}{A} = \frac{p.l.D}{2.l.t} = \frac{p.D}{2.t}$$

This stress, known as **hoop** stress**, is the tensile stress in the material of the shell set up in the tangential direction, all the way along length, l, of the

---

* We shall use the term "seam" to mean an imaginary line of separation between two halves of a shell, forming a cross-section considered for the purpose of stress analysis. The presence, or otherwise, of an actual butt-welded joint along that line is immaterial to our discussion, provided that the joint efficiency is 100 per cent.

** The term originates from a classic problem of stress analysis in a thin circular ring, or "hoop", subjected to uniformly distibuted radial forces, which produce uniform enlargement of the ring. Hoop stress is sometimes referred to as circumferential stress due to its tangential direction. However, this often creates confusion in the learner's mind between circumferential stress, i.e. hoop stress, and stress on a circumferential seam which, to make things worse, is sometimes called longitudinal stress. To avoid ambiguity and confusion we shall only use the terms "hoop stress" and "axial stress" in this chapter.

longitudinal seam. One should note very carefully that the direction of this stress is perpendicular to the cross-section of the shell material made by the imaginary plane of separation between the two halves.

Therefore, hoop stress $$f_H = \frac{p.D}{2.t}$$

Example 18.1

Determine the hoop stress in the material of a cylindrical air receiver 1.2 m long and 350 mm in diameter, with a wall thickness of 6 mm, subjected to a pressure of 1 MPa.

Solution

Keeping in mind that in this context pressure is similar to stress, we can use pressure in megapascals and dimensions in millimetres for convenience.

$$\text{Hoop stress, } f_H = \frac{p.D}{2.t} = \frac{1\ \text{MPa} \times 350\ \text{mm}}{2 \times 6\ \text{mm}} = 29.2\ \text{MPa}$$

Stress on circumferential seam: Axial stress

Axial stress in a shell of a pressure vessel is tensile stress in the direction of the principal axis of the cylinder.

To establish the magnitude of this stress it is necessary to consider a section along the plane of a circumferential seam as shown in Figure 18.4.

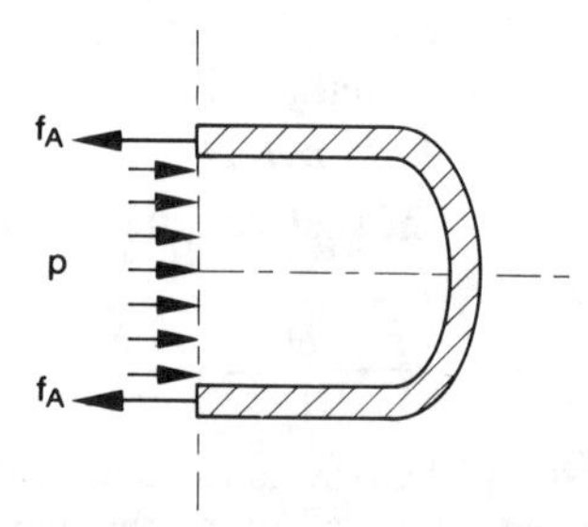

Fig. 18.4

This time the pressure p can be regarded as acting in the axial direction, i.e. perpendicular to the projected area of the cylinder end, equal to $\pi D^2/4$.

This gives rise to a total axial force:

$$F = p.\frac{\pi D^2}{4}$$

The area of the shell material resisting this force can be estimated approximately as the product of the circumference based on the nominal diameter of the shell and the wall thickness:

$$A = \pi.D.t$$

The accuracy of this approximation is quite acceptable provided that the cylinder is thin, i.e. its diameter is at least ten times greater than the wall thickness, and the nominal diameter is somewhere between the inside and outside diameters of the shell.

The stress resulting from the force acting on the cross-section of the circumferential seam can be found from

$$f = \frac{F}{A} = \frac{p.\frac{\pi D^2}{4}}{\pi D.t} = \frac{p.D}{4.t}$$

This is axial stress:

$$\boxed{f_A = \frac{p.D}{4.t}}$$

Example 18.2

Calculate the axial stress in the material of the air receiver in the previous example.

Solution

$$f_A = \frac{p.D}{4t} = \frac{1\ \text{MPa} \times 350\ \text{mm}}{4 \times 6\ \text{mm}} = 14.6\ \text{MPa}$$

The first step in the design of a pressure vessel is to determine the thickness of plate required. A simple comparison shows that a given pressure in a cylindrical pressure vessel causes hoop stress which is twice as high as the axial stress. This means that a longitudinal seam is more vulnerable to rupture. Therefore if other conditions are equal, the strength of a longitudinal seam is the limiting factor in the design of cylindrical vessels, and the hoop stress formula should be used for calculating required plate thickness.

Example 18.3

Determine the minimum plate thickness required for a steam boiler drum 1.2 m diameter, if the maximum allowable stress in the material is 75 MPa and the pressure is 1.5 MPa.

Solution

Hoop stress $\quad f_H = \frac{p.D}{2t}$ is the limiting factor,

from which $\quad t = \frac{p.D}{2f_H}$

Substitute $\quad t = \frac{1.5\ \text{MPa} \times 1200\ \text{mm}}{2 \times 75\ \text{MPa}} = 12\ \text{mm}$

Effect of joint efficiency

For general purposes it may be assumed that good quality butt-welded joints in mild steel have the same strength as the plates being joined, i.e. a joint efficiency of 100 per cent is assumed. However, under certain conditions of workmanship or service it would be safer to assume an efficiency of the joint of 90 per cent or less.

Another, more obvious, example is when the joint is a riveted joint. Riveted joints are never as strong as the solid plate. Consequently, an allowance must be made for the weakness of the joint. Depending on the actual type of the riveted connection used and the arrangement and size of the rivets, the efficiency of such joints may be as low as 50 per cent.

It is not our purpose here to discuss various boiler and pressure vessel codes and specifications. It is sufficient, at this stage, to understand that if a joint is only a certain percentage as strong as the solid plate, the thickness of the plate to allow for the weakness of the joint must be increased accordingly.

This in effect means that, to allow for joint efficiency, the required plate thickness must be equal to the calculated minimum thickness, t, divided by the joint efficiency.

Example 18.4

Determine the actual plate thickness that would be required in the previous example if an allowance for joint efficiency of 80 per cent was to be made.

Solution

Calculated thickness, $t = 12$ mm
Required thickness, to allow for joint efficiency of 80 per cent is:

$$\frac{12 \text{ mm}}{0.8} = 15 \text{ mm}$$

18.2 Stress in a spherical shell

A pressure vessel of spherical shape is symmetrical in all directions, suggesting that stresses in the material of its wall induced by internal pressure are the same at all points and in all directions.

If we consider any diametral section of the sphere, as in Figure 18.5, the similarity with stresses on circumferential seams can easily be seen.

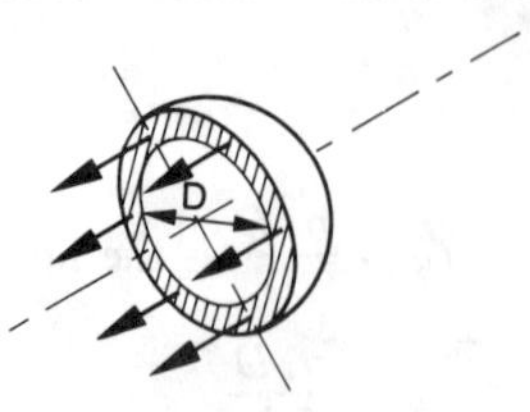

Fig. 18.5

The force due to pressure on a circular area $\pi D^2/4$ is equal to $p.\pi D^2/4$.

This force is distributed over the cross-sectional area of the material equal to $\pi D.t$, again assuming "thin-wall" approximation.

Therefore stress in the wall material is:

$$f = \frac{F}{A} = \frac{p.\dfrac{\pi D^2}{4}}{\pi.D.t} = \frac{p.D}{4t}$$

Thus, for the same p, D and t stress in a spherical shell is equivalent to axial stress on a circumferential seam in a cylindrical vessel, given by:

$$f = \frac{p.D}{4.t}$$

It should be understood, however, that a cylindrical shell of equal diameter and wall thickness, subjected to the same pressure will also have hoop stress in its longitudinal seams of twice the magnitude of axial stress.

It follows, therefore, that since a spherical shell does not have longitudinal seams, it is not subjected to hoop stress, and can be said to be twice as strong as a corresponding cylindrical container. The following example should illustrate this point.

Example 18.5

Given the ultimate strength of steel plate of 380 MPa and using a factor of safety of 5, compare the maximum allowable pressure in (a) spherical and (b) cylindrical pressure vessels of 1 m diameter and 10 mm wall thickness, assuming 100 per cent joint efficiency in fully welded construction.

Solution

Allowable stress, $f = \dfrac{380 \text{ MPa}}{5} = 76 \text{ MPa}$

Stress in a spherical shell is given by $f = \dfrac{p.D}{4t}$, from which maximum allowable pressure in a spherical pressure vessel is:

$$p = \frac{4tf}{D} = \frac{4 \times 10 \text{ mm} \times 76 \text{ MPa}}{1000 \text{ mm}} = 3.04 \text{ MPa}$$

Critical stress in a cylindrical shell is hoop stress, $f = \dfrac{p.D}{2t}$, from which maximum allowable pressure in a cylindrical pressure vessel is:

$$p = \frac{2tf}{D} = \frac{2 \times 10 \text{ mm} \times 76 \text{ MPa}}{1000 \text{ mm}} = 1.52 \text{ MPa}$$

It is apparent that within the specified parameters the spherical vessel can withstand twice as large a pressure as would be allowed in the cylindrical vessel.

Although theoretically quite valid, this comparison should not be overemphasised. Other comparisons on the basis of equal volume or of equal cost may be more useful. Practical considerations, such as ease of manufacture and resulting costs may dictate a search for other alternatives. For example, it could be easier and cheaper to manufacture a cylindrical vessel. The resultant saving may allow the use of heavier plate for extra strength.

Problems

18.1 Determine the stress in the material of a spherical container, 600 mm diameter with 5 mm wall thickness if the pressure is 1 MPa.

18.2 If the maximum allowable stress in the material is 75 MPa, calculate the minimum allowable wall thickness for the container in the previous problem.

18.3 Determine the hoop stress and the axial stress in the wall of a cylindrical air receiver of 300 mm diameter and 3 mm wall thickness, when the working pressure is 720 kPa.

18.4 Find the minimum thickness of the boiler plate, with ultimate tensile strength of 375 MPa, required for the construction of a 600 mm diameter cylindrical drum for the working pressure of 1 MPa, assuming a safety factor of 5. Compare with the answer to problem 18.2.

18.5 A cylindrical air receiver 800 mm in diameter is to be designed to withstand safely a working pressure of 900 kPa. Boiler plate with an allowable stress of 80 MPa will be used. Efficiency of all welded joints is assumed to be 90 per cent. Determine the required wall thickness.

18.6 Calculate the safe working pressure for a 2 m diameter spherical container, with 15 mm wall thickness if the ultimate tensile strength of its material is 420 MPa, the factor of safety is 6 and the efficiency of welded joints is 90 per cent.

18.7 What is the maximum volume capacity of a spherical container that can be made from 8 mm thick plate, having an allowable stress of 80 MPa and 100 per cent joint efficiency, designed to withstand a pressure of 1.8 MPa?

18.8 Determine the factor of safety when a 1 m diameter cylindrical pressure vessel operates at 950 kPa internal pressure. Wall thickness is 10 mm and the ultimate tensile strength is 380 MPa.

18.9 A spherical tank 12 m in diameter is used to contain gas under pressure. The wall thickness is 10 mm with joint efficiency of 90 per cent. If the allowable stress in the wall material is 76 MPa, determine the maximum safe pressure.

18.10 A boiler drum, 1800 mm in diameter, is to have a riveted longitudinal joint with an efficiency of 75 per cent. The operating pressure is expected to be 1.35 MPa.

Calculate the required thickness of boiler plate with a maximum allowable tensile stress of 67.5 MPa.

Review questions

1. What is the most common shape of a pressure vessel?
2. Sketch a cylindrical pressure vessel showing circumferential and longitudinal seams.
3. What is the stress on a longitudinal seam called?
4. What is the stress on a circumferential seam called?
5. State the expressions for calculating stresses in cylindrical shells.
6. In a cylindrical shell which is higher, hoop stress or axial stress?
7. State the expression for calculating stress in a spherical shell.
8. Explain how joint efficiency affects the wall thickness required to withstand a specified pressure.

19

Bending of beams

In Chapter 6 we introduced beams as structural members, or machine components, used to support transverse loads due to weight or other causes. In particular, we learned how to calculate reactions at beam supports if all loads are known. We are about to discuss the effects of the applied loads on the beam and the relationship between the conditions of loading, the cross-sectional dimensions of the beam and the internal resistance in the material of the beam.

In order to understand the behaviour of beams under load we must introduce the concepts of **shear force** and **bending moment** produced in a beam by the applied load, and must learn to draw the shear force and bending moment diagrams for beams subjected to various loading conditions.

The last part of this chapter will then be concerned with determination of the bending stress and radius of curvature in a beam subjected to bending.

As before, our discussion in this chapter will be limited to horizontal simply supported and cantilever beams subjected to vertical point loads. Distributed loads and loads which are not vertical will not be considered.

19.1 Shear force

We are now turning our attention to internal forces that exist within the material of a beam subjected to loads.

Consider two portions of a beam cut by an imaginary section transverse to the beam as shown in Figure 19.1(a). The effect of the unbalanced forces on each part of the beam is to move the left-hand portion upwards relative to the right-hand portion as shown in Figure 19.1(b). The reason that we never actually observe such movement between parts of a solid beam is because internal forces exist within the material of the beam in the plane of the imaginary cross-section which resist the tendency for such movement. The magnitude of the internal force at any given cross-section depends on the sum of the external forces acting on each portion of the beam to one side of the cross-section. This internal resistance force is called **shear force**.

The magnitude of the shear force at any cross-section of a beam is equal to the algebraic sum of all external forces, i.e. loads and reactions, acting on either portion of the beam to one side of the section only.

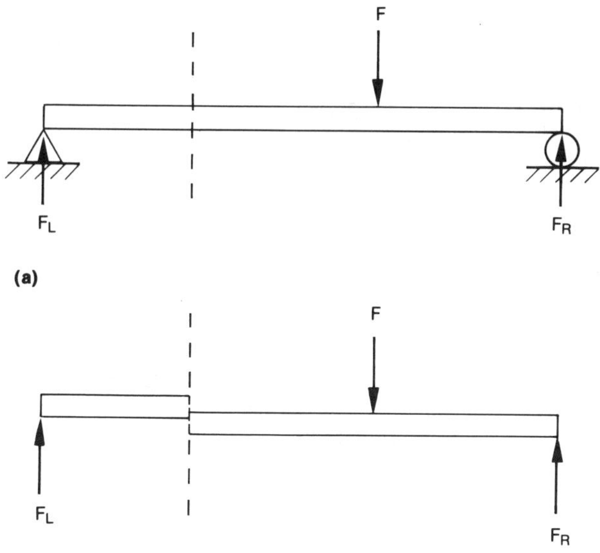

Fig. 19.1

The sign convention commonly adopted is that shear force at a cross-section is positive if it tends to push the left-hand portion of the beam upwards in relation to the right-hand portion, as shown in Figure 19.2(a). The opposite effect would then be regarded as negative, Figure 19.2(b).

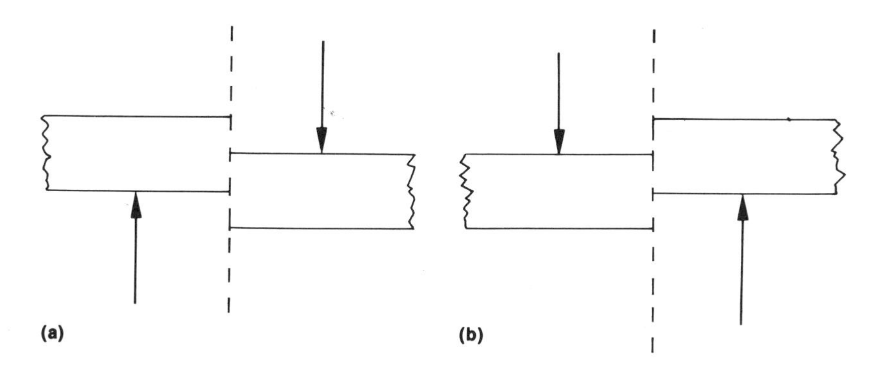

Fig. 19.2(a) *Positive shear* **(b)** *Negative shear*

Example 19.1

Determine the shear forces at the three cross-sections for the beam and loading shown in Figure 19.3(a).

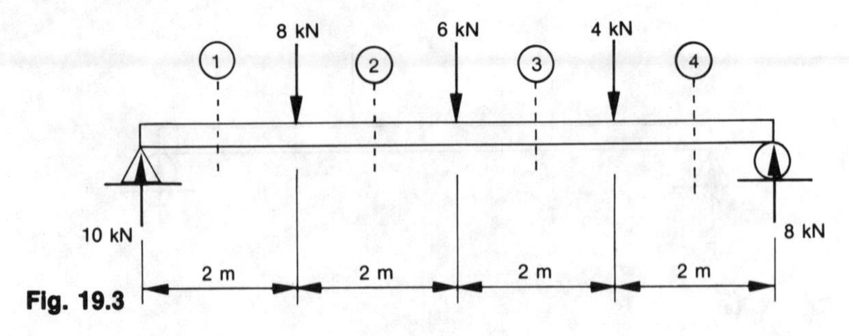

Fig. 19.3

Solution

It is usually convenient to start from the left-hand side and consider all forces to the left of the respective cross-sections.

Shear force at cross-section 1:

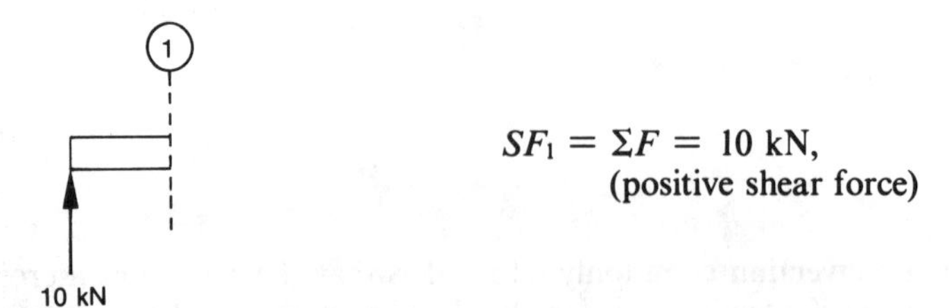

$SF_1 = \Sigma F = 10$ kN,
(positive shear force)

Shear force at cross-section 2:

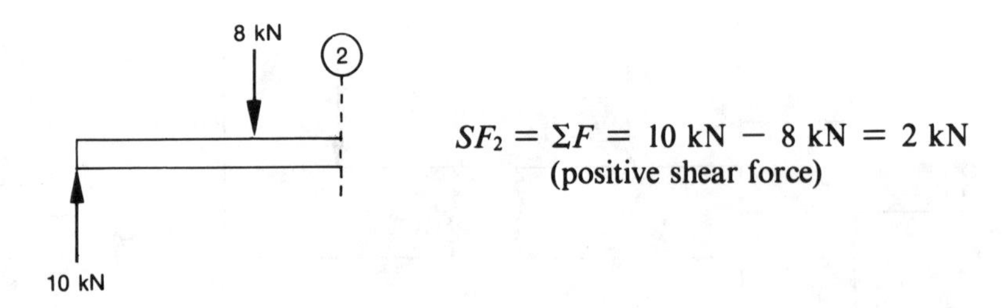

$SF_2 = \Sigma F = 10 \text{ kN} - 8 \text{ kN} = 2 \text{ kN}$
(positive shear force)

Shear force at cross-section 3:

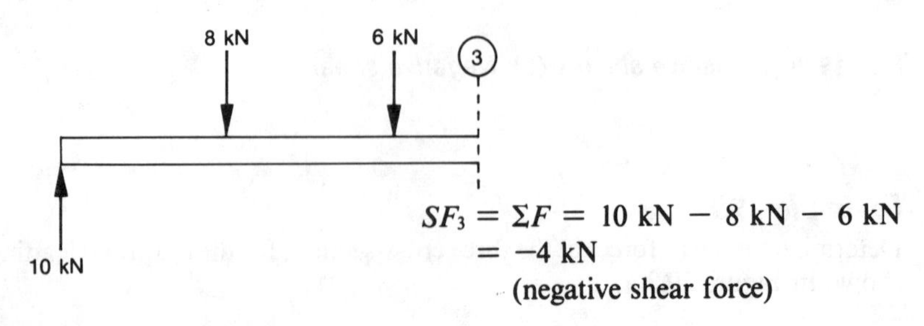

$SF_3 = \Sigma F = 10 \text{ kN} - 8 \text{ kN} - 6 \text{ kN}$
$= -4 \text{ kN}$
(negative shear force)

Shear force at cross-section 4:

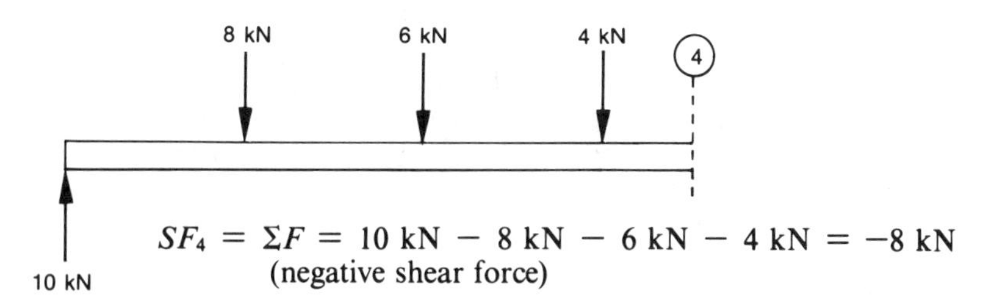

$$SF_4 = \Sigma F = 10 \text{ kN} - 8 \text{ kN} - 6 \text{ kN} - 4 \text{ kN} = -8 \text{ kN}$$
(negative shear force)

It is useful to recognise that if we always work consistently from left to right, i.e. consider forces on the left-hand portion of the beam only, and use the usual sign convention of forces, i.e. positive up and negative down, the answers obtained automatically give the correct sign for the shear force.

19.2 Shear force diagrams

To avoid long and unwieldy statements of results obtained in problems similar to Example 19.1, particularly if a large number of forces is involved, a graphical method of representing results called **shear force diagram** has been developed. It consists simply of plotting the values obtained by calculation against distance measured along the beam.

It should be noted that in the space between two adjacent external forces, shear force is constant irrespective of the actual position of the reference cross-section. The shear force diagram will therefore consist of horizontal straight lines, with step changes under each load as illustrated by the following example.

Example 19.2

Draw the shear force diagram for the beam in Example 19.1.

Solution

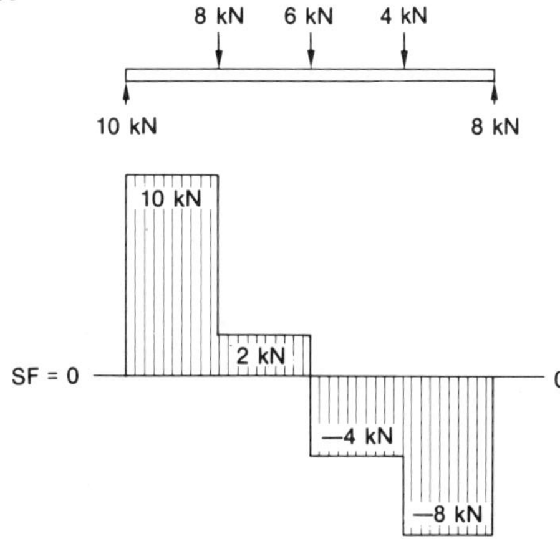

Fig. 19.4

The shear force diagram is plotted directly below the space diagram of the beam, the horizontal distance representing the length of the beam. Vertical distances are drawn to some convenient scale, up from the zero line to represent positive shear force and down for negative shear force. Horizontal lines represent constant shear force between loads. A step change in magnitude equal to the load occurs under each load. Vertical cross-hatching is common practice, but not absolutely necessary. Magnitudes can be shown along the lines for clarity. (See Fig. 19.4.)

Problem

19.1 In each of the following diagrams, Figure 19.5(a) to (j), beams are shown with all external forces, i.e. loads and reactions. Do all necessary calculations and draw the shear force diagram for each beam shown.

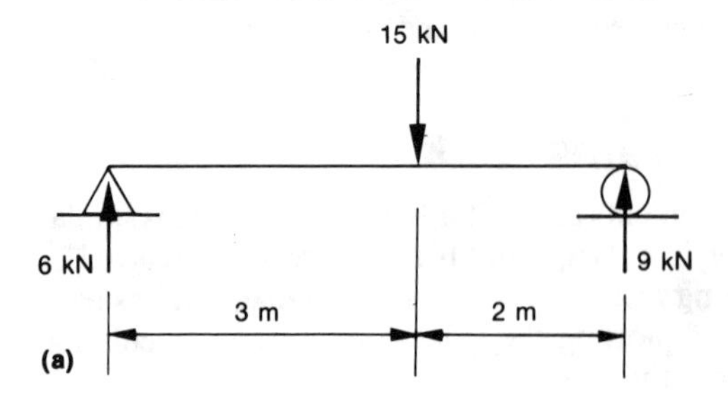

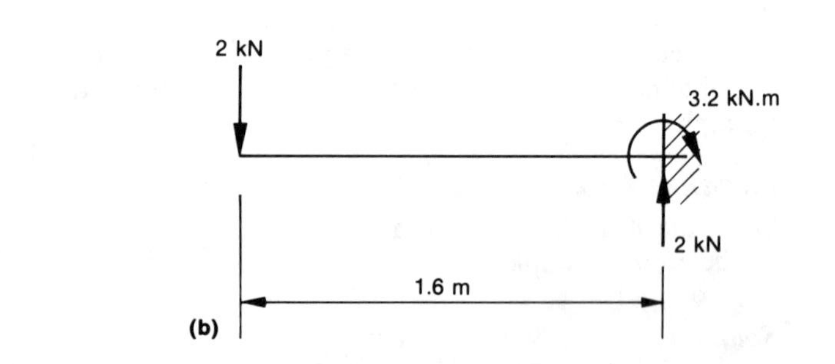

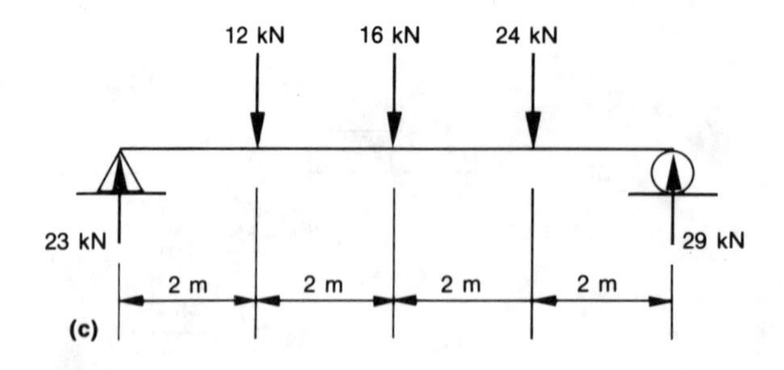

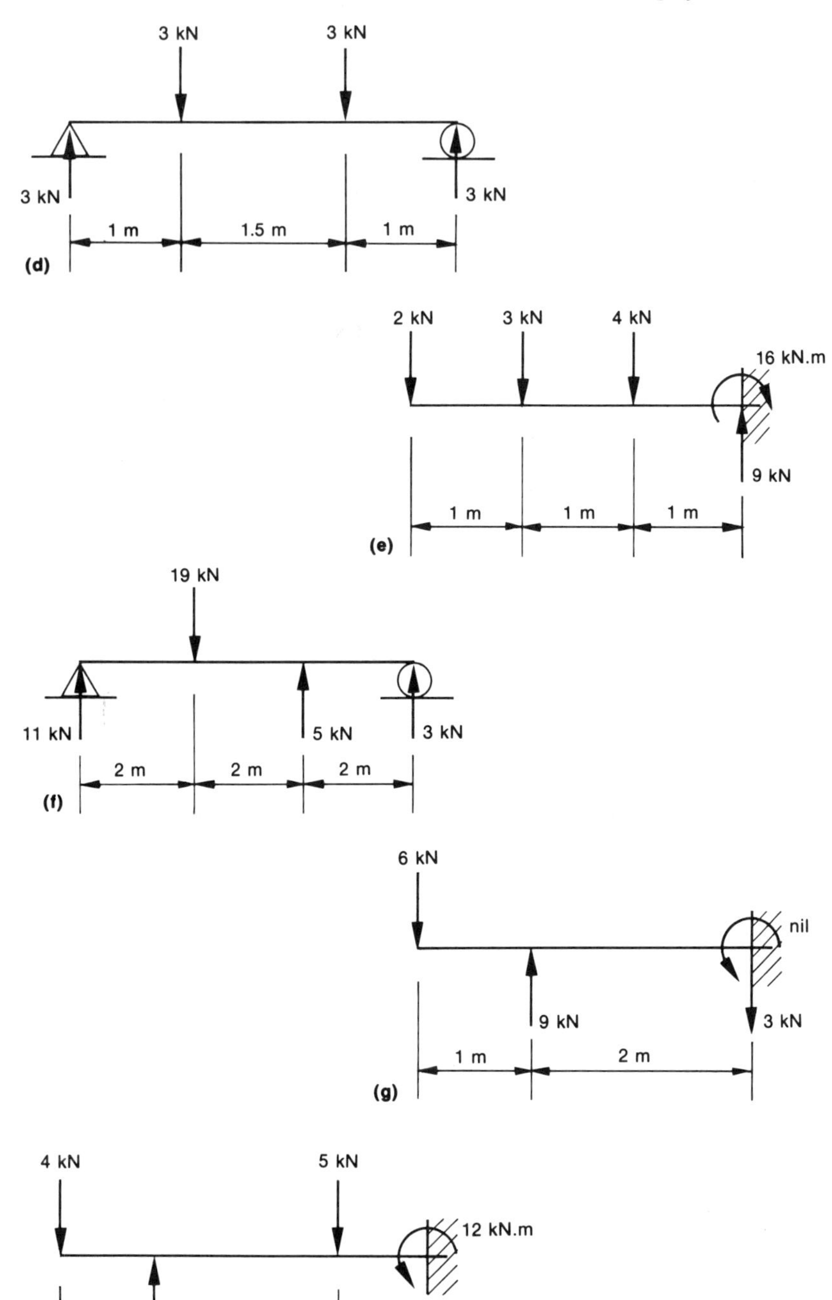
3 kN
3 kN
3 kN
3 kN
1 m
1.5 m
1 m
(d)
2 kN
3 kN
4 kN
16 kN.m
9 kN
1 m
1 m
1 m
(e)
19 kN
11 kN
5 kN
3 kN
2 m
2 m
2 m
(f)
6 kN
nil
9 kN
3 kN
1 m
2 m
(g)
4 kN
5 kN
12 kN.m
9 kN
2 m
4 m
2 m
(h)

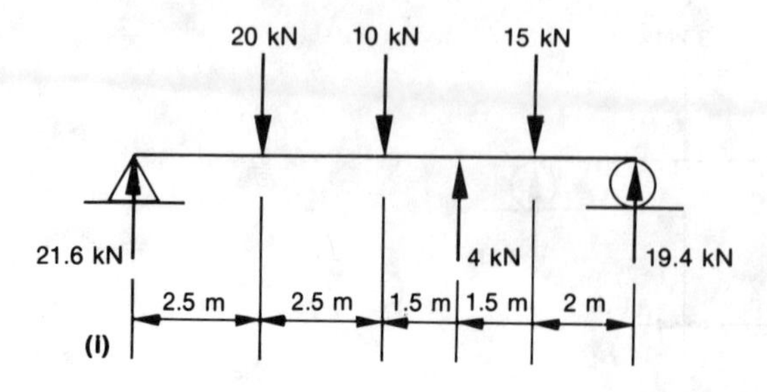

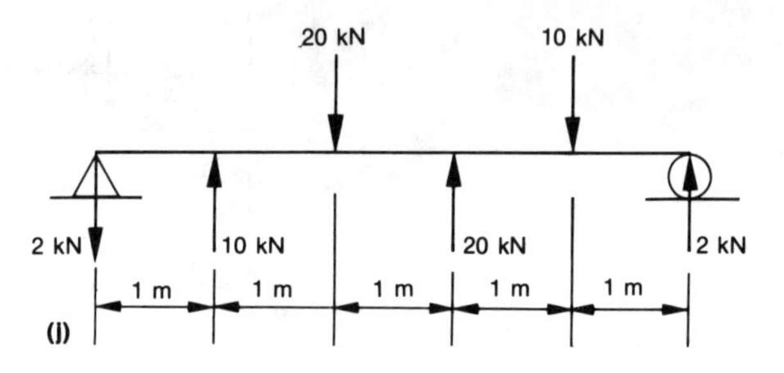

Fig. 19.5

19.3 Bending moment

In addition to internal shear forces, every cross-section in a beam may also experience an internal moment called **bending moment**. Consider the same two portions of a beam, as before, cut by an imaginary reference cross-section. (See Fig. 19.6(a)).

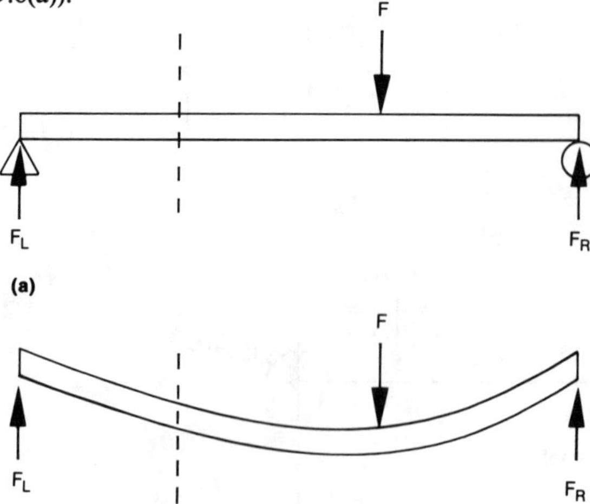

Fig. 19.6

Another effect of the unbalanced forces on each part of the beam, not considered previously, is to bend the beam, as shown in Figure 19.6(b).

The amount of bending tendency at the reference cross-section is measured by the summation of the moments about the cross-section of all external forces on a portion of the beam to one side of the cross-section. The sum of the moments is called **bending moment**.

The magnitude of the bending moment at any cross-section of a beam is equal to the algebraic sum of the moments about the cross-section of all external forces, i.e. loads and reactions, acting on either portion of the beam to one side of the section only.

The usual sign convention is that bending moment is positive if it produces bending in a beam which is convex downwards, as in Figure 19.7. The opposite effect is negative.

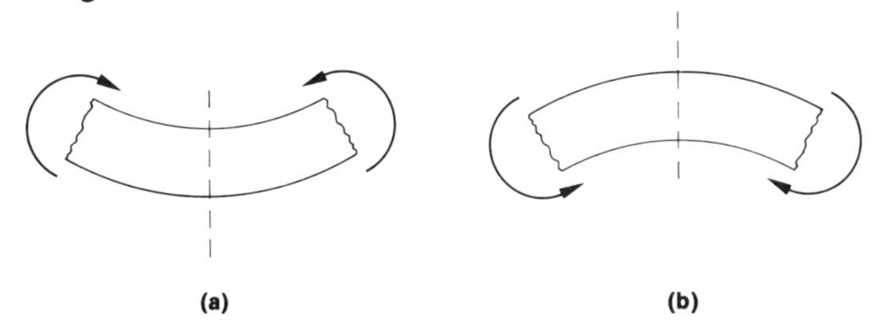

Fig. 19.7(a) *Positive bending* **(b)** *Negative bending*

Example 19.3

Determine the bending moment at the three points under the applied forces, for the beam and loading shown in Figure 19.8.

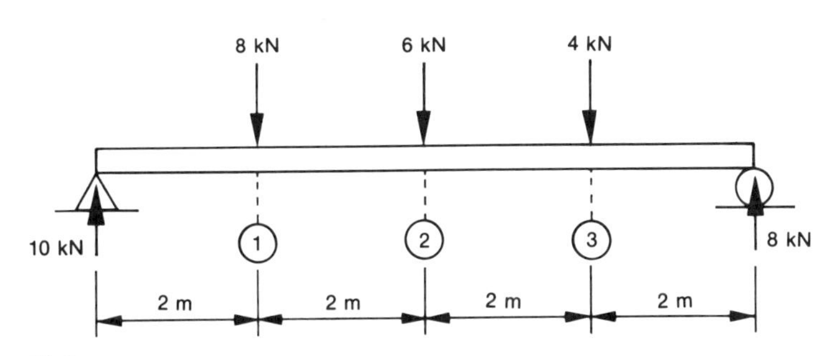

Fig. 19.8

Solution

In the case of bending moment calculations, it is convenient to select the reference cross-sections at the point of application of external loads. Again, we will start from the left-hand side and consider all forces to the left of the respective cross-sections.

Bending moment at cross-section 1:

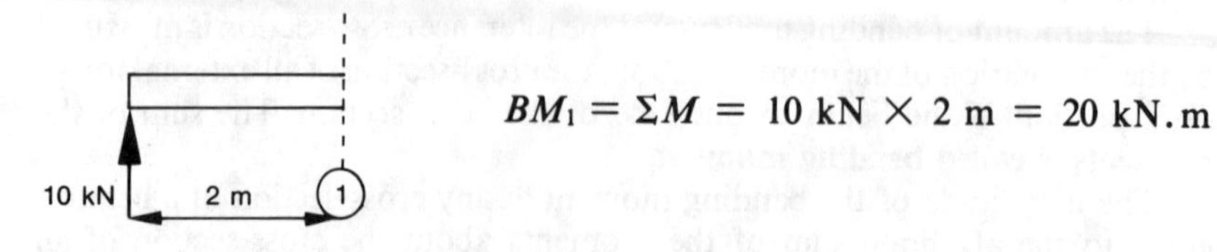

$$BM_1 = \Sigma M = 10 \text{ kN} \times 2 \text{ m} = 20 \text{ kN.m}$$

Bending moment at cross-section 2:

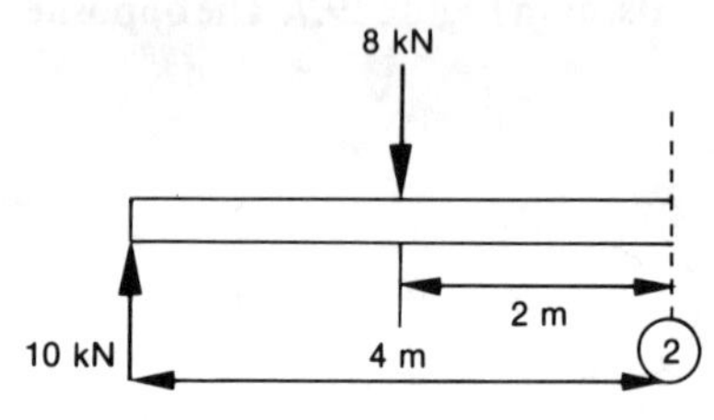

$$BM_2 = \Sigma M = 10 \text{ kN} \times 4 \text{ m} - 8 \text{ kN} \times 2 \text{ m}$$
$$= 24 \text{ kN.m}$$

Bending moment at cross-section 3:

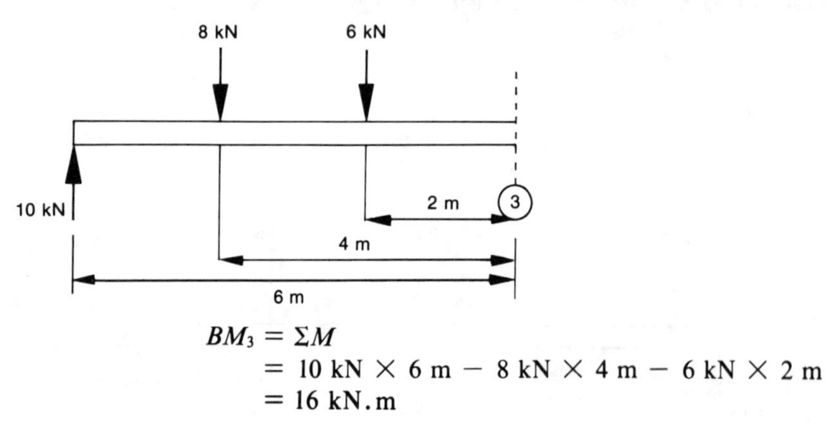

$$BM_3 = \Sigma M$$
$$= 10 \text{ kN} \times 6 \text{ m} - 8 \text{ kN} \times 4 \text{ m} - 6 \text{ kN} \times 2 \text{ m}$$
$$= 16 \text{ kN.m}$$

Summation of moments about each of the two supports will show that bending moment at the supports of a simply supported beam is always zero. It is left to the student to verify this statement by actual calculations.

Here again, it is useful to recognise that if we always work consistently from left to right, i.e. consider moments due to forces acting on the left-hand portion of the beam only, and use the usual sign convention for moments, i.e. positive-clockwise, negative-anticlockwise, the answers obtained automatically give the correct sign for the bending moments.

19.4 Bending moment diagrams

Information about bending moments can be conveniently represented by means of a **bending moment diagram**, which is a plot of the values obtained by calculation, against distance measured along the beam.

It can be shown mathematically that in the spaces between two adjacent external forces, bending moment varies directly with the distance measured along the beam. The bending moment diagram will therefore consist of straight lines joining the points representing the computed magnitude of the bending moment at each load-bearing point.

Example 19.4

Draw the bending moment diagram for the beam in Example 19.3.

Solution

The bending moment diagram is usually plotted directly below the shear force diagram using the same horizontal scale along the length of the beam. Vertical distances are drawn to some convenient scale, up from the zero line for positive bending moments and down for negative bending moments, under each of the external forces acting on the beam. The points obtained in this manner are joined by straight lines. Magnitudes of bending moments can be indicated on the diagram and vertical cross-hatching used for clarity. (See Fig. 19.9.)

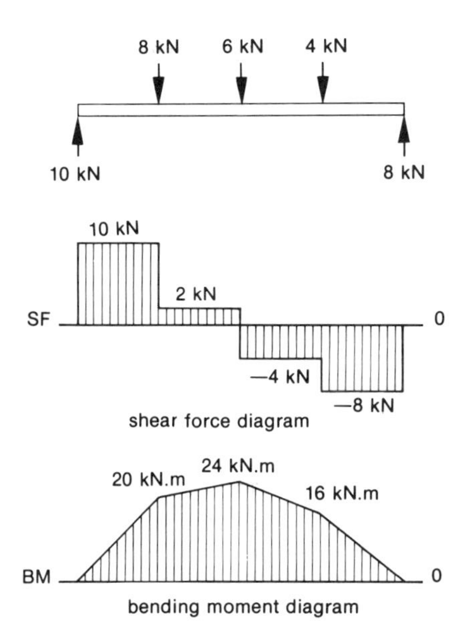

Fig. 19.9

Problem

19.2 For the beams and loadings shown in Figure 19.5 do all necessary calculations and draw the bending moment diagrams.

Compare with answers to problems 19.1, (a) to (j), by drawing *BM* diagram directly below *SF* diagram for each beam.

In each case note the position of the maximum bending moment.

19.5 Position of maximum bending moment

The problem of beam design consists essentially in selecting the cross-section which will offer the most effective resistance to shear forces and bending moments produced by the loads, without excessive stresses or bending of the beam.

Shear forces in a long slender beam by themselves are seldom a major criterion in beam design. Bending moments, on the other hand, can be responsible for very high stresses in the material of the beam and contribute to the curvature of the beam distorted by bending, which can lead to unacceptably high deflections.

In many engineering applications, only the maximum value of the bending moment needs to be known corresponding to the point where the beam is most likely to fail under bending. However, if a beam carries more than two or three loads, all of different magnitude, it is not always possible to tell at a glance where the maximum bending moment occurs.

It has already been observed that when a beam is subjected to several concentrated loads, the shear force is of constant value between the loads and changes abruptly at each load, by an amount equal to the load. This property greatly facilitates the construction of shear force diagrams.

A further observation shows that the bending moment is maximum at points where the shear force diagram changes from positive to negative, i.e. passes through zero axis.* The beam in Figure 19.9, for example, has a maximum bending moment of 24 kN.m at the point under the 6 kN load, where the shear force changes its value from +2 kN to −4 kN.

This relationship between the shear force diagram and the position of maximun bending moment often allows the designer to locate and determine the maximum bending moment in a beam without constructing the bending moment diagram.

---

* Those familiar with differential calculus will be interested to know that shear force is equal to the first derivative of bending moment with respect to distance − *x* measured along the beam,

$SF = \frac{d(BM)}{dx}$.

Hence where $SF = \frac{d(BM)}{dx} = 0$, *BM* is a maximum.

19.6 Bending stress

This part of the chapter deals with the relationship between bending moment in a beam, the stress produced in the material of the beam and certain geometrical properties of its cross-section.

Let us consider a particular cross-section in a beam at which a bending moment (M) is applied. It can be shown* that the magnitude of bending stress (f_b) produced in the material of the beam at the cross-section is directly proportional to the bending moment.

On the other hand, the size and shape of the cross-section of the beam give rise to a resistance which balances the applied bending moment. A geometrical measure of this resistance is known as the **moment of inertia** (I) of the cross-section. Appendix A at the end of the book deals with the concept of moment of inertia and the methods of calculating it for a variety of possible cross-sectional shapes.**It can be shown that bending stress is inversely proportional to the moment of inertia of the cross-section.

One important difference between bending stress and other types of stress considered previously is that, unlike direct tension or compression, bending stress is not distributed uniformly over the cross-sectional area of the beam. Every beam can be considered as being made up of a large number of horizontal layers or fibres held together by the internal adhesion between them. When a beam is subjected to bending, the fibres on the convex side are extended, while those on the concave side are compressed, as in Figure 19.10(a). Somewhere between the stretched and the compressed fibres there is a longitudinal plane along which there is no deformation of length. This plane is known as the **neutral plane**.

It can be shown mathematically that the neutral plane passes through the geometrical centre of the cross-section of the beam, called the centroid. Appendix A also deals with the methods of locating centroids of various geometrical shapes which may be encountered in practice.

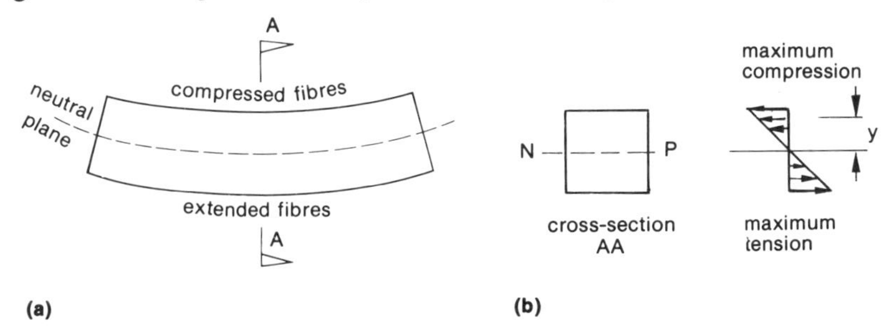

Fig. 19.10(a) *Beam in bending* **(b)** *Stress distribution*

* The full mathematical proof of this and following propositions, which is omitted here, can be found in more advanced texts on Strength of Materials.

** For the purposes of this chapter we shall note that the moment of inertia of a rectangular cross-section is given by $I = b.h^3/12$, where b is the width and h is the depth of the beam. (See Table A1 in the Appendix.)

The further a particular fibre is from the neutral plane the greater will be the amount of elongation or compression experienced by the fibre. We also know that according to Hooke's Law tensile and compressive stress in a material is proportional to the amount of change in length. Therefore, the stress in any one fibre is proportional to its distance (y) from the neutral plane (Fig. 19.10(b)).

The important conclusions which follow from the distribution of stress in a cross-section of a beam can be summarised as follows:

1. There is no stress at the neutral plane.
2. The maximum tensile stress occurs in the extreme fibre on the convex side of the beam.
3. The maximum compressive stress occurs in the extreme fibre on the concave side of the beam.

The formula which embodies the relationships discussed above is:

$$f_b = \frac{M.y}{I}$$

where f_b is bending stress in MPa
M is bending moment at a given cross-section in N.mm
y is distance from the neutral plane to a particular fibre in mm
I is moment of inertia of the cross-section in mm^4

This formula is applicable to any cross-section in a beam and any fibre within the beam. However, in most cases, the designer is only interested in the maximum values of stress. Therefore, he would use the maximum value of M and the distance to the extreme fibres, i.e. maximum value of y, for his calculations.

If the cross-section is not symmetrical about its centroidal (neutral) axis, a distinction has to be made between tension and compression sides by using different values of y.

Example 19.3

A beam of rectangular cross-section, 300 mm deep by 100 mm wide is subjected to a positive bending moment of 67.5 kN.m. Determine the maximum value of bending stress.

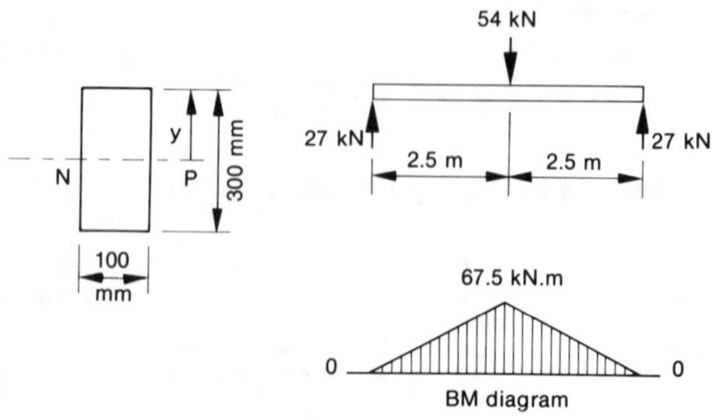

Fig. 19.11

Solution

The cross-section of the beam is as shown in Figure 19.11.

The moment of inertia $I = \frac{bh^3}{12} = \frac{100 \times 300^3}{12} = 225 \times 10^6 \text{ mm}^4$

The distance to the extreme fibre $y = 150$ mm

Bending moment $M = 67.5$ kN.m $= 67.5 \times 10^6$ N.mm

Bending stress $f_b = \frac{My}{I} = \frac{67.5 \times 10^6 \text{ N.mm} \times 150 \text{ mm}}{225 \times 10^6 \text{ mm}^4}$

$= 45$ MPa

In this case, because of the symmetrical cross-section, the extreme fibres on the tension and compression sides are at the same distance from the neutral plane. Therefore, the answer represents the maximum compressive stress in the top fibre as well as the maximum tensile stress in the bottom fibre.

Example 19.4

For the cantilever beam shown in Figure 19.12, determine the maximum value of stress.

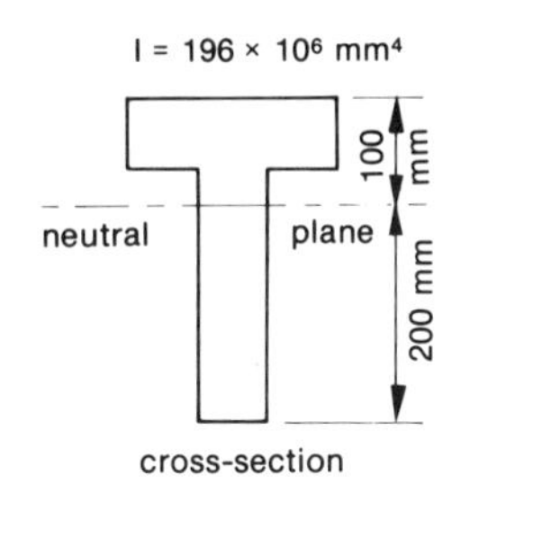

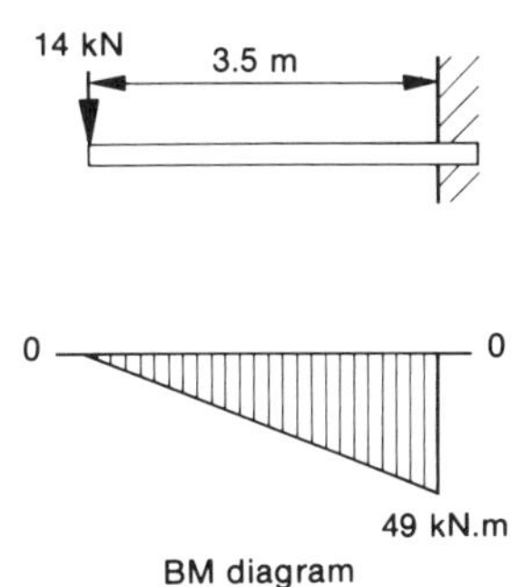

Fig. 19.12

Solution

Maximum bending moment occurs at the fixed end of the beam.

$M = 14 \text{ kN} \times 3.5 \text{ m} = 49 \text{ kN.m} = 49 \times 10^6 \text{ N.mm}$

This is negative bending moment. Therefore, the convex or tension fibre is on top and concave or compression fibre is on the bottom.

Hence, for maximum tension $y_t = 100$ mm and for maximum compression $y_c = 200$ mm.

Maximum stresses can now be calculated:

Tensile stress in the top fibre is:

$$f_t = \frac{My_t}{I} = \frac{49 \times 10^6 \text{ N.mm} \times 100 \text{ mm}}{196 \times 10^6 \text{ mm}^4} = 25 \text{ MPa}$$

Compressive stress in the bottom fibre is:

$$f_c = \frac{My_c}{I} = \frac{49 \times 10^6 \text{ N.mm} \times 200 \text{ mm}}{196 \times 10^6 \text{ mm}^4} = 50 \text{ MPa}$$

Such non-symmetrical cross-sections are sometimes used for beams made from materials which are stronger in compression than they are in tension, such as concrete.

19.7 Radius of curvature

When a beam is subjected to bending, its shape is distorted into a curve. The radius of curvature, R, is not constant along the beam, but depends on the magnitude of bending moment which exists at each cross-section. The radius also depends on the moment of inertia (I) of the cross-section and on Young's modulus of elasticity (E) of the material of the beam.

The expression relating these variables is:

$$R = \frac{EI}{M}$$

where R is radius of curvature in mm
M is bending moment in N.mm
E is Young's modulus in MPa
I is moment of inertia in mm^4

It should be understood that the radius of curvature is the inverse measure of distortion in bending, i.e. the greater the curvature the smaller the radius. For the undistorted, i.e. straight, beam the radius is infinite.

Example 19.5

Determine the radius of curvature at the point of maximum bending moment:

(a) for the beam in Example 19.3, if the material is steel, $E = 200\ 000$ MPa (N/mm^2);
(b) for the beam in Example 19.4, if the material is concrete, $E = 23\ 000$ MPa (N/mm^2).

Solution

(a) $R = \dfrac{EI}{M} = \dfrac{200\ 000 \text{ MPa} \times 225 \times 10^6 \text{ mm}^4}{67.5 \times 10^6 \text{ N.mm}} = 666\ 700 \text{ mm}$

$R = 666.7$ m

(b) $R = \dfrac{EI}{M} = \dfrac{23\ 000 \text{ MPa} \times 196 \times 10^6 \text{ mm}^4}{49 \times 10^6 \text{ N.mm}} = 92\ 000 \text{ mm}$

$R = 92$ m

As expected, the radii are large, because under moderate conditions of loading, the curvature produced in a solid beam is relatively small.

Problems

19.3 Determine the maximum stress for the beam and loading shown in Figure 19.13, if the beam is rectangular, 250 mm deep by 75 mm wide.

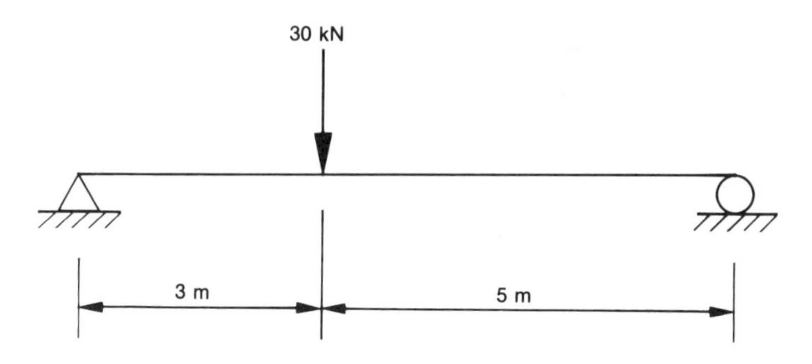

Fig. 19.13

19.4 If the material of the beam in problem 19.3 is steel, E = 200 000 MPa, determine the radius of the maximum curvature.

19.5 Determine the maximum load in kilonewtons that can be applied at the free end of a 3 m cantilever beam of universal rolled steel beam cross-section, 356 mm deep, with a moment of inertia of 142×10^6 mm^4, if the allowable stress is 76 MPa.

19.6 How wide should a rectangular cross-section of a beam be if its depth is 150 mm and it is to carry a load of 15 kN in the middle of an 8 m span with an allowable stress of 80 MPa?

19.7 Determine the stress and the radius of curvature at mid-span of the beam in problem 19.6, if the load is reduced to 12 kN. Take E = 200 000 MPa.

19.8 A beam of T-section has a moment of inertia of 350×10^6 mm^4 and distances from the neutral plane to the extreme fibre on the compression side of 150 mm and on the tension side 200 mm. The material is steel, E = 200 000 MPa.

Determine the stresses and the radius of curvature at mid-span, if the beam is 14 m long and carries a load of 40 kN in the middle.

19.9 For any given beam, what is the relationship between bending stress and radius of curvature?

Determine the radius of curvature of a 10 mm × 10 mm steel bar, E = 200 000 MPa, if during an experiment it is subjected to a maximum bending stress of 100 MPa.

19.10 Determine the maximum allowable bending moment and the corresponding radius of curvature for a timber beam, 200 mm deep by 100 mm wide, if the allowable stress is 20 MPa and the modulus of elasticity of timber is 12 000 MPa.

19.11 Repeat problem 19.10, for the beam placed so that it is 100 mm deep by 200 mm wide. Compare the answers. Which is the better way to place the beam in order to get the maximum load-carrying capacity?

19.12 For the loading as in problem 19.1(c) (Fig. 19.5(c)) select a suitable size of a rectangular timber beam if the allowable stress is 25 MPa and the cross-section is to be twice as deep as it is wide.

Review questions

1. Explain the concept of shear force in a loaded beam.
2. State the sign convention for shear forces.
3. Explain the concept of bending moment in a loaded beam.
4. State the sign convention for bending moments.
5. How can the shear force diagram be used to locate the maximum bending moment?
6. State the formula for calculating bending stress.
7. Explain the significance of the neutral plane.
8. Describe the pattern of stress distribution on a beam cross-section.
9. State how the radius of curvature of a beam can be calculated.

PART SIX

Fluid mechanics

We live, submerged at the bottom of an ocean of air, which by experiment, is known to have weight.

On the surface of the liquid in the bowl, a fifty-mile column of air presses down. It is not to be wondered, that in the tube, with nothing to oppose its entry, it rises until it balances the weight of the air on the outside, which supports it.

Evangelista Torricelli,
on the principle of mercury barometer

20

Properties of fluids

In this chapter we introduce some concepts and definitions related to the properties and applications of fluids in engineering. In particular **density** and **specific volume** of fluids will be discussed.

20.1 Fluid mechanics

Fluid mechanics is a branch of engineering science concerned with the properties and behaviour of fluids at rest or in motion. It deals with forces required to hold a fluid at rest, or with the interaction of forces that cause motion of the fluid. Fluid mechanics supplies the theoretical foundation for many engineering applications where the use of fluids is involved.

A fluid can generally be defined as a substance which can offer no permanent resistance to change of shape, i.e. a substance which can flow. This definition includes liquids and gases.

A **liquid** is a fluid which is practically incompressible, i.e. it retains constant volume regardless of the size or shape of its container. Water, oil and mercury are common examples of liquids used in engineering.

A **gas** is a fluid which expands to completely fill all the available space in a vessel in which it is contained. Air and steam are two examples of gaseous substances in common use.

Fluid mechanics and its applications can be subdivided into a number of different headings. For our purposes it is sufficient to define fluid statics, hydraulics and pneumatics as three major areas.

Fluid statics is the study of fluids at rest, concerned with pressure and its effects on submerged surfaces. It begins with manometry or the study of pressure and its measurement. Major applications of fluid statics are in the design of dams and storage tanks for the containment of liquids, and in ship design.

The theory of fluid flow and engineering applications which involve fluids in motion are further subdivided into hydraulics and pneumatics.

Hydraulics is concerned with **liquids in motion**, especially the pumping and flow of water in pipes and open channels. An important branch of hydraulics, called fluid-power engineering, covers the field of engineering applications in which the energy of pressurised or moving liquids is made to do mechanical work. Fluid-power technology finds its uses in the operation and control of machinery and transmission of power. The elements of fluid-power systems include pumps, hydraulic cylinders and controls. Water and oil are the main liquids used in fluid-power systems.

Examples of hydraulic equipment include: a motor vehicle brake in which the shoes are expanded by small pistons operated by oil pressure and supplied by a pedal-operated master cylinder and piston; pumping plants and pipelines; the hydraulic jack in which the lifting head is carried on a plunger working in a cylinder to which oil is supplied under pressure from a small hand-operated pump; the hydraulic press for exerting large forces for metal pressing and hydraulic riveters.

Pneumatics is concerned with applications involving **compressible fluids**, especially air, and can generally be described as the study of mechanical properties of air and other gases. Applications include automotive tyres, pneumatic brakes, paint sprayers and construction equipment, such as pneumatic drills. Compressed air power is flexible, economic and safe.

20.2 Mechanical properties of fluids

The main characteristic of a fluid is its ability to flow, i.e. its ability to be deformed continuously under the action of forces which cause the fluid to flow. Apart from mass-related properties discussed in the following sections, compressibility and viscosity require the greatest consideration in the study of fluid flow.

Compressibility of a fluid is a measure of its ability to change volume when subjected to variations in pressure. Compressibility is measured in terms of relative change in volume for a given change of pressure. The mathematical treatment of compressibility is outside the scope of this book, but a general understanding of the concept is useful. In particular, it is essential to understand that one of the main differences between liquids and gases is that for most practical purposes **all liquids are incompressible**, while gases will decrease in volume when subjected to increased pressure.

Viscosity is the property of fluids by which they resist forces which tend to change their shape, i.e. viscosity is the resistance to flow. The nature and characteristics of viscosity are not discussed in this book. It is sufficient to say that viscosity is a result of cohesive forces and of transfer of momentum between molecules of a fluid in motion, i.e. viscosity is analogous to friction. Gases and water have relatively low viscosity, while oils are highly viscous liquids.

In a fluid at rest there are no forces related to motion and viscosity has no relevance. This greatly simplifies the study of fluid statics.

When solving fluid flow problems it is possible to obtain an approximate theoretical answer by completely ignoring fluid viscosity. A fluid which is assumed to have negligible viscosity is referred to as **ideal fluid**.

20.3 Density and relative density

The **density** (ρ) of a fluid is defined as its mass per unit volume. The SI unit of density is derived from the units of mass (kg) and volume (m^3) and is kilogram per cubic metre (kg/m^3).

Table 20.1 Densities of liquids

Liquid	*Density* *(kg/m³)*	*Relative density* *(water = 1.0)*
Petrol	700	0.70
Alcohol (pure)	790	0.79
Turpentine	870	0.87
Oil (petroleum)	880	0.88
Oil (vegetable)	930	0.93
Water (pure)	1000	1.00
Beer, milk	1030	1.03
Water (sea)	1030	1.03
Hydrochloric acid	1200	1.20
Glycerine	1280	1.28
Mercury	13590	13.59

$$\rho = \frac{\text{mass}}{\text{volume}}$$

It is sometimes convenient to re-define the unit of density in terms of tonnes for mass, i.e. tonne per cubic metre (t/m^3) or litres for volume, i.e. kilogram per litre (kg/L).

Water is the most common and useful substance. It is helpful to memorise the density of liquid water at normal conditions, which is approximately equal to:

$$\rho_{water} = 1000 \text{ kg/m}^3 = 1 \text{ kg/L} = 1 \text{ t/m}^3$$

Other liquids have different densities as can be seen from Table 20.1.

Unlike liquids, gases are compressible fluids. It is therefore more important in the case of gases to specify the exact conditions for which densities are given. In Table 20.2 densities of some common permanent gases are listed for normal atmospheric conditions at sea level.

Table 20.2 Densities of gases (at 15° C and 101.3 kPa)

Gas	*Density* *(kg/m³)*	*Relative density RD* *(air = 1.0)*
Hydrogen	0.085	0.07
Helium	0.169	0.14
Neon	0.846	0.69
Nitrogen	1.185	0.97
Carbon monoxide	1.185	0.97
Air	1.226	1.00
Oxygen	1.354	1.10
Argon	1.692	1.38
Carbon dioxide	1.861	1.52
Sulphur dioxide	2.708	2.21

Relative density (RD) is a comparative measure of density. Liquids are usually compared to pure water and gases to air.

Relative density of a liquid is the ratio of its density to the density of water, and relative density of a gas is the ratio of its density to the density of air. For example, relative density of mercury is:

$$RD_{merc} = \frac{\rho_{merc}}{\rho_{water}} = \frac{13\ 590\ kg/m^3}{1000\ kg/m^3} = 13.59$$

Similarly, relative density of carbon dioxide is:

$$RD_{\substack{carb\\dioxide}} = \frac{\rho_{carb\ diox}}{\rho_{air}} = \frac{1.861\ kg/m^3}{1.226\ kg/m^3} = 1.518$$

Note that relative density is a dimensionless ratio.

Example 20.1

Determine the mass of air in a room 10 m × 6 m × 2.5 m at normal conditions.

Solution

From Table 20.2, $\rho = 1.226\ kg/m^3$

The volume of the room $= 10\ m \times 6\ m \times 2.5\ m = 150\ m^3$

Mass of air $= 1.226\ kg/m^3 \times 150\ m^3 = 183.9\ kg$

Example 20.2

600 mL of sulphuric acid has a mass of 1.11 kg. What is the density and relative density of sulphuric acid?

Solution

$$\text{Density} = \frac{\text{mass}}{\text{volume}} = \frac{1.11\ kg}{0.6\ L} = 1.85\ kg/L = 1850\ kg/m^3$$

$$\text{Relative density} = \frac{1850\ kg/m^3}{1000\ kg/m^3} = 1.85$$

The fact that liquids are incompressible means that a volume already occupied by a liquid cannot accommodate any additional mass of liquid. On the other hand, if a gaseous substance occupies a vessel, additional quantity of gas can be pumped into the vessel, thus increasing the mass contained in the same volume, i.e. increasing its density. Density of a gas can also be increased by compressing a given quantity of gas to a higher pressure. The properties of gases are treated in more detail in Chapters 26 and 27.

Example 20.3

A pressure vessel having a volume of 0.5 m^3 contains 0.677 kg of oxygen at normal atmospheric conditions. If additional 5 kg of oxygen are pumped into the vessel, determine the initial and final densities of oxygen in the vessel.

Solution

$$\text{Initial density} = \frac{0.677 \text{ kg}}{0.5 \text{ m}^3} = 1.354 \text{ kg/m}^3$$

$$\text{Final density} = \frac{0.677 \text{ kg} + 5.0 \text{ kg}}{0.5 \text{ m}^3} = 11.35 \text{ kg/m}^3$$

Example 20.4

An air receiver, 1.5 m^3 volume, contains 9.2 kg of compressed air. Determine the relative density of air in the receiver.

Solution

$$\text{Density} = \frac{9.2 \text{ kg}}{1.5 \text{ m}^3} = 6.133 \text{ kg/m}^3$$

$$\text{Relative density} = \frac{6.133 \text{ kg/m}^3}{1.226 \text{ kg/m}^3} = 5$$

20.4 Specific volume

Specific volume (v) of a fluid is defined as its volume per unit mass, i.e. the volume occupied by one unit of mass.

$$v = \frac{\text{volume}}{\text{mass}}$$

The SI unit of specific volume is cubic metre per kilogram (m^3/kg). Alternatively, cubic metre per tonne (m^3/t) and litre per kilogram (L/kg) may be used.

It follows from the definitions that specific volume is the reciprocal of density, or vice versa.

$$v = \frac{1}{\rho} \quad \text{or} \quad \rho = \frac{1}{v}$$

Example 20.5

Determine specific volume of air at sea level and normal temperature (15° C).

Solution

Density of air (from Table 20.2) is 1.226 kg/m^3.

$$\text{Specific volume } v = \frac{1}{\rho} = \frac{1}{1.226 \text{ kg/m}^3} = 0.8157 \text{ m}^3/\text{kg}$$

Example 20.6

Relative density of liquid bitumen is 0.85. What is its specific volume?

Solution

$$\text{Density} = 0.85 \times 1000 \text{ kg/m}^3 = 850 \text{ kg/m}^3 = 0.85 \text{ t/m}^3$$

$$\text{Specific volume, } v = \frac{1}{\rho} = \frac{1}{0.85 \text{ t/m}^3} = 1.176 \text{ m}^3\text{/t}$$

Problems

20.1 A bottle having a volume equal to 750 mL contains 637 g of kerosene when full. Determine density, relative density and specific volume of kerosene.

20.2 Determine the weight of one litre of mercury.

20.3 A swimming pool is 8 m long × 3 m wide with an average depth of 2 m. Determine the mass of water in the pool.

20.4 A cylindrical tank 1 m diameter × 1.5 m high is half filled with petrol. Determine the mass of petrol in the tank.

20.5 Determine the compressive force in each of the four legs of a platform supporting a water storage tank 2 m diameter and 1 m deep, when the tank is filled with water.

20.6 What is the mass of air in a room, 6 m × 3 m × 2.5 m, at normal atmospheric condition?

20.7 An air compressor takes air in at normal atmospheric conditions and compresses it to one-sixth of its initial volume. Determine the specific volume of air after compression.

20.8 A cylindrical air receiver is 1.5 m long × 0.5 m diameter and contains 1 kg of air. What mass of air must be pumped into the receiver in order to increase the air density in it to 10 kg/m^3?

20.9 A cylindrical bottle 200 mm diameter × 750 mm long contains 10 g of hydrogen. What is the density of hydrogen in the bottle? What mass of hydrogen can be removed from the bottle before the density of gas remaining in the bottle decreases to 0.085 kg/m^3?

20.10 A balloon is filled with 10 m^3 of helium at normal atmospheric conditions at sea level, where it is not fully extended. After the balloon reaches a certain height it extends to its full spherical shape with a diameter of 6 m. Determine the mass of helium in the balloon and its specific volume at the height reached.

Review questions

1. Define a fluid.
2. Describe the essential difference between liquids and gases.
3. What is fluid mechanics?
4. Explain the difference between fluid statics, hydraulics and pneumatics.
5. What is the meaning of compressibility?
6. Which fluids are incompressible?
7. What is meant by fluid viscosity?
8. What is an ideal fluid?
9. Define and state units of
   (a) density,
   (b) relative density,
   (c) specific volume.
10. What is the relationship between specific volume and density?
11. What is the density of water?

21

Pressure and its measurement

When dealing with liquids and gases we ordinarily speak of pressure without too much concern for its exact meaning. Thus we speak of pressure cookers, inflate automotive tyres to prescribed pressure, complain that there is insufficient pressure behind a water tap, etc. In engineering, however, the concept and units of pressure must be defined, and the means of measurement established before further progress can be made.

In common experience, pressure is usually associated with either a fluid confined in a container, i.e. fluid pressure, or pressure due to the weight of a liquid.

21.1 Fluid pressure

When a fluid is confined in a container, e.g. air in a pressure vessel, it exerts a force on every part of the inside surface of the container. This force results from the average effect of the multitude of minute forces produced by the rapid and repeated bombardment of the container walls by the moving molecules of the fluid.

The force on the inside surface of the container is a distributed force, which is everywhere perpendicular to the surface of the container. It is convenient to describe the distributed force acting on the surface as **pressure** due to the fluid. However, it should also be understood that pressure exists at every point within the fluid and not only at the surface.

Pressure can be defined as the normal force acting on each unit of area of the inside surface of the container. If the total force acting on the area A is F, the pressure p is given by

$$p = \frac{F}{A}$$

The SI unit of pressure is the **pascal**, with the symbol, Pa. The definition of the pascal follows from the definition of pressure, i.e. force per unit area, or newton per square metre. The pascal is a small unit suitable only for measuring small pressures or pressure differences. For the majority of engineering applications larger prefixed units called kilopascals (kPa) and megapascals (MPa) are used.*

The pressure corresponding to a perfect vacuum, i.e. completely empty space, is zero, as there can be no force exerted by empty space. Any pressure measured relative to this zero is called **absolute pressure** and is always positive.

Example 21.1

Gas is contained in a cylinder with a close fitting piston of 200 mm diameter. If the absolute pressure of the gas in the cylinder is 500 kPa determine the force exerted by the gas on the piston.

Solution

The area of the piston

$$A = \frac{\pi D^2}{4} = \frac{\pi \times 0.2^2}{4} = 0.0314 \text{ m}^2$$

From $p = \dfrac{F}{A}$,

$$\begin{aligned} F = p \times A &= 500 \text{ kPa} \times 0.0314 \text{ m}^2 \\ &= 500 \text{ kN/m}^2 \times 0.0314 \text{ m}^2 \\ &= 15.7 \text{ kN} \end{aligned}$$

Note that although the pressure unit in this case is the kilopascal, it is helpful to express it as kN/m^2, in order to cancel out the square metre, so that the answer is in kilonewtons, i.e. in units of force. It also makes clear that the homogeneity of units demands that the area be expressed in square metres.

Absolute pressure due to a gas is practically the same at all points within a container. On the other hand, if the fluid is a liquid we can only say that the pressure is the same at all points at the same level within the liquid, due to variation of pressure with depth in a liquid. We shall examine this in more detail in Chapter 22.

21.2 Atmospheric pressure

Here on the surface of the Earth we are at the bottom of an ocean of atmospheric air, the weight of which is pushing down on each unit area of the Earth's surface and all over the surface of all objects on the Earth.

Atmospheric pressure varies with altitude, geographical location and weather conditions. However, its average value at sea level can be taken as standard, equal to 101.3 kPa absolute.

---

* A further variation of the metric unit is the **bar** equal to 100 kPa. Although it has the advantage of being approximately equal to standard atmospheric pressure, the bar is specifically excluded from Australia's metric system. However, its derivative, the millibar, equal to 100 Pa, has been allowed in this country for use in meteorology, i.e. the study of the Earth's atmosphere in its relation to weather and climate. In this book we shall use only pascals, kilopascals and megapascals as the units of pressure.

The existence of atmospheric pressure was not realised until the seventeenth century when, in 1643, Evangelista Torricelli (1608–47), pursuing a suggestion by Galileo, inverted a long glass tube filled with mercury into a dish and observed that some of the mercury did not flow out and that above the mercury in the tube was a vacuum. He concluded that the column of mercury was supported by atmospheric pressure, and that the changes of the height of the mercury from day to day were due to variations in atmospheric pressure. The space above the mercury in the tube is called a Torricellian vacuum.

Torricelli did not live to see his explanation of atmospheric pressure generally accepted among scientists. After his death, Blaise Pascal (1623–62) repeated and extended Torricelli's experiments, and in particular demonstrated the variation of atmospheric pressure with altitude. Another famous demonstration of the effect of atmospheric pressure was performed in 1654 by a German scientist, Otto van Guericke (1602–86), who is credited with the invention of the air pump which made creation of a partial vacuum possible. For his experiment, Guericke made two hollow bronze hemispheres ("Magdeburg hemispheres"), which were placed together to form a hollow sphere about 350 mm in diameter with an airtight ring of greased leather. After the air was pumped out from the sphere, the external atmospheric pressure acting on the hemispheres prevented two teams of eight horses, each pulling in opposite directions, from separating them, until air had been re-admitted.

The mercury barometer

The most common device used for accurate measurements of atmospheric pressure is the mercury barometer. In its simplest form it is the basic Torricelli's tube, just under one metre in length, in which the atmospheric pressure balances a column of mercury, the height of which can be measured with accuracy (see Fig. 21.1(a)).

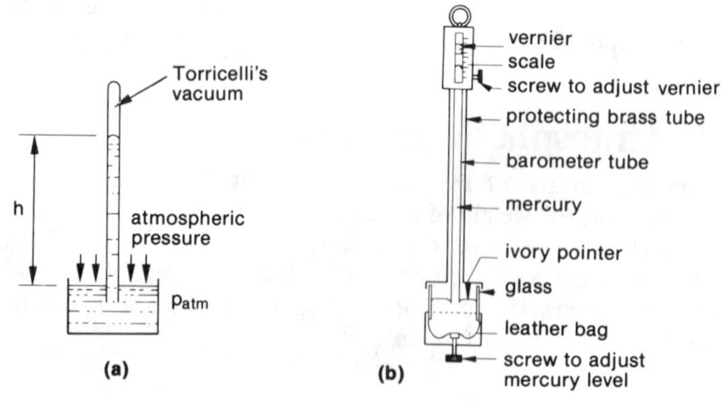

Fig. 21.1

If the cross-sectional area of the tube is A and the height of the column of mercury is h, the weight of the column is

$$F_w = m \times g = (\rho . A . h) \times g = 13\,590 \times A \times h \times 9.81$$

where ρ is density of mercury equal to 13 590 kg/m^3.

The weight of the column is exactly balanced by the force due to atmospheric pressure, acting at the foot of the column, equal to:

$$F = p_{atm} \times A = 101\ 300 \text{ Pa} \times A$$

Equating the two forces yields:

$$13\ 590 \times A \times h \times 9.81 = 101\ 300 \times A$$

Hence,
$$h = \frac{101\ 300}{13\ 590 \times 9.81} = 0.760 \text{ m}$$

Thus the column of mercury supported by standard atmospheric pressure is 760 mm high.

It can therefore be stated that standard atmospheric pressure is:

$$p_{atm} = 101.3 \text{ kPa} = 760 \text{ mm Hg}$$ *

For any other value of atmospheric pressure, the column of mercury will be proportional to pressure.

Example 21.2

At a particular altitude the barometer reading is 742 mm. What is the atmospheric pressure?

Solution

$$p_{atm} = 742 \text{ mm Hg} \times \frac{101.3 \text{ kPa}}{760 \text{ mm Hg}} = 98.9 \text{ kPa}$$

Most variations of the many different types of mercury barometers arise from different techniques for measuring the height of the mercury column. The Fortin barometer, shown in Figure 21.1(b), employs a vernier scale for accurate readings of the mercury column. It also incorporates an adjustable leather bag which replaces the dish of mercury which is used in the simple barometer. The adjusting screw allows the open surface of mercury to be raised or lowered to coincide with the zero of the vertical millimetre scale.

The aneroid and the altimeter

The aneroid barometer is a mechanical device for measuring atmospheric pressure. It consists of a partially evacuated cylindrical capsule or bellows connected to a pointer by a system of levers, so that the movements of the bellows resulting from changes in atmospheric pressure are transmitted to the pointer moving over a suitably calibrated scale.

Aneroid barometers are less accurate than mercury barometers, but are smaller in size and more convenient in use, especially as portable instruments.

---

* In this expression the chemical symbol for mercury, Hg, is used.

Table 21.1 Standard atmospheric pressure for different altitudes above sea level

Altitude (m)	*Atmospheric pressure (kPa)*
500	95.2
1000	89.6
1500	84.3
2000	79.3
2500	74.5
3000	69.9
4000	61.7
5000	54.1

Aneroid barometers are also used as altimeters for aircraft, in which case the scale is calibrated in metres of ascent. Roughly speaking, the pressure falls by 1.1 kPa for every 100 m of ascent in the lower atmosphere. A more accurate summary of the variation in standard atmospheric pressure with altitude is given in Table 21.1.

The highest location in Australia, the top of Mount Kosciusko, is at 2230 m above sea level, where standard atmospheric pressure is 77.0 kPa, while that on top of Mount Everest is 31.6 kPa.

21.3 Gauge pressure

In most theoretical calculations, especially those concerning compressed gases, we are concerned with absolute pressure. However, absolute pressure is not easy to measure directly. Most pressure gauges read the difference between the absolute pressure and the atmospheric pressure existing at the time of measurement, and this is referred to as **gauge pressure**.

The relationship between gauge pressure (p_{ga}), absolute pressure (p_{abs}) and atmospheric pressure (p_{atm}) is as follows:

$$p_{ga} = p_{abs} - p_{atm}$$

or alternatively:

$$p_{abs} = p_{atm} + p_{ga}$$

It should be noted that, unlike absolute pressure, gauge pressure can be positive or negative, depending on whether it measures above or below atmospheric pressure. Negative gauge pressure, i.e. pressure below atmospheric, is sometimes referred to as vacuum.

The relationship between gauge, absolute and atmospheric pressure is illustrated graphically in Figure 21.2.

The most common type of pressure gauge is the Bourdon-tube gauge, invented in 1850 by Eugene Bourdon. As shown in Figure 21.3, it has a working element in the form of a hollow tube of elliptical cross-section, shaped

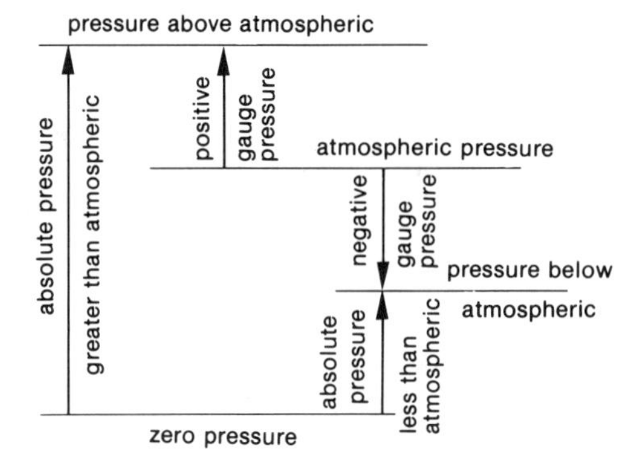

Fig. 21.2

like a question mark and connected directly to the source of pressure, e.g. air receiver or steam boiler. As the pressured fluid enters the tube, it tends to straighten it out, thus moving the tube tip. This movement actuates the pointer spindle and pressure is read on a graduated circular scale.

The selection of a suitable gauge depends on the type of fluid used, expected pressure levels, size, accuracy and safety considerations. Alternatives include pressure gauges incorporating flexible diaphragms or bellows as principal working elements, with direct mechanical linkage attached to an indicator.

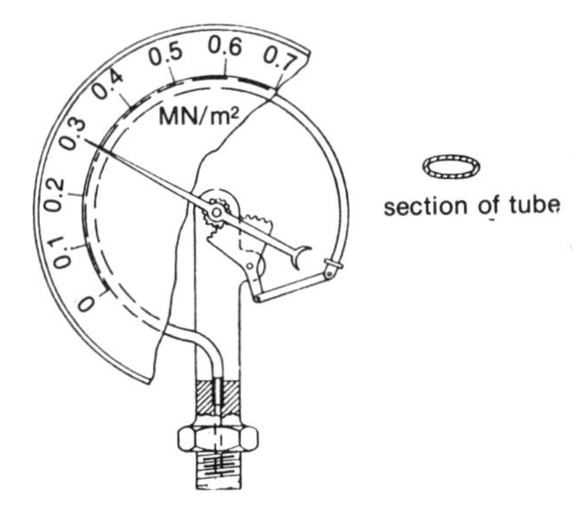

Fig. 21.3

Example 21.3

A pressure gauge on an air receiver reads 200 kPa when the mercury barometer reading is 750 mm Hg. What is the absolute pressure of air in the receiver?

Solution

$$p_{ga} = 200 \text{ kPa}$$

$$p_{atm} = 750 \text{ mm Hg} \times \frac{101.3 \text{ kPa}}{760 \text{ mm Hg}} = 100 \text{ kPa}$$

$$p_{abs} = p_{ga} + p_{atm} = 200 \text{ kPa} + 100 \text{ kPa} = 300 \text{ kPa}$$

Problems

21.1 Express a pressure of 500 kPa in (a) pascals, (b) megapascals and (c) newtons per square metre.

21.2 A gas exerts a uniform pressure of 1.5 MPa on a piston of 150 mm diameter. Determine the force due to gas pressure.

21.3 If a 100 mm diameter piston exerts a force of 4 kN on liquid in a reciprocating pump, what is the pressure due to this force?

21.4 When a Fortin barometer reading is 767.5 mm Hg, what is the atmospheric pressure?

21.5 What will the height of the mercury column be when the atmospheric pressure is 95 kPa?

21.6 What will the height of the mercury barometer column be when measuring atmospheric pressure on top of Mount Kosciusko?

21.7 A gauge on a steam boiler reads 700 kPa, atmospheric pressure is 101.3 kPa. What is the absolute pressure of steam?

21.8 A vacuum gauge reads 25 kPa, atmospheric pressure is 101.3 kPa. What is the absolute pressure?

21.9 An air receiver contains air at 800 kPa (gauge) while atmospheric pressure is 100 kPa. If the air at this pressure operates a piston of 10 mm diameter, determine the force exerted by the compressed air on the piston.

Fig. 21.4(a) *Dead weight pressure gauge test*

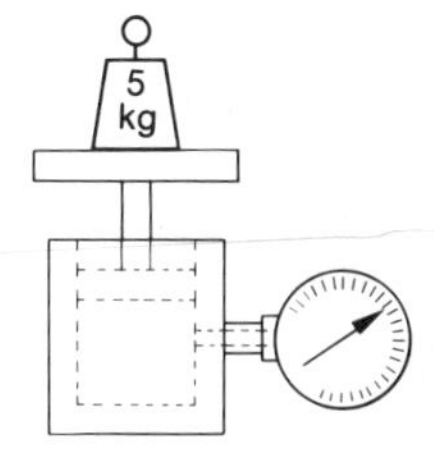

Fig. 21.4(b)

21.10 A pressure gauge is tested as shown in Figure 21.4(b), by placing a 5 kg mass on a piston of 30 mm diameter. The atmospheric pressure is 101.3 kPa. What should the gauge reading be?

21.4 Manometry

Apart from mechanical pressure gauges described in the previous section, pressure measurement equipment includes a range of instruments which can generally be described as **manometers**. Hence, pressure-measuring techniques, particularly those which involve the use of manometers, are commonly described as **manometry**.

In general, a manometer can be described as an instrument used to measure the pressure of a fluid, composed of a transparent tube of suitable shape, containing a liquid, such as mercury, water, oil or alcohol, with one end connected to the pressure source and the other open to the atmosphere or connected to a source of reference pressure for differential reading.

The operation of a manometer depends on the principle that pressure can be balanced by a column of liquid so that:

$$p = \rho . g . h$$ *

where p is pressure corresponding to the column of liquid
ρ is density of the liquid
g is 9.81 N/kg, gravitational constant
h is the vertical height of the column of liquid

* For the proof of this relationship see Chapter 22.

Depending on the purpose and the arrangement of the limbs of the manometer tube, all manometers can be divided into a number of types which include piezometers, simple U-tube manometers and differential manometers. The scale can be vertical or inclined. The inclined scale provides better accuracy of measurement, especially when small pressure differences are involved. The types are summarised diagrammatically in Figure 21.5.

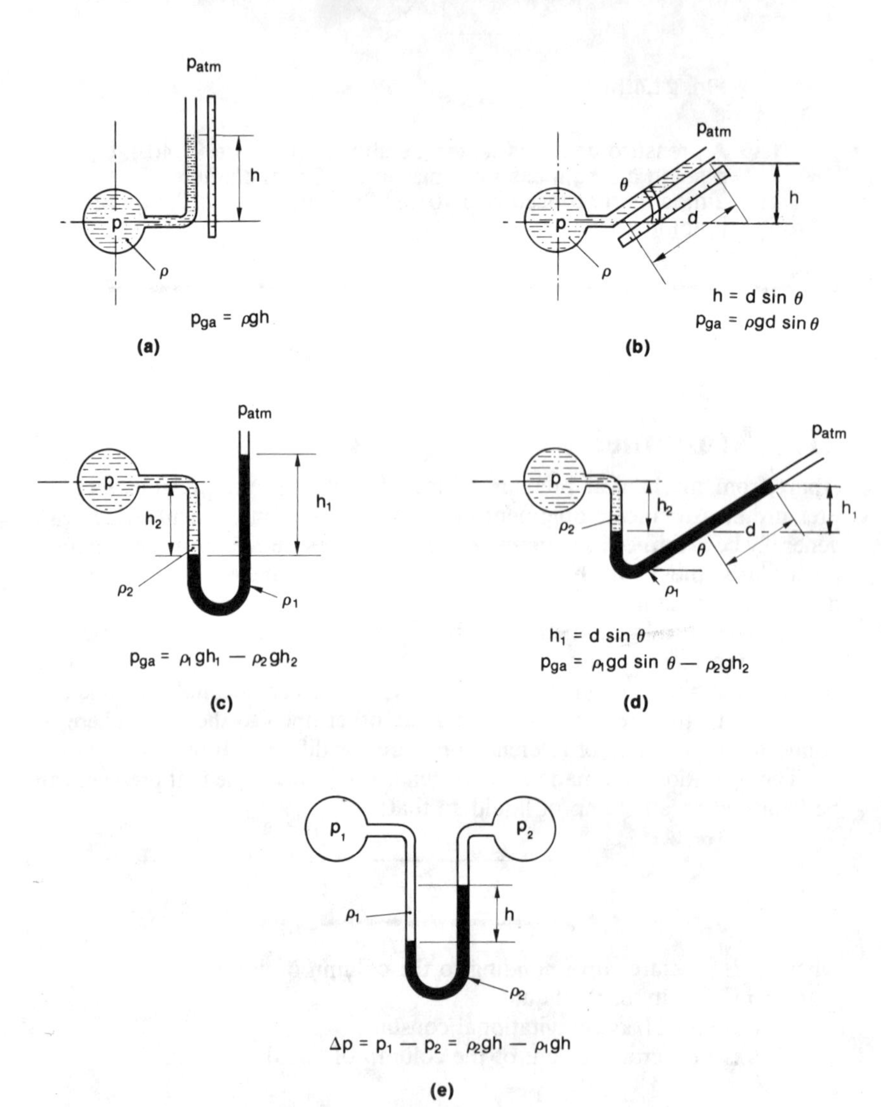

Fig. 21.5 **(a)** *Vertical piezometer* **(b)** *Inclined piezometer* **(c)** *Simple U-tube manometer* **(d)** *Inclined manometer* **(e)** *Differential manometer*

Solution of each individual problem involving manometers depends on the arrangement of a particular instrument. However, the steps for determining pressures in complicated gauges can be summarised as follows:

1. Start at the lowest level, below which there is complete symmetry, and calculate pressure due to each column of liquid using $p = \rho.g.h$.
2. Equate the sum of all pressures on one side to the sum of all pressures on the other side.
3. Solve for absolute pressure, if required, or for $(p_{abs} - p_{atm})$, i.e. gauge pressure.

The following examples should help to illustrate these steps.

Example 21.4

A simple U-tube manometer using mercury ($\rho = 13\,590$ kg/m^3) is used to measure pressure in a hot water boiler.

The readings are as shown in Figure 21.6.

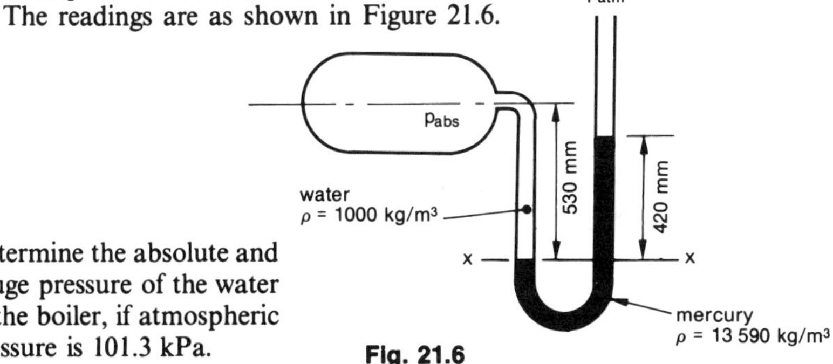

Determine the absolute and gauge pressure of the water in the boiler, if atmospheric pressure is 101.3 kPa.

Fig. 21.6

Solution

Starting from level *X-X*, the pressures are:

Pressure due to a column of water* 530 mm high, equal to:

$$p_{wat} = \rho.g.h = 1000 \text{ kg/m}^3 \times 9.81 \text{ N/kg} \times 0.53 \text{ m}$$
$$= 5199 \text{ Pa} = 5.2 \text{ kPa}$$

Pressure due to a column of mercury, 420 mm high, equal to:

$$p_{merc} = \rho.g.h = 13\,590 \text{ kg/m}^3 \times 9.81 \text{ N/kg} \times 0.42 \text{ m}$$
$$= 55\,994 \text{ Pa} = 56.0 \text{ kPa}$$

Equating pressures on two sides yields:

$$p_{abs} + 5.2 = p_{atm} + 56.0$$
$$p_{abs} + 5.2 = 101.3 + 56.0$$

Hence, absolute pressure,

$$p_{abs} = 152.1 \text{ kPa}$$

Gauge pressure is therefore:

$$p_{ga} = p_{abs} - p_{atm} = 152.1 - 101.3 = 50.8 \text{ kPa}$$

* Note that if the fluid being measured is a gas, e.g. steam, its density is relatively low and can be neglected.

Example 21.5

An inclined piezometer with an angle of 25° to the horizontal is measuring pressure in a horizontal pipe carrying oil of relative density 0.78, as shown in Figure 21.7.

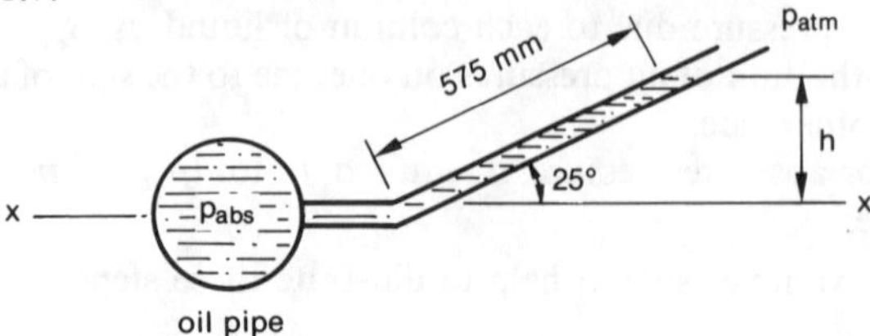

Fig. 21.7

The reading along the incline is 575 mm. What is the gauge pressure of oil in the pipe?

Solution

The vertical height $h = 0.575 \text{ m} \times \sin 25° = 0.243 \text{ m}$

$$\text{Oil density } \rho = 0.78 \times 1000 \text{ kg/m}^3$$
$$= 780 \text{ kg/m}^3$$

Pressure due to the column of oil:

$$p = \rho.g.h = 780 \text{ kg/m}^3 \times 9.81 \text{ N/kg} \times 0.243 \text{ m}$$
$$= 1859 \text{ Pa} = 1.86 \text{ kPa}$$

This is the gauge pressure of the oil, as can readily be seen from the summation of pressure above *X-X*:

$$p_{abs} = 1.86 + p_{atm}$$

Hence, $$p_{ga} = p_{abs} - p_{atm} = 1.86 \text{ kPa}$$

Fig. 21.8 *U-tube and inclined manometers*

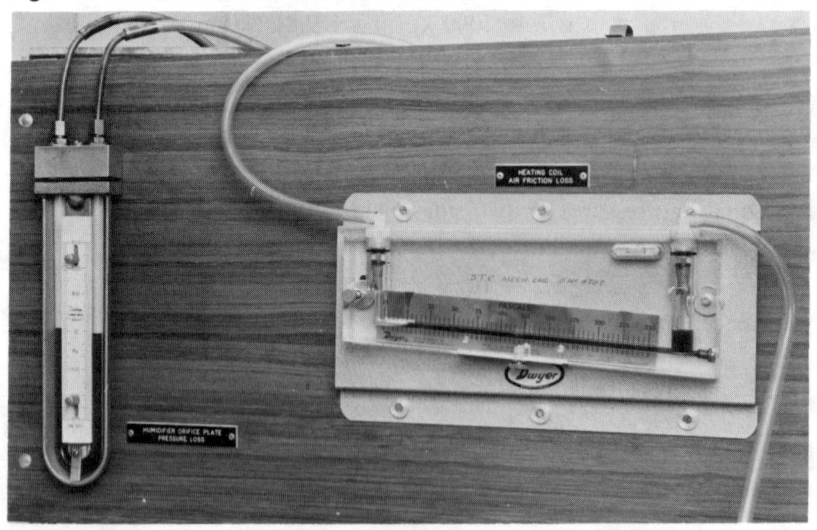

Problems

In the following problems, refer to Figure 21.5 if necessary. Atmospheric pressure may be taken as standard at 101.3 kPa.

21.11 Determine the gauge and absolute pressures in a water pipe, if the water level in a vertical piezometer is 300 mm above the centreline of the pipe.

21.12 An inclined piezometer reads 530 mm along a 35° incline when the fluid of relative density 0.83 is in the vessel. Determine the corresponding gauge pressure at the point of measurement.

21.13 A simple U-tube manometer, using manometer fluid of relative density 1.78, reads 246 mm below atmospheric when attached to an air duct. Determine the absolute pressure of air in the duct.

21.14 An inclined mercury manometer is used to measure the pressure in a gas main. The reading along a 25° incline is 120 mm. What is the gauge pressure of the gas?

21.15 A U-tube water manometer is used to measure pressure rise across an air fan. The reading is 20 mm. What is the pressure rise in kilopascals?

21.16 A differential manometer using mercury is connected between two water pipes as shown in Figure 21.5(e).

If pressure p_1 is 150 kPa (absolute) and the reading h = 350 mm, what is the absolute and gauge pressure in the second pipe?

21.17 A U-tube manometer using mercury is connected to a pipe carrying oil of relative density 0.75 as shown in Figure 21.5(c). The readings are h_1 = 255 mm and h_2 = 200 mm.

What is the absolute pressure of the oil?

21.18 A simple U-tube manometer is to be used to measure gauge pressure of air within the range of 0–5 kPa. If the height of the column is not to exceed 300 mm and accuracy is important, which manometer fluid would you recommend? mercury, water, or a special fluid of RD = 1.78?

Review questions

1. State the definition of pressure.
2. What is the SI unit of pressure?
3. Explain the meaning of hydrostatic pressure.
4. Explain the difference between absolute, atmospheric and gauge pressure.
5. What is the numerical value of standard atmospheric pressure at sea level?
6. Explain the principle of operation of the mercury barometer.
7. What is an aneroid barometer?
8. With the aid of a sketch, explain the operation of a Bourdon-tube pressure gauge.
9. With the aid of sketches explain the operation of
   (a) piezometer
   (b) U-tube manometer and
   (c) differential manometer.
10. Explain the advantage of the inclined limb manometer.

22

Applications of fluid pressure

In this chapter we consider fluid pressure in action. First, we look at applications of fluid statics, i.e. the study of fluids at rest. These include transmission of pressure and its conversion to a useful force in hydraulic jacks and presses, buoyancy and flotation. Secondly, the work done by fluid pressure and work done during pumping are considered.

22.1 Pressure in a liquid

It can be shown that pressure within a body of liquid depends only on two factors, its density and the depth. To prove this proposition let us consider a tank of uniform cross-sectional area, A, h units deep, holding a liquid of density ρ, as shown in Figure 22.1.

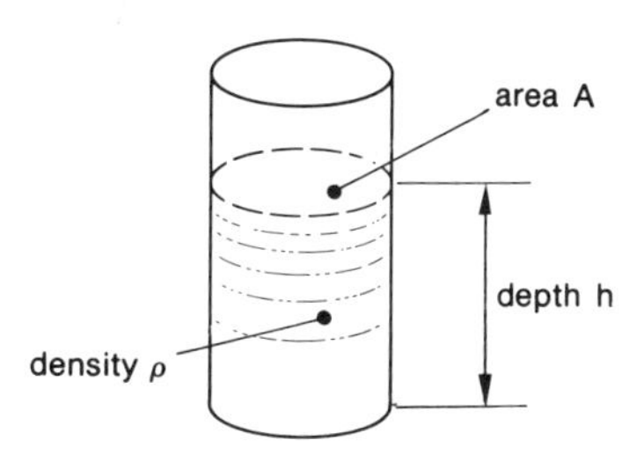

Fig. 22.1

We can write the following series of statements:

Volume of the liquid $V = A \times h$

Mass of the liquid $m = \rho \times V = \rho \times A \times h$

Weight of the liquid $F_w = m \times g = \rho \times A \times h \times g$

Pressure on the bottom of container,

$$p = \frac{F_w}{A}$$

$$= \frac{\rho \times A \times h \times g}{A}$$

The cross-sectional area cancels out and we are left with the now familiar

$$p = \rho \times g \times h$$

where h is the depth of the liquid.

In this context several basic principles of fluid statics must be mentioned and clearly understood:

1. Pressure exists at every point within the liquid.
2. Pressure at a point is the same in all directions, i.e. pressure is a scalar.
3. The pressure is directly proportional to the depth below the surface.
4. The pressure is the same at all points at the same level within a liquid.

Students must satisfy themselves that the above-mentioned principles are in fact valid. Of these, the last could present difficulties and is sometimes referred to as the "hydrostatic paradox", particularly when presented in the form of several vessels of different shape filled to the same depth with identical liquid (see Fig. 22.2(a)). According to the equation, $p = \rho . g . h$, the pressure at the base is the same in all the vessels, regardless of their shape or the amount of liquid they contain, so long as the depth is the same. The complete proof of the validity of this principle is outside the scope of this book. Could it help our understanding if the vessels were connected at the base to provide an obvious level of equal pressure, as shown in Figure 22.2(b)?

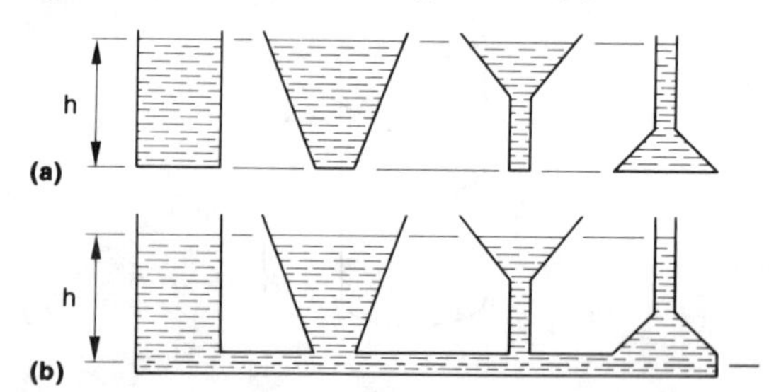

Fig. 22.2

In addition, we must note that pressure given by the equation $p = \rho . g . h$ does not include the effect of atmospheric pressure acting on the open surface of the liquid, i.e. it is a measure of gauge pressure.

Example 22.1

A water tank of 2 m diameter contains 1.5 m depth of water. What is the pressure exerted by the water on the base of the tank?

Solution

$$p = \rho \times g \times h = 1000 \text{ kg/m}^3 \times 9.81 \text{ N/kg} \times 1.5 \text{ m}$$
$$= 14\,715 \text{ Pa} = 14.7 \text{ kPa}$$

22.2 Forces on submerged surfaces

Since pressure is distributed force, it should be possible to calculate the total force acting on a submerged surface. It follows from the definition of hydrostatic pressure (see Chap. 21) that the force on any plane surface submerged in liquid is perpendicular to the surface. This makes calculation of forces acting on horizontal submerged surfaces very easy, as the following example illustrates.

Example 22.2

What is the force due to water pressure acting on the base of a water tower having dimensions as shown in Figure 22.3?

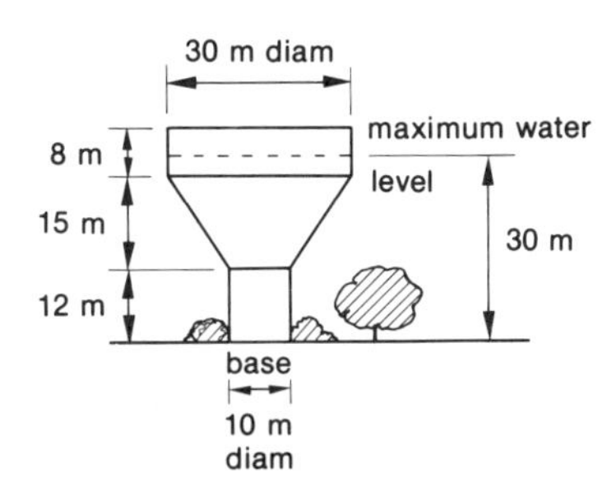

Fig. 22.3

Solution

Pressure at the base, $p = \rho.g.h$

$= 1000 \text{ kg/m}^3 \times 9.81 \text{ N/kg} \times 30 \text{ m}$

$= 294\,300 \text{ Pa} = 294.3 \text{ kPa}$

Area of the base, $A = \dfrac{\pi D^2}{4}$

$= 78.54 \text{ m}^2$

Force on the base, $F = p \times A$

$= 294.3 \text{ kN/m}^2 \times 78.54 \text{ m}^2$

$= 23\,114 \text{ kN} = 23.1 \text{ MN}$

When a rectangular* submerged surface is vertical, pressure distribution over the surface is not uniform but varies linearly from zero gauge at the free surface to $p = \rho.g.h$ at the bottom. In this case the force can be calculated using average pressure.

Example 22.3

Determine the force due to water pressure on a dam 20 m long if the depth of water is 12 m from the water surface to the base of the dam (Fig. 22.4).

* Non-rectangular surfaces are not considered in this book.

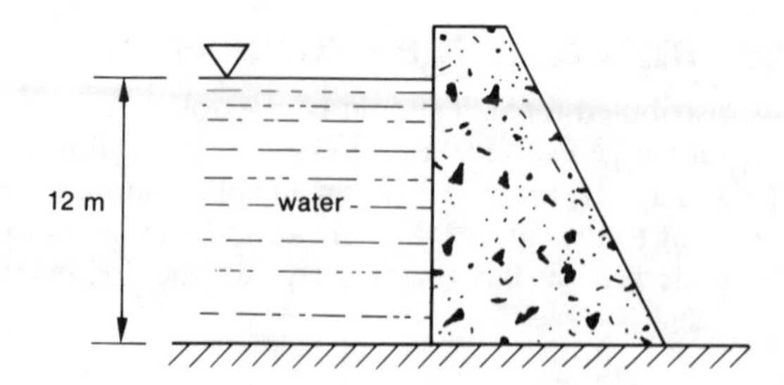

Fig. 22.4

Solution

Pressure at the base,
$$p = \rho . g . h$$
$$= 1000 \text{ kg/m}^3 \times 9.81 \text{ N/kg} \times 12 \text{ m}$$
$$= 117.7 \text{ kPa}$$

Average pressure,
$$\frac{p_{\text{surf}} + p_{\text{base}}}{2} = \frac{0 + 117.7}{2} = 58.86 \text{ kPa}$$

Area,
$$A = 12 \text{ m} \times 20 \text{ m} = 240 \text{ m}^2$$

Force,
$$F = p \times A$$
$$= 58.86 \text{ kN/m}^2 \times 240 \text{ m}^2$$
$$= 14\,126 \text{ kN} = 14.13 \text{ MN*}$$

22.3 Transmission of pressure by fluids

Some time around 1650, French scientist–philosopher, Blaise Pascal (1623–62), formulated the law which bears his name, although the principles behind it had been stated earlier by Benedette and also independently by Stevin.

Pascal's Law states that **pressure impressed at any place on a confined fluid is transmitted undiminished to all other portions of it.**

This means that if a force is applied anywhere on a fluid contained in a closed system, thereby increasing its pressure, that pressure will be transmitted without loss in every direction. The law does not imply that pressures are equal at all points throughout the confined fluid, but rather that when pressure is increased by applying an external force, the amount of increase is the same at all points.

One of the most common applications of Pascal's law is the hydraulic press, which is a machine used to multiply force. In a hydraulic press, a small force is applied to a small piston in a small cylinder of a reciprocating pump,

---

* It can be shown that this force is not acting at the centre of the submerged area, but at a point known as the "centre of pressure" which lies at a distance of $\frac{2}{3}h$ from the free water surface (e.g. at a depth of $\frac{2}{3} \times 12 = 8$ m in the above example). For a full mathematical treatment of the concept of centre of pressure, one must refer to more advanced texts.

thus generating pressure. This pressure is transmitted throughout the oil or water used as the hydraulic fluid, and presses on a large piston in a large cylinder producing a large force.

Example 22.4

In a hydraulic press (Fig. 22.5), a force of 30 N is applied at the end of the operating lever as shown. The diameter of the small piston is 50 mm and that of the large piston 250 mm. Determine the force on the large piston exerted by fluid pressure.

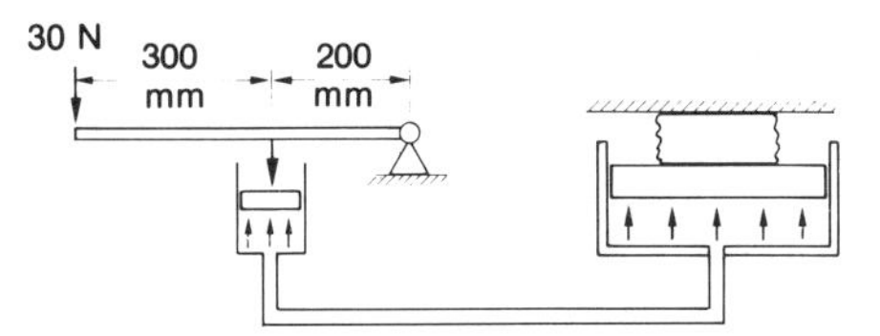

Fig. 22.5

Solution

Taking moments about the pivot point, the force applied to the small piston is:

$$F_1 = \frac{30 \times 500}{200} = 75\ \text{N}$$

The pressure of the fluid due to this force is:

$$p = \frac{F_1}{A_1} = \frac{75}{\left(\dfrac{\pi \times 0.05^2}{4}\right)} = 38.2\ \text{kPa}$$

This pressure transmitted to the large cylinder exerts a force on the large piston equal to:

$$F_2 = p \times A_2 = 38.2 \times \left(\frac{\pi \times 0.25^2}{4}\right) = 1.875\ \text{kN}$$

Thus in a simple hydraulic device like this the force is magnified:

$$\frac{1875\ \text{N}}{30\ \text{N}} = 62.5\ \text{times.}$$

Problems

22.1 Determine the pressure at the base due to water contained in a tank of 1 m diameter filled to a depth of 1.2 m.

22.2 If water was used as a fluid in a barometer, instead of mercury, what would the height of the water column be, when atmospheric pressure is standard at 101.3 kPa?

22.3 Pressure due to a particular liquid increases with depth at the rate of 7.85 Pa per mm. What is the relative density of the liquid?

22.4 A cylinder and piston 500 mm diameter are provided with a side tube as shown in Figure 22.6. If the cylinder contains oil of relative density 0.88 and a 100 kg load is placed on the piston, how high will the oil be forced in the tube?

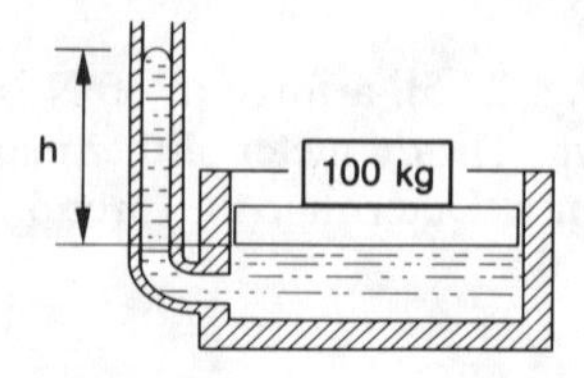

Fig. 22.6

22.5 A cylindrical tank, 500 mm diameter and 1.5 m high, contains glycerine of relative density 1.28 to a depth of 1.3 m. The air space above the liquid surface is at a pressure of 30 kPa gauge. Determine the maximum pressure in the tank and the total force on the inside surface of its base.

22.6 A water tank, 2 m diameter and 1.5 m deep, has a horizontal opening 300 mm diameter in its base, covered by a watertight cover plate held in position by 8 bolts. Determine the force in each bolt when the tank is full.

22.7 Determine the force on a water dam 10 m long, if the water depth at the dam is 1.8 m.

22.8 A hydraulic hoist has a main cylinder of 350 mm diameter and a pump cylinder of 20 mm diameter. If the force applied by the pump piston is 100 N, determine the load it can raise in tonnes, assuming no losses.

22.9 A hydraulic press has a ram cylinder of 200 mm diameter and a pump cylinder of 25 mm diameter. If a ram force of 50 kN is required, determine the necessary oil pressure and the force exerted by the pump piston.

22.10 A hydraulic jack is to be designed so that a force of 60 N applied at a small piston of 10 mm diameter would lift a load of 0.5 tonne. What should be the diameter of the large piston?

22.4 Buoyancy and flotation

When an object is placed in a fluid, it receives an upward force or upthrust. Archimedes, who lived in the third century BC in Syracuse, Sicily, and is considered to be one of the greatest scientists of all times, was the first to carry out practical experiments to measure the upthrust, otherwise known as the buoyant force, on objects submerged in liquids. The discovery he made, now called Archimedes' principle, was contained in one of his famous works "On Floating Bodies", which did not survive in its original form.

The principle states that, **when a body is wholly or partially immersed in a fluid, it experiences an upthrust equal to the weight of the fluid displaced.**

$$F_b = V.\rho.g$$

where F_b is buoyant force (upthrust on submerged body)
V is volume of the fluid displaced
ρ is fluid density
g is gravitational constant, 9.81 N/kg

Example 22.5

A solid block, 200 mm × 150 mm × 100 mm, is immersed in water. What is the buoyant force acting on the block?

Solution

The volume of the water displaced is:

$$V = 0.2 \times 0.15 \times 0.1 = 0.003 \text{ m}^3$$

The buoyant force, equal to the weight of the water displaced, is:

$$F_b = V \times \rho \times g$$
$$= 0.003 \text{ m}^3 \times 1000 \text{ kg/m}^3 \times 9.81 \text{ N/kg}$$
$$= 29.43 \text{ N}$$

If an object is immersed in a liquid as shown in Figure 22.7, while its apparent weight is being measured, the reading on the spring balance will be affected by the magnitude of the buoyant force acting on the object.

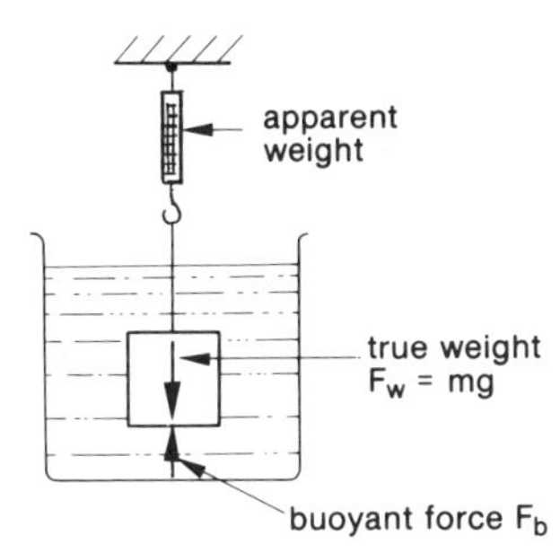

Fig. 22.7

Example 22.6

If the block in Example 22.5 is made of steel, $\rho = 7800 \text{ kg/m}^3$, what will be the apparent weight of the block when it is immersed in water?

Solution

The true weight of the block, i.e. the downward force due to gravity is:

$$F_w = \rho.V.g$$
$$= 7800 \text{ kg/m}^3 \times 0.003 \text{ m}^3 \times 9.81 \text{ N/kg}$$
$$= 229.6 \text{ N}$$

The buoyant force, calculated previously, i.e. the upthrust due to immersion, is:

$$F_b = 29.4 \text{ N}$$

The net force or the apparent weight of the block is the difference between F_w and F_b.

Apparent weight, $F_w - F_b = 229.6 - 29.4 = 200.2$ N

The concept of apparent weight can be used as a convenient way of determining unknown density of a solid substance by immersing an object in a liquid of known density and measuring the true and the apparent weight of the body.

We know that true weight is given by $F_w = \rho_s . V.g$ and the buoyant force is $F_b = \rho_l . V.g$, where ρ_s and ρ_l are densities of the substance and the liquid respectively.

Dividing one of these equations by the other yields:

$$\frac{F_w}{F_b} = \frac{\rho_s}{\rho_l}$$

or

$$\rho_s = \rho_l \frac{F_w}{F_b}$$

Example 22.7

When a piece of alloy metal having a mass of 200 g is immersed in water, its apparent weight is 1.66 N. What is its density?

Solution

The true weight,

$$F_w = m \times g$$
$$= 0.2 \text{ kg} \times 9.81 \text{ N/kg}$$
$$= 1.96 \text{ N}$$

Since the apparent weight $= F_w - F_b$

$$1.66 = 1.96 - F_b$$

Hence,

$$F_b = 1.96 - 1.66 = 0.30 \text{ N}$$

Density of the metal is found from

$$\rho_s = \rho_l \times \frac{F_w}{F_b} = 1000 \text{ kg/m}^3 \times \frac{1.96 \text{ N}}{0.3 \text{ N}}$$
$$= 6533 \text{ kg/m}^3$$

When an object floats on a liquid, it is in equilibrium under the action of two vertical forces: the weight of the object and the buoyant force, which is equal to the weight of the liquid displaced.

For equilibrium, the weight must be exactly equal to the buoyant force.

Example 22.8

Determine the volume of sea water of relative density 1.03, displaced by a ship, mass 2000 tonnes.

Solution

The weight of the ship, $F_w = m.g$

$$= 2 \times 10^6 \text{ kg} \times 9.81 \text{ N/kg}$$

$$= 19.62 \text{ MN}$$

Therefore, the buoyant force = 19.62 MN

Since, $F_b = \rho_l . V.g$ for displaced liquid,

$$19.62 \times 10^6 \text{ N} = 1030 \text{ kg/m}^3 \times V \times 9.81 \text{ N/kg}$$

Hence, volume of sea water displaced:

$$V = \frac{19.62 \times 10^6}{1030 \times 9.81} = 1942 \text{ m}^3$$

Problems

22.11 Determine the buoyant force acting on an object of 1 litre volume when it is fully immersed in (a) water and (b) mercury.

22.12 A cube, 100 mm side, is held under water so that its top face is parallel to the surface of the water and is 150 mm under it.

Calculate the force due to water pressure acting on each face of the cube and hence the net force on the cube in each of the three directions perpendicular to its faces.

22.13 For the cube in problem 22.12, calculate the buoyant force and compare the answer with that obtained in 22.12.

22.14 What will be the apparent weight of an aluminium cylinder, 200 mm diameter and 300 mm long, when immersed in water? Density of aluminium is 2780 kg/m^3.

22.15 A log of wood, relative density 0.52, 300 mm diameter and 1.5 m long, is held under water. Determine the force required to hold it.

22.16 In an experiment to determine relative density of a material, a block of mass 1.2 kg is fully immersed in water and an apparent weight of the block while immersed is observed to be 10 N. What is the relative density of the block?

22.17 A flat-bottomed barge is 20 m long, 5 m wide and has a mass of 150 tonnes. Determine the position of its waterline above its bottom when floating empty in a river.

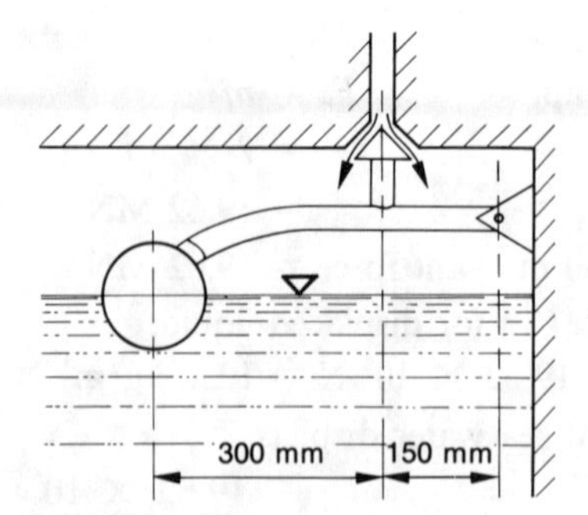

Fig. 22.8

22.18 A float-valve shown in Figure 22.8 is just in equilibrium when the spherical float of 125 mm diameter and 0.17 kg mass is half submerged in liquid of relative density 0.83. What is the force at the valve? The weight of the arm may be neglected.

22.19 A weather balloon is inflated with 15 cubic metres of hydrogen and is anchored in still air by a cable in preparation for release. If the total mass of the payload, fabric and attachments, is 7.5 kg, determine the force in the cable. Take densities of air and hydrogen as 1.226 and 0.085 kg/m³ respectively.

22.20 A spherical balloon of 10 m diameter is inflated with helium. Calculate the number of people of average mass 70 kg each that can be lifted by the balloon, if the mass of its fabric, gondola and attachments is 200 kg. Take densities of air and helium as 1.226 and 0.169 kg/m³ respectively.

22.5 Work done by fluid pressure

In many engineering applications, such as hydraulic presses and reciprocating internal combustion engines, work is done by fluid pressure acting on a reciprocating piston.

Let us consider a fluid being admitted at constant pressure p into a cylinder of cross-sectional area A for the full length of stroke l, as shown in Figure 22.9.

The force on the piston due to constant fluid pressure is constant throughout the stroke and equal to

$$F = p.A$$

By definition, work is equal to force times displacement, measured in the direction of the force. Therefore, work done by fluid pressure on the piston is given by:

$$\boxed{W = F.l = p.A.l}$$

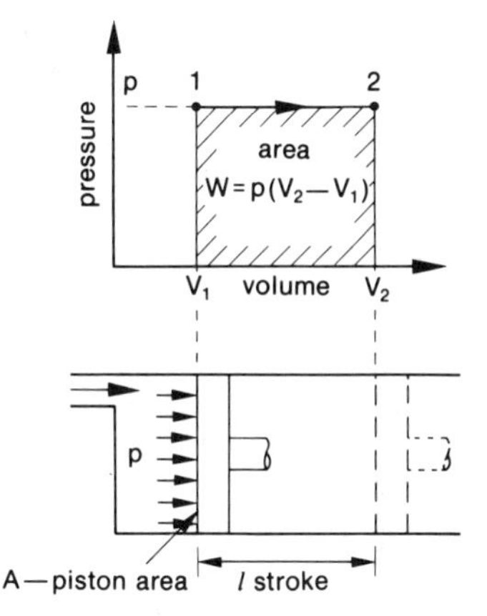

Fig. 22.9

The product $A.l$ of the piston area and the length of stroke is the volume swept out by the piston, called **swept volume**.

Example 22.9

Given the bore and stroke of a hydraulic cylinder of 100 mm and 250 mm respectively, determine the work done by oil at constant pressure of 1.5 MPa absolute.

Solution

Piston area, $$A = \frac{\pi D^2}{4} = \frac{0.1^2}{4} = 7.85 \times 10^{-3} \text{ m}^2$$

The work done during the full stroke:

$$W = p.A.l = 1.5 \times 10^6 \text{ N/m}^2 \times 7.85 \times 10^{-3} \text{ m}^2 \times 0.25 \text{ m}$$
$$= 2945 \text{ J} = 2.95 \text{ kJ}$$

If the operation during which pressure is acting on a moving piston is shown as a graph of pressure plotted against volume, such a graph, known as a pressure-volume diagram, can be used to represent or evaluate the work done.

In case of pressure remaining constant, the graph appears as a horizontal straight line whose height is at pressure p and whose length is from original volume V_1 to final volume V_2, as shown in Figure 22.9 above the diagram of the piston and cylinder. It can easily be seen that for a full stroke, l, the difference between the final and original volume of fluid in the cylinder is equal to the swept volume:

$$A.l = V_2 - V_1$$

from which it follows that:

$$\boxed{W = p\,(V_2 - V_1)}$$

We see, therefore, that the shaded area under the p-V line represents the work done, which can be easily confirmed by an analysis of the units:

$$(\text{pressure}) \times \begin{pmatrix}\text{difference}\\ \text{of volumes}\end{pmatrix} = \frac{\text{N}}{\text{m}^2} \times \text{m}^3 = \text{N.m} = \text{J},$$

i.e. unit of work

We have only considered here a case of constant pressure expansion or admission of fluid. In the majority of practical cases the pressure would change simultaneously with changes in volume and the p-V line, instead of being horizontal, is a curve resembling a hyperbola. Furthermore, practical applications usually involve repeatable cycles consisting of several processes in a sequence. The general principles outlined above, particularly the relationship between work done by the fluid and the area of the p-V diagram, still apply, provided that we find some way of averaging the variable pressure. How this is done is not discussed here. It is sufficient to say that an average pressure, known as mean effective pressure* can in fact be determined and used as if it was a constant pressure for $W = p(V_2 - V_1)$.

Example 22.10

One litre of air is allowed to expand to five times that volume in a cylinder against a moving piston. The average pressure during expansion is 300 kPa absolute.

Determine the work done by the air on the piston.

Solution

Initial volume, $V_1 = 0.001 \text{ m}^3$

Final volume, $V_2 = 0.001 \times 5 = 0.005 \text{ m}^3$

Work done, $W = p(V_2 - V_1)$

$= 300 \text{ kN/m}^2 \times (0.005 - 0.001)\text{m}^2$

$= 1.2 \text{ kJ}$

Example 22.11

A single-cylinder engine has 87.5 mm bore and 100 mm stroke and makes 900 cycles per minute. If the mean effective pressure is 500 kPa, determine the power, i.e. work per unit time, developed by the gases within the cylinder.

Solution

Area of piston, $A = \dfrac{\pi \times 0.0875^2}{4} = 6.01 \times 10^{-3} \text{ m}^2$

Work per cycle, $W = p.A.l$

$= 500 \text{ kN/m}^2 \times 6.01 \times 10^{-3} \text{ m}^2 \times 0.1 \text{ m}$

$= 0.3 \text{ kJ}$

---

* Mean effective pressure is defined as that constant pressure which, if acting on the piston for the whole length of the stroke, would do the same amount of work as is actually done by the fluid during a complete cycle.

Every second the cycle is repeated:

$$\frac{900}{60} = 15 \text{ times}$$

Therefore, work done per second, i.e. power, is:

$$P = 0.3 \times 15 = 4.5 \text{ kJ/s} = 4.5 \text{ kW}$$

Problems

22.21 If a constant force of 5 kN due to fluid pressure acts on a piston having a stroke of 120 mm, calculate the work done during the stroke.

22.22 If the diameter of the piston in the previous question is 100 mm, determine the swept volume and the pressure acting on the piston.

22.23 A lift truck has a rated capacity of 1.5 tonnes and can raise the load to a height of 1 m in 5 seconds. Its hydraulic cylinder diameter is 150 mm. Determine the oil pressure in the cylinder and the work done per second.

22.24 Two pistons of a hydraulic press have diameters of 200 mm and 20 mm. If the stroke of the small piston is 30 mm, through what distance will the large piston have moved after 50 strokes?

22.25 If the hoist in the previous problem is 80 per cent efficient as a simple machine, determine:
(a) the velocity ratio,
(b) the mechanical advantage,
(c) the input work required to raise a 1 tonne vehicle 2 m above floor level.

22.26 A gas in a cylinder is at a pressure of 0.8 MPa absolute. It is heated and expanded at constant pressure from a volume of 1.3 litres to a volume of 3.7 litres. Determine the work done by the gas on the piston and draw a pressure-volume diagram representing the process.

22.27 One litre of air is held in a cylinder by means of a spring-loaded piston, as shown in Figure 22.10, so that the pressure is 200 kPa absolute. The air is then heated and expands to a volume of 3 litres. Due to the action of the spring, the pressure increases linearly to 350 kPa absolute. Draw a pressure-volume diagram to illustrate this process and hence calculate the work done by the air.

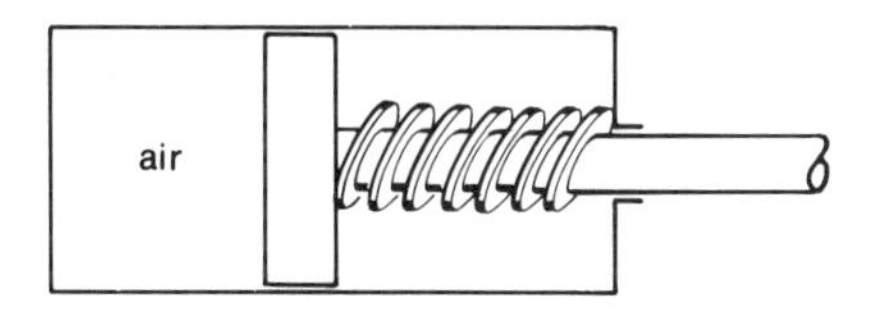

Fig. 22.10

22.28 The following results were tabulated during a test on a pneumatic cylinder.

Air pressure	kPa abs	200	250	300	350	400	450
Air volume	L	0.500	0.465	0.430	0.395	0.360	0.325

Plot the pressure-volume graph and determine the work done by the piston on the air in the cylinder.

22.29 Determine the power developed in the cylinder of a steam engine having a stroke of 450 mm and a piston diameter of 300 mm, if the mean effective pressure is 275 kPa, when the engine makes 90 cycles per minute.

22.30 In a test on a single cylinder reciprocating internal combustion engine, the following observations were recorded:

Bore, 200 mm
Stroke, 250 mm
Mean effective pressure, 320 kPa
Cycles per second, 6

Calculate the power developed within the cylinder.

Review questions

1. State the formula for calculating pressure due to a column of liquid.
2. Is the pressure calculated using this formula absolute or gauge pressure?
3. Explain how you would calculate the force exerted by liquid on:
   (a) a horizontal submerged surface,
   (b) a vertical submerged surface.
4. State Pascal's Law.
5. Explain the principle of operation of a hydraulic jack.
6. State Archimedes Principle.
7. Explain the concept of apparent weight.
8. Describe how apparent weight can be used experimentally to determine density of a solid substance.
9. Can the same principle be used to determine density of a liquid?
10. State the condition under which a body will float.
11. Explain how work can be done by fluid pressure acting on a piston.
12. State the relationship between work, pressure and swept volume.
13. With the aid of a sketch explain how work done by a fluid can be related to the pressure-volume diagram.

23

Fluid flow

In this chapter we introduce two of the most fundamental equations of fluid dynamics: the continuity equation and Bernoulli's equation. The former is derived from the general law of conservation of mass and the latter is a special statement of conservation of energy applied to the steady flow of a fluid.

23.1 Fluid flow measurement

When a fluid is flowing inside a pipe or a duct its flow can be described in terms of several related parameters. One of these is the velocity (v) with which the fluid is crossing a particular cross-section of the duct or pipe. Another is a measure known as flow rate, which can be expressed either as mass flow rate ($\dot{m}$) or as volume flow rate ($\dot{V}$).*

Fig. 23.1 *Air ducts and water pipes*

* A dot above the symbol of mass or volume denotes mass or volume flow rate.

Mass flow rate is defined as the mass of fluid flowing across a given cross-section per unit time. The SI unit of mass flow rate is kilogram per second.

Volume flow rate is defined as the volume of fluid flowing across a given cross-section per unit time. SI units of volume flow rate are cubic metres per second and litres per second. There is no hard and fast rule to suggest a preference between these units, except when volume flow rate is to be mathematically related to other quantities, such as velocity, in which case all units must be consistent, i.e. m^3/s must be used.

Of course, mathematical conversion between flow rate and velocity, and between mass and volume flow rates, is always possible and is the subject of the continuity principle and its mathematical expressions.

The flow of liquids is most commonly measured with variable area flow meters called rotameters or with positive displacement meters. A rotameter is a device, consisting of a transparent tapered tube in which a float is supported by the liquid flowing up through the tube (Fig. 23.2(a)). As the flow increases, the float is lifted higher in the tube, thus increasing the area between the float and the tube wall. Usually rotameters are used for a single type of liquid with the tube-face calibrated accordingly in kg/s or L/s.

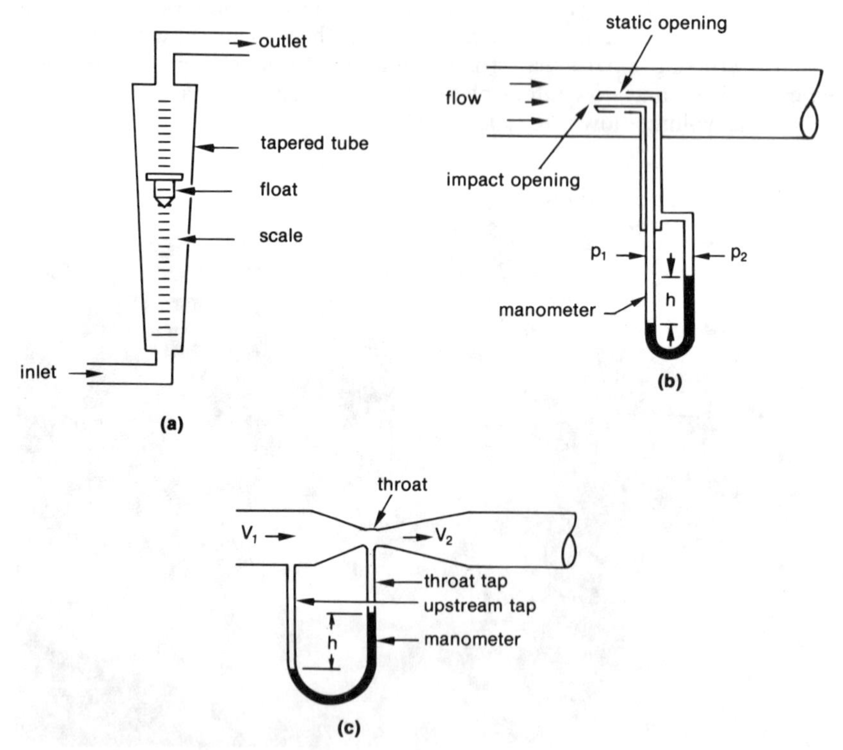

Fig. 23.2(a) *Variable-area flowmeter—a rotameter*
(b) *Pitot tube with manometer*
(c) *Venturi tube with manometer*

A positive-displacement meter consists of tight fitting vanes or gears rotated by the liquid as it passes through. These meters can be very accurate but require some interference with the flow stream.

The flow of gases is usually measured in terms of flow velocity by means of Pitot or Venturi tubes. The former is a tube (Fig. 23.2(b)) facing into the flow, measuring the velocity-related impact pressure and the latter is a tube (Fig. 23.2(c)) which when inserted into the flow line acts as a calibrated restriction that creates a velocity and pressure differential. The measurements of pressure obtained with the aid of U-tube manometers attached to the Pitot or Venturi tubes can be converted mathematically into velocity and hence to volume flow rate.

23.2 The continuity equation

Continuity equation is a mathematical expression of one of the most basic and important laws of nature, that of conservation of mass.

Under steady state conditions of flow of a fluid in a pipe or tube of varying cross-sectional area, as in Figure 23.3, the mass of fluid that passes cross-section 2-2 must be the same as that which passes cross-section 1-1 per unit time.

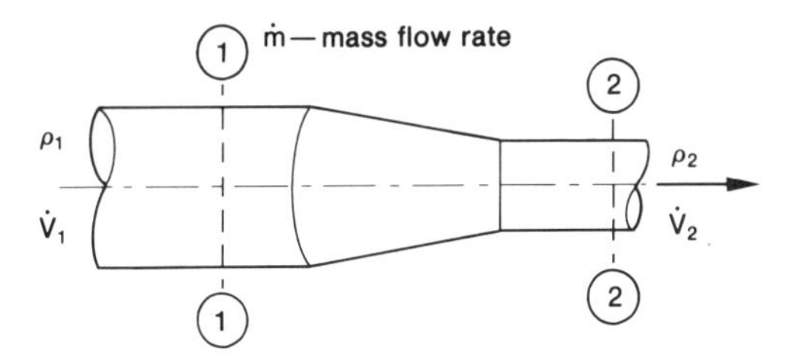

Fig. 23.3

If we let ρ_1 and ρ_2 be density of the fluid at cross-sections 1-1 and 2-2, and $\dot{V}_1$ and $\dot{V}_2$ the volume flow rates at these cross-sections respectively, we can write:

$$\dot{V}_1 . \rho_1 = \dot{m} = \dot{V}_2 \rho_2$$

where $\dot{m}$ is the mass flow rate, which is the same for each cross-section.

Example 23.1

The volume flow rate of air in a duct at point 1 is 30 litres per second and density is 1.26 kg/m^3, and, due to heat gains between 1 and 2, density at point 2 is 1.05 kg/m^3. Determine (a) the mass flow rate in the duct and (b) the volume flow rate at point 2.

Solution

Mass flow rate, which is the same for both cross-sections, is:

$$\dot{m} = \dot{V}_1 \times \rho_1 = 0.03 \text{ m}^3/\text{s} \times 1.26 \text{ kg/m}^3 = 0.0378 \text{ kg/s}$$

The volume flow rate at the second cross-section is:

$$\dot{V}_2 = \frac{\dot{m}}{\rho_2} = \frac{0.0378 \text{ kg/s}}{1.05 \text{ kg/m}^3} = 0.036 \text{ m}^3/\text{s} = 36 \text{ L/s}$$

If we consider an element of fluid flowing across a given cross-section of area A with a velocity v, it can be seen that the volume flow rate is related to both area and velocity:

$$\boxed{\dot{V} = A.v}$$

When substituted into the previous equation this relationship yields

$$\boxed{A_1 v_1 \rho_1 = \dot{m} = A_2 v_2 \rho_2}$$

This is the equation which is taken to express the continuity principle. This form is particularly useful if we realise that cross-sectional area can readily be related to the dimensions, usually diameter, of the pipe.

It is also convenient to remember that for liquids, which are practically incompressible, there can be little, if any, change in density between the two cross-sections.*

If $\rho_1 = \rho_2$, the continuity equation reduces to

$$A_1.v_1 = A_2.v_2$$

Example 23.2

A centrifugal water pump is connected to a 75 mm diameter inlet pipe and discharges through a 50 mm diameter outlet pipe. The inlet velocity is 1.5 m/s. Determine the volume and mass flow rates and the discharge velocity.

Solution

Using inlet conditions, the volume flow rate is:

$$\dot{V} = A_1 \times v_1 = \frac{\pi D^2}{4} \times v_1 = \left(\frac{\pi \times 0.075^2}{4}\right) \text{m}^2 \times 1.5 \text{ m/s}$$

$$= 0.00663 \text{ m}^3/\text{s} = 6.63 \text{ L/s}$$

Hence, mass flow rate is:

$$\dot{m} = \dot{V} \times \rho = 0.00663 \text{ m}^3/\text{s} \times 1000 \text{ kg/m}^3 = 6.63 \text{ kg/s}$$

The discharge velocity is related to the cross-sectional area of the outlet pipe:

$$v_2 = \frac{\dot{V}}{A_2} = \frac{\dot{V}}{\left(\frac{\pi D^2}{4}\right)} = \frac{0.00663 \text{ m}^3/\text{s}}{\left(\frac{\pi \times 0.05^2}{4}\right) \text{m}^2} = 3.38 \text{ m/s}$$

---

* This also applies to gases, if there is no appreciable change in temperature and pressure between the cross-sections.

The conservation of mass principle applies also to branching pipes and ducts. Furthermore, if the assumption of constant density is justified, a satisfactory solution is obtained by considering volume flow rates only.

Example 23.3

In an air conditioning system, branching ducts are as shown in Figure 23.4. If the flow rates in branches B and C are 3 m^3/s and 2 m^3/s respectively, and air velocity is to be 3 m/s in the main duct A and 2.5 m/s in each of the branches, determine the required cross-sectional area of each duct.

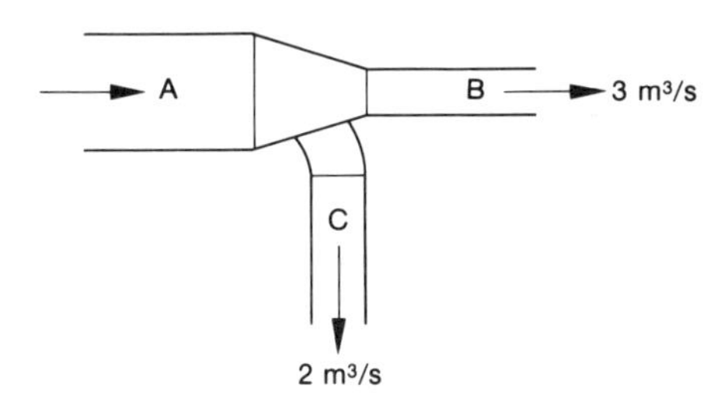

Fig. 23.4

Solution

Volume flow rate in the main A is:

$$\dot{V}_A = \dot{V}_B + \dot{V}_C = 3\ m^3/s + 2\ m^3/s = 5\ m^3/s$$

Hence, the cross-sectional areas required are:

$$A_A = \frac{5\ m^3/s}{3\ m/s} = 1.67\ m^2$$

$$A_B = \frac{3\ m^3/s}{2.5\ m/s} = 1.20\ m^2$$

$$A_C = \frac{2\ m^3/s}{2.5\ m/s} = 0.8\ m^2$$

Problems

23.1 Determine the mass and volume flow rate of oil of relative density 0.83 flowing in a 100 mm diameter pipe with a velocity of 2 m/s.

23.2 Determine the side dimensions of a square duct required to supply 4.5 kg/s of air (ρ = 1.226 kg/m³) with a flow velocity of 7.5 m/s.

23.3 The suction line to an air compressor is 75 mm diameter and carries air at atmospheric conditions (ρ = 1.226 kg/m³).

The discharge line is 25 mm diameter and carries compressed air (ρ = 6.13 kg/m³).

If the flow rate through the compressor is 0.1 kg/s, determine the volume flow rate and velocity in the suction and discharge pipes.

23.4 A circular swimming pool 3.5 m diameter is being filled through a garden hose 15 mm diameter in which water velocity is 2.26 m/s. Determine how long it will take to fill the pool to a depth of 1.5 m.

23.5 Kerosene (RD = 0.85) is being taken out of a 2 m diameter storage tank at the rate of 35 litres per minute through a 25 mm diameter pipe. Determine:
(a) the mass flow rate of kerosene,
(b) the velocity in the pipe,
(c) the rate at which the level in the tank drops.

23.6 Chilled water is flowing through a cooling coil consisting of a parallel arrangement of 24 tubes 10 mm diameter each. The total flow rate is 75 litres per minute. Determine velocity of flow in each tube.

If the velocity in the inlet pipe is to be 4 times that in the coil, determine the required inlet diameter.

23.3 Bernoulli's equation

The equation known as Bernoulli's equation was developed by Daniel Bernoulli (1700–82), one in a famous family of Swiss mathematicians and physicists. His most important work, dealing with the behaviour of fluids, the "Hydrodynamica", was published in 1738 and contained the theorem expressing the relationships between pressure, velocity and elevation of a moving fluid. The theorem states that the total mechanical energy of an incompressible fluid in steady flow remains constant. The theorem is, therefore, a statement of the law of conservation of energy applied to the flow of ideal fluids. It recognises, however, that total energy of the fluid stream comprises three components, namely the energy associated with fluid pressure, the gravitational potential energy of elevation and the kinetic energy of fluid motion. Although the total energy remains constant, it is possible for changes to take place between the component parts, e.g. a decrease of pressure energy can be compensated by a corresponding increase in kinetic and/or potential energy.

Considering each component separately, it can be shown that for a mass, m, of the fluid in steady, incompressible flow:

$$\text{Pressure energy} = m.\frac{p}{\rho}$$

$$\text{Kinetic energy} = m.\frac{v^2}{2}$$

$$\text{Potential energy} = m.g.h$$

According to Bernoulli's equation the sum of these three components of energy* is constant, i.e.

$$m.\frac{p}{\rho} + m\frac{v^2}{2} + mgh = \text{constant}$$

If we apply this to any two cross-sections of a pipe or duct, we can write:

$$m\frac{p_1}{\rho} + m\frac{v_1^2}{2} + mgh_1 = m\frac{p_2}{\rho} + m\frac{v_2^2}{2} + mgh_2$$

where: p_1 and p_2 = pressures at points 1 and 2
v_1 and v_2 = velocities at points 1 and 2
h_1** and h_2 = elevations at points 1 and 2
ρ = fluid density (constant)

If the mass is equal to one unit, e.g. kilogram, the equation becomes:

$$\frac{p_1}{\rho} + \frac{v_1^2}{2} + gh_1 = \frac{p_2}{\rho} + \frac{v_2^2}{2} + gh_2$$

This is the Bernoulli equation which is widely used to analyse fluid flow problems, as illustrated by the following examples.

Example 23.4

Frictionless flow in a pipe of varying cross-section.

500 litres per minute of oil of relative density 0.8 flow in the vertical pipeline shown in Figure 23.5.

The pressure gauge at 1 reads 30 kPa. What does the gauge at 2 read?

---

* Note the homogeneity of units:

(a) $m.\frac{p}{\rho} \rightarrow \text{kg} \times \frac{\text{N/m}^2}{\text{kg/m}^3} = \text{N.m} = \text{J}$, i.e. unit of energy

(b) $m\frac{v^2}{2} \rightarrow \text{kg}.\left(\frac{\text{m}}{\text{s}}\right)^2 = \frac{\text{kg.m}}{\text{s}^2} \times \text{m} = \text{N.m} = \text{J}$, also energy

(c) $m.g.h \rightarrow \text{kg} \times \frac{\text{N}}{\text{kg}} \times \text{m} = \text{N.m} = \text{J}$

If the mass is given per unit time, i.e. as mass flow rate in kg/s, then the energy will also be per unit time, i.e. joules per second.

** Alternatively the symbol z is often used for elevation, to avoid confusion with another concept called enthalpy, introduced in Chapter 25, for which h is a universally accepted symbol. However, problems involving both elevation and enthalpy are not likely to be encountered at this level, and it is therefore safe to use h as the most logical symbol for elevation, i.e. height.

Solution

Consider all variables for Bernoulli's equation.

Pressures:

$p_1 = 30\,000$ Pa*

$p_2 =$ unknown, to be determined

Density:

$\rho = 0.8 \times 1000 \text{ kg/m}^3 = 800 \text{ kg/m}^3$

Velocities:

$$v_1 = \frac{\dot{V}}{A_1} = \frac{\left(\frac{0.5}{60}\right) \text{m}^3/\text{s}}{\left(\frac{\pi \times 0.1^2}{4}\right) \text{m}^2} = 1.061 \text{ m/s}$$

$$v_2 = \frac{\dot{V}}{A_2} = \frac{\left(\frac{0.5}{60}\right) \text{m}^3/\text{s}}{\left(\frac{\pi \times 0.05^2}{4}\right) \text{m}^2} = 4.244 \text{ m/s}$$

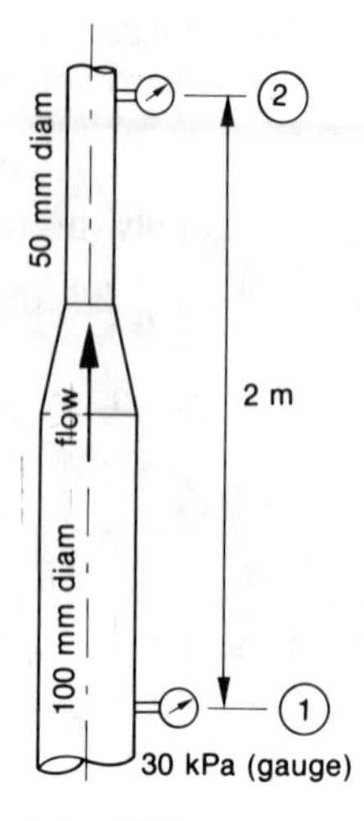

Fig. 23.5

Elevation:

$h_1 = 0$ (assuming datum level at this point)

$h_2 = 2$ m

Substitute into Bernoulli's equation:

$$\frac{p_1}{\rho} + \frac{v_1^2}{2} + gh_1 = \frac{p_2}{\rho} + \frac{v_2^2}{2} + gh_2$$

$$\frac{30\,000}{800} + \frac{1.06^2}{2} + 9.81 \times 0 = \frac{p_2}{800} + \frac{4.24^2}{2} + 9.81 \times 2$$

$$37.5 + 0.56 + 0 = \frac{p_2}{800} + 9.01 + 19.62$$

Solving for the unknown pressure at cross-section 2 yields

$$p_2 = 7540 \text{ Pa} = 7.54 \text{ kPa gauge.}$$

Example 23.5

Reduction of pressure at a constriction of a tube.

Air of density 1.226 kg/m³ flows with a velocity of 4 m/s inside a horizontal round duct of 300 mm diameter. The gauge pressure is 500 Pa.

If the diameter tapers to 150 mm, determine the pressure at the reduced diameter.

---

* Gauge or absolute pressures may be used here, so long as both pressures are expressed in the same way. Needless to say, all units must be homogeneous.

Solution

Since the duct is horizontal, $h_1 = h_2$, and there is no change in potential energy, the equation reduces to:

$$\frac{p_1}{\rho} + \frac{v_1^2}{2} = \frac{p_2}{\rho} + \frac{v_2^2}{2}$$

where:

$p_1 = 500$ Pa (gauge)

p_2 is unknown

$v_1 = 4$ m/s

$$\dot{V} = v_1 A_1 = 4\ \text{m/s}\left(\frac{\pi\, 0.3^2}{4}\right)\text{m}^2 = 0.2827\ \text{m}^3/\text{s}$$

$$v_2 = \frac{\dot{V}}{A_2} = \frac{0.2827\ \text{m}^3/\text{s}}{\left(\frac{\pi\, 0.15^2}{4}\right)\text{m}^2} = 16\ \text{m/s*}$$

Substitute and solve:

$$\frac{500}{1.226} + \frac{4^2}{2} = \frac{p_2}{1.226} + \frac{16^2}{2}$$

$$p_2 = 353\ \text{Pa gauge}$$

Note that as velocity increases pressure decreases with the change in diameter, i.e. pressure energy is converted into kinetic energy.

Example 23.6

Liquid flowing out of a small orifice.

Water flows out of a tank shown in Figure 23.6 through a small orifice 1.2 m below the water surface. What is the velocity of the emergent jet of water? If the diameter of the orifice is 10 mm, what is the flow rate out of the orifice? **

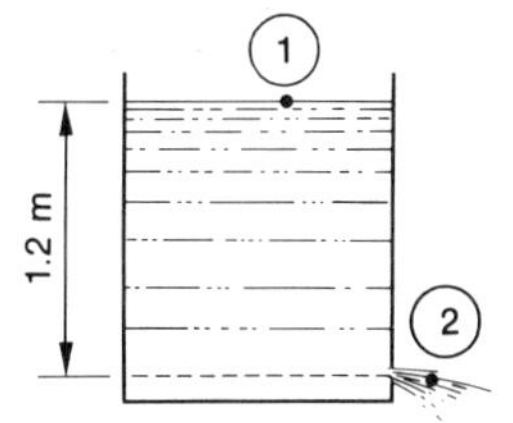

Fig. 23.6

---

* Alternatively, from continuity $v_1 A_1 = v_2 A_2$, hence

$$v_2 = v_1 \frac{A_1}{A_2} = v_1 \frac{\frac{\pi}{4} \times D_1^2}{\frac{\pi}{4} \times D_2^2} = v_1 \left(\frac{D_1}{D_2}\right)^2 = 4 \times \left(\frac{300}{150}\right)^2 = 16\ \text{m/s}$$

i.e. velocity is inversely proportional to the square of the diameter.

** In practice, a jet of fluid discharged from an orifice usually contracts to a throat of smaller diameter just outside the orifice. This effect, known as "vena contracta", depends on the shape and sharpness of the orifice edge and its practical significance is unimportant at this introductory level of study.

Solution

In this case we can consider the free surface of the water and a point just outside the orifice as points 1 and 2 for reference.

This time it is the pressure that is the same at points 1 and 2 and consequently cancels out of the equation. It can also be seen that the velocity at point 1, i.e. the surface of water at the top of the tank, is negligible.

Mathematically $p_1 = p_2$ and $v_1 = 0$.

If we take the orifice as datum level, then $h_2 = 0$ and the equation reduces to:

$$0 + 0 + gh_1 = 0 + \frac{v_2^2}{2} + 0$$

Substituting,
$$9.81 \times 1.2 = \frac{v_2^2}{2}$$

hence
$$v_2 = \sqrt{2 \times 9.81 \times 1.2} = 4.85 \text{ m/s}$$

The flow rate can now be found using the continuity equation:

$$\dot{V} = A_2 v_2 = \frac{\pi}{4} D_2^2 \times v_2 = \frac{\pi}{4}\ 0.01^2 \times 4.85$$

$$= 0.381 \times 10^{-3} \text{ m/s}$$

$$= 0.381 \text{ L/s}$$

Looking back at each of the examples given here, it can be seen that the method of solution consists of considering pressure, velocity and elevation of the fluid at two reference points along the flow stream. In many instances, one or more of these variables are negligible or exactly equal to zero, which simplifies the solution. Sometimes continuity equation is also used to relate some variables together or to convert velocity to flow rate.

The following situations occur frequently in fluid flow systems and, if recognised, simplify the solution of Bernoulli's equation by reducing the number of terms required.

1. Elevation
   (a) For any pipe or other system element located at datum level, elevation is zero, $h = 0$.
   (b) For any two points at the same level, e.g. along a horizontal pipe, $h_1 = h_2$. Potential energy terms cancel out.
2. Velocity
   (a) For fluid surface in a tank or reservoir, velocity, if any, is negligible, $v = 0$.
   (b) For a pipe of uniform diameter, velocity of flow is constant, $v_1 = v_2$. Kinetic energy terms cancel out.
3. Pressure
   For open fluid surface in a tank or reservoir or a discharge from a pipe or an orifice into atmospheric conditions, gauge pressure is zero, $p = 0$.

It should be remembered that each of the terms in Bernoulli's equation may be thought of as a description of energy. It must also be understood that Bernoulli's equation is a theoretical relationship between these energy levels and does not take into account losses of energy due to fluid friction or the presence of a pump or a turbine as part of a hydraulic system. These will be dealt with in the next chapter.

Problems

23.7 A water main, 75 mm diameter, is located 2.5 m above floor level and carries 1300 litres per minute at a pressure of 150 kPa gauge. For each unit mass of water, calculate:
(a) pressure energy, based on gauge pressure,
(b) kinetic energy,
(c) potential energy, relative to floor level.

23.8 Oil of relative density 0.83 flows in a tapered pipe at the rate of 25 litres per second, from a cross-section A of 100 mm diameter to a cross-section B of 50 mm diameter, which is 1.5 m higher than A.
If the pressure at A is 57 kPa gauge, what is the pressure at B?

23.9 The water mains of a city are laid over a hill 30 m above the level of the pumping station. If the water pressure in the main at the top of the hill is 50 kPa gauge, calculate the pressure required at the station. Pipe size is constant throughout.

23.10 At one point in a pipeline carrying water at the rate of 100 litres per second, the diameter is 150 mm and gauge pressure is 75 kPa. At another point where the diameter is 200 mm, the pressure is 20 kPa gauge. What is the difference in level between these points?

23.11 Water is maintained at a height of 5 m in a tank, above a circular opening of 20 mm diameter.
Determine the flow rate of water out of the opening.

23.12 A funnel is used to pour petrol into a tank as shown in Figure 23.7. If the outlet diameter is 15 mm and the level of petrol in the funnel is constant, how long will it take to fill a tank of 50 litre capacity?

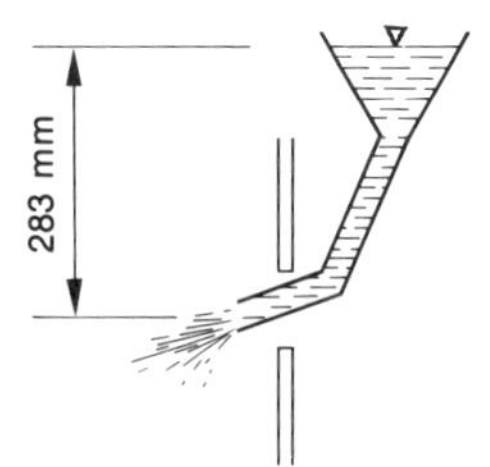

Fig. 23.7

23.13 Air in a 400 mm diameter duct flows through a Venturi meter with a 250 mm diameter throat. The pipe-to-throat pressure difference measured with a U-tube manometer is equivalent to 55 mm of manometer fluid of RD = 1.78.
What is the volume flow rate of air, if its density is 1.226 kg/m^3?

23.14 A siphon arrangement is used to empty a water tank over a 2 m wall with a 20 mm diameter hose as shown in Figure 23.8. Determine:
(a) velocity of discharge from the hose,
(b) volume flow rate,
(c) pressure at the highest point of the hose.

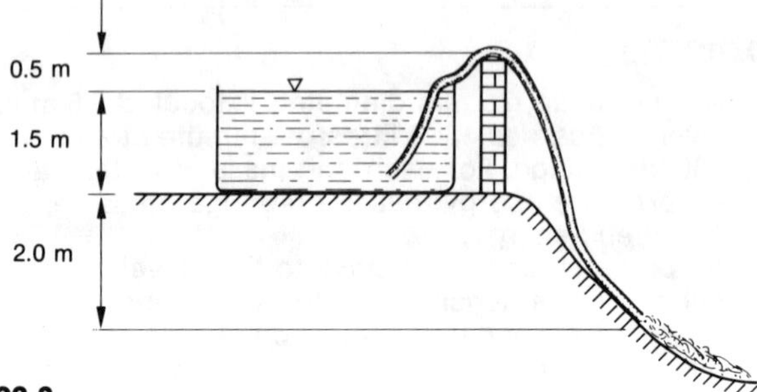

Fig. 23.8

Review questions

1. Define mass flow rate and volume flow rate.
2. Describe (a) rotameter, (b) Pitot tube, (c) Venturi tube.
3. State the conservation of mass principle applied to flowing fluids.
4. State the continuity equation.
5. Show that in a tube of varying cross-section velocity is inversely proportional to the tube diameter.
6. State Bernoulli's equation.
7. Explain the significance of each term in Bernoulli's equation.
8. What are the limitations of Bernoulli's equation?

24

Hydraulic systems

The scope of this book does not permit extensive study of diverse applications of fluid flow and fluid power engineering. This chapter will serve only as an introduction to the principles behind hydraulic systems, i.e. systems of pumps, turbines, tanks and pipework for the storage and distribution of liquids and for conversion of hydraulic energy into mechanical power.

24.1 Hydraulic head

You will recall that the form of Bernoulli's equation used in Chapter 23 was:

$$\frac{p_1}{\rho} + \frac{v_1^2}{2} + gh_1 = \frac{p_2}{\rho} + \frac{v_2^2}{2} + gh_2$$

in which each term represented energy per kilogram of fluid flowing in a pipe, expressed in joules per kilogram.

If we divide this equation by g, we will obtain an alternative form of Bernoulli's equation, which is widely used by engineers, particularly where the flow of liquids is involved:

$$\boxed{\frac{p_1}{\rho g} + \frac{v_1^2}{2g} + h_1 = \frac{p_2}{\rho g} + \frac{v_2^2}{2g} + h_2}$$

It can easily be shown that each term in this equation will now have to be expressed in units of length, i.e. metres. When expressed in this way, each term is referred to as "hydraulic head". Hydraulic head should be understood as an indirect measure of energy level in a flowing fluid expressed in metres. Obviously there are three components of the total hydraulic head, corresponding to the three components of energy, as follows:

1. **pressure head** $= \frac{p}{\rho g}$, which is related to the energy level of the fluid associated with its pressure;
2. **velocity head** $= \frac{v^2}{2g}$, related to the level of kinetic energy of the fluid; and
3. **static head** $= h$, also called potential head, which is the actual elevation of the fluid above some reference level.

Just as three separate components of energy add up to a constant, or total energy, the three components of hydraulic head constitute a constant total head.

$$\textit{Total head} = \begin{cases} \text{pressure head} \\ + \\ \text{velocity head} \\ + \\ \text{static head} \end{cases}$$

Example 24.1

A pipe, 150 mm diameter, located 3 m above ground level, carries petrol (RD = 0.72) with a velocity of 5.5 m/s. If pressure in the pipe is 75 kPa gauge, calculate the total hydraulic head, relative to ground level.

Solution

Pressure head, $\dfrac{p}{\rho g} = \dfrac{75\,000 \text{ N/m}^2}{720 \text{ kg/m}^3 \times 9.81 \text{ N/kg}} = 10.62 \text{ m}$

Velocity head, $\dfrac{v^2}{2g} = \dfrac{(5.5 \text{ m/s})^2}{2 \times 9.81 \text{ N/kg}} = 1.54 \text{ m}$*

Static head, $h = 3.00 \text{ m}$

Total hydraulic head $= 15.16 \text{ m}$

The concept of hydraulic head can quite conveniently be used for relating hydraulic conditions at more than one point in a system.

Example 24.2

If the pipe in the previous example rises to a level 7 m above ground, determine the pressure in the elevated part of the pipe.

Solution

The total hydraulic head does not change, i.e. total head is 15.16 m.

The components of total head are:

Pressure head $= \dfrac{p}{\rho g}$ (unknown)

Velocity head = 1.54 m (this does not change since the pipe diameter is the same)

Static head = 7 m (the actual elevation above ground level)

Therefore, $15.16 = \dfrac{p}{\rho g} + 1.54 + 7$

* From the definition of the newton it follows that N/kg = m/s² and the units reduce to metres of velocity head.

Hence
$$\frac{p}{\rho g} = 6.62 \text{ m}$$

and
$$p = 6.62 \times \rho g$$
$$= 6.62 \text{ m} \times 720 \text{ kg/m}^3 \times 9.81 \text{ N/kg}$$
$$= 46\ 760 \text{ N/m}^2$$

or
$$p = 46.76 \text{ kPa (ga)}$$

24.2 Pumping head and power

In the majority of hydraulic systems, the flow of fluids through pipes is produced and maintained by means of mechanical pumps. It is therefore necessary to be able to relate the pumping action of a pump to the characteristics of the flow in a system of pipework.

A pump is a mechanical device which uses force and motion to transport liquids.* Energy is imparted to the liquids by means of volumetric displacement in the case of reciprocating piston and rotary pumps, or by means of addition of kinetic energy by a rapidly rotating impeller.

Pumps are classified according to their mechanical moving elements. The major types include reciprocating piston, centrifugal impeller, rotary vanes and gear types. The centrifugal pump is the most common type for general use in pumping most liquids. Pumps are used extensively for irrigation and drainage, in oil and chemical industry, boiler plants, steelmaking and many other industrial applications.

Fig. 24.1 *Centrifugal pump*

* We limit our study to pumps which are used to move liquids. However, there are some similarities in performance and design between pumps and fans, i.e. devices for handling air and other gases.

Let us consider a pump with a flow rate of $\dot{m}$ kilograms per second. If all of the energy supplied by the pump to the fluid was used to lift the fluid through a vertical height H, without any increase in pressure or velocity, the energy per unit time, or power P, would be equal to the increase in potential energy of the fluid, or:

$$P = \dot{m}.g.H$$

When a pump operates as part of a hydraulic system, its effect is more likely to be a combination of vertical lift, an increase of pressure and velocity of the fluid pumped. It is therefore common practice not to think of H as elevation only, but rather as a term which embraces a combination of effects. By analogy with other similar terms, it is referred to as the "pumping head", identified by the use of an appropriate subscript H_p

$$H_p = \frac{P}{\dot{m}.g}$$

It should be noted that P is the power, i.e. energy per unit time supplied to the flowing fluid by the pump, and not the power required to drive the pump.

The latter can be determined if the mechanical efficiency of the pump is known:

$$\text{Shaft input power} = \frac{P}{\eta} = \frac{\dot{m}.g.H_p}{\eta}$$

where η is the mechanical efficiency of the pump.

Example 24.3

Determine the pumping head developed when a pump delivers 600 litres per minute of oil of relative density 0.83, using 10 kW of shaft power input, while its mechanical efficiency is 75 per cent.

Solution

Mass flow rate is:

$$\dot{m} = \left(\frac{0.6}{60}\right) \text{m}^3/\text{s} \times 830 \text{ kg/m}^3 = 8.3 \text{ kg/s}$$

Power supplied to the fluid:

$$P = 75 \text{ per cent of } 10 \text{ kW} = 7500 \text{ W}$$

Pumping head:

$$H_p = \frac{P}{\dot{m}.g} = \frac{7500 \text{ N.m/s}}{8.3 \text{ kg/s} \times 9.81 \text{ N/kg}} = 92.1 \text{ m}$$

As has been explored previously, hydraulic head terms can be interpreted as a measure of energy levels in a flowing fluid. It is therefore appropriate to regard the pumping head as a measure of the energy input into the fluid stream. The Bernoulli equation can be modified accordingly.

total head before the pump + pumping head = total head after the pump

$$\frac{p_1}{\rho g} + \frac{v_1^2}{2g} + h_1 + H_p = \frac{p_2}{\rho g} + \frac{v_2^2}{2g} + h_2$$

Example 24.4

A feed pump supplies water at the rate of 5 litres per second from a hot-well at atmospheric pressure into a boiler which is at a pressure of 200 kPa gauge, also involving lifting the water through a vertical height of 3.5 m (Fig. 24.2).

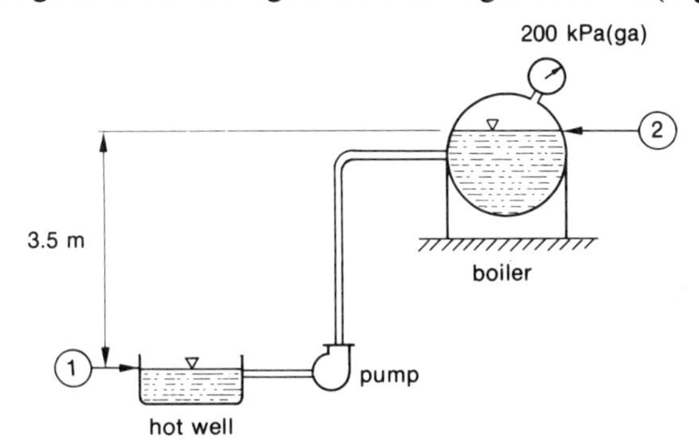

Fig. 24.2

Determine the power required to drive the pump, if its mechanical efficiency is 78 per cent.

Solution

Consider Bernoulli's equation between the two surfaces, 1 and 2.

For the hot-well, $p_1 = 0$ gauge, $v_1 = 0$ and $h_1 = 0$

For the boiler, $p_2 = 200\ 000$ Pa gauge, $v_2 = 0$ and $h_2 = 3.5$ m

Therefore, substitution yields:

$$(0 + 0 + 0) + H_p = \frac{200\ 000\ \text{N/m}^2}{1000\ \text{kg/m}^3 \times 9.81\ \text{N/kg}} + 0 + 3.5\ \text{m}$$

$$\text{Pumping head,}\quad H_p = 20.4\ \text{m} + 3.5\ \text{m}$$

$$H_p = 23.9\ \text{m}$$

Hence, power supplied to the water:

$$P = \dot{m}.g.H_p = 5\ \text{kg/s} \times 9.81\ \text{N/kg} \times 23.9\ \text{m}$$

$$= 1172\ \text{W} = 1.17\ \text{kW}$$

Shaft input required to drive the pump is:

$$\frac{P}{\eta} = \frac{1.17}{0.78} = 1.5\ \text{kW}$$

24.3 Turbines in hydraulic systems

We have seen how pumps add energy to fluids, i.e. increase total hydraulic head by an amount H_p.

Turbines, on the other hand, extract energy from a fluid stream and convert that energy into useful mechanical power. Conversion is accomplished by passing the fluid through a system of fixed passages on to moving blades attached to a rotor, causing the latter to rotate.* Apart from some minor uses, water turbines are mostly employed for generation of electric power.

Let us compare turbines and pumps as elements of a hydraulic system. It can be said that turbines are "reversed pumps". Unlike pumps, which increase total hydraulic head of a fluid, turbines remove energy from the fluid, thus reducing its total head by an amount which we will designate by H_t.

Bernoulli's equation becomes:

$$\frac{p_1}{\rho g} + \frac{v_1^2}{2g} + h_1 - H_t = \frac{p_2}{\rho g} + \frac{v_2^2}{2g} + h_2$$

Just as it is for pumps, turbine head, H_t, should not simply be regarded as a difference in height, but rather as a composite term, indicative of the energy extracted from the fluid and reflecting changes in pressure, velocity and potential head of the fluid stream before and after the turbine.

Turbine head can be related to the amount of power, i.e. energy per unit time removed from the fluid.

$$H_t = \frac{P}{\dot{m}.g}$$

Since turbines can never be 100 per cent efficient mechanically, the shaft output power is given by:

$$\text{shaft output} = \eta.P = \dot{m}.g.H_t.\eta$$

where η is the mechanical efficiency.

Example 24.5

A hydroelectric power generating plant is shown diagrammatically in Figure 24.3. If water is discharged from the outlet 1 m in diameter with a velocity of 10 m/s and the efficiency of the plant is 65 per cent, determine the power output.

---

* Latin *turbo* = a whirling object.

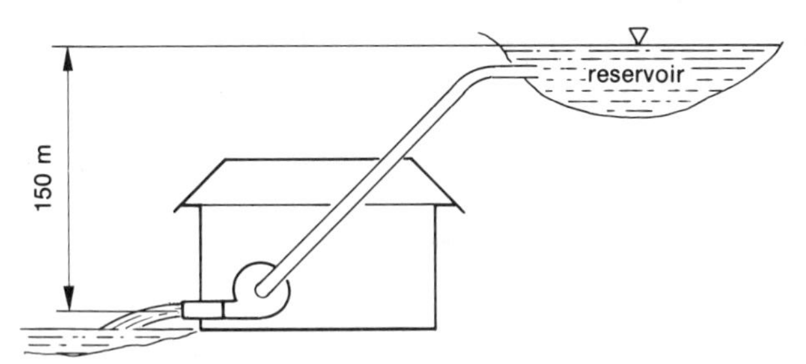

Fig. 24.3

Solution

At the free water surface in the reservoir, $p_1 = 0$, $v_1 = 0$ and $h_1 = 150$ m.

At the outlet, $p_2 = 0$, $v_2 = 10$ m/s and $h_2 = 0$.

Bernoulli's equation yields:

$$(0 + 0 + 150) - H_t = \left(0 + \frac{10^2}{2 \times 9.81} + 0\right)$$

Hence, turbine head $\quad H_t = 144.9$ m

Using continuity equation at outlet conditions, we can determine the mass flow rate of water.

$$\dot{m} = \rho . v . A = 1000 \text{ kg/m}^3 \times 10 \text{ m/s} \times \frac{\pi \times 1^2}{4} \text{ m}^2$$

$$= 7854 \text{ kg/s}$$

The energy extracted from the flow can now be calculated:

$$P = \dot{m} . g . H_t = 7854 \text{ kg/s} \times 9.81 \text{ N/kg} \times 144.9 \text{ m}$$

$$= 11.16 \text{ MW}$$

Allowing for plant efficiency, the power output is 65 per cent of 11.16 MW, or: Power output = 11.16 MW × 0.65 = 7.26 MW

24.4 Head loss due to fluid friction

It was mentioned at the end of the previous chapter that Bernoulli's equation in its original form is only a theoretical relationship between the energy levels in a flowing fluid.

In actual flow of real viscous fluids, considerable losses of energy may occur due to fluid friction in a pipe. The concept and study of viscosity of fluids is outside the scope of this book. However, the student should be aware of the presence of frictional energy losses, which result in a gradual reduction of the total energy level along the pipe.

When energy levels are expressed in terms of hydraulic head, it can be said that a head loss H_L occurs along the pipe, proportional to the energy loss.

Accordingly, Bernoulli's equation can be modified by inclusion of the head loss term as follows:

$$\frac{p_1}{\rho g} + \frac{v_1^2}{2g} + h_1 - H_L = \frac{p_2}{\rho g} + \frac{v_2^2}{2g} + h_2$$

Example 24.6

A 100 mm diameter pipe carries oil of relative density 0.78 along a horizontal distance of 560 m. The head loss is known to be 8 m per 100 metres length of the pipe. Determine the pressure drop along the pipe.

Solution

Substitute the given information into modified Bernoulli's equation, where total head loss for the entire length of the pipe is $H_L = 8\text{ m} \times \frac{560\text{ m}}{100\text{ m}} = 44.8\text{ m}$, $h_1 = h_2$ and $v_1 = v_2$.

$$\frac{p_1}{780 \times 9.81} + \frac{v_1^2}{2g} + h_1 - 44.8 = \frac{p_2}{780 \times 9.81} + \frac{v_2^2}{2g} + h_2$$

Hence,

$$\frac{p_1}{780 \times 9.81} - 44.8 = \frac{p_2}{780 \times 9.81}$$

Solving for the pressure difference between two ends of the pipe yields:

$$\text{Pressure drop } \Delta p = p_1 - p_2 = 342.8\text{ kPa*}$$

At this point we will refer the student to more advanced texts for further study of fluid mechanics, and turn our attention in the following chapters to the effects of temperature and heat on the behaviour of gases and other fluids used in engineering.

Problems

24.1 Convert the following values into pressure head, velocity head and static head, if water flows in a pipe where p = 55 kPa gauge, v = 8 m/s and h = 6.5 m.

24.2 Calculate the pumping head if the energy supplied to water pumped at the rate of 30 litres per second is 8 kW.

24.3 A pump having an efficiency of 75 per cent pumps water from a lake into an open tank through an elevation of 25 m at the rate of 0.4 m³ per minute.

Determine the shaft input power required to drive the pump. The difference between suction and discharge velocities is negligible.

* Δ– "delta", the fourth letter of the Greek alphabet, is a symbol often used in mathematical expressions to signify a difference or a change in some quantity.

24.4 A turbine which is 81 per cent efficient is driven by a water flow of 700 litres per second. If the available turbine head is 90 m, determine the power output from the turbine.

24.5 During a test on a centrifugal pump using water, inlet and outlet pressures were found to be minus 15 kPa gauge and 53 kPa gauge respectively. The flow rate was 12.6 L/s and shaft input power was 1.17 kW.

If the inlet and outlet diameters are the same, determine the pump efficiency.

24.6 A fire-hose is supplied with water from an open tank in which the minimum level of water is 8 m above ground. The maximum height at which the nozzle may have to operate is 3.5 m above ground. The diameter of the nozzle is 20 mm and the hose and pipe diameter is 50 mm throughout.

If there is no pump in the system, determine the flow rate from the fire-hose for these conditions.

24.7 If it is necessary to ensure a minimum flow of 10 litres per second from the fire-hose in problem 24.6, a pump may be installed as shown in Figure 24.4.

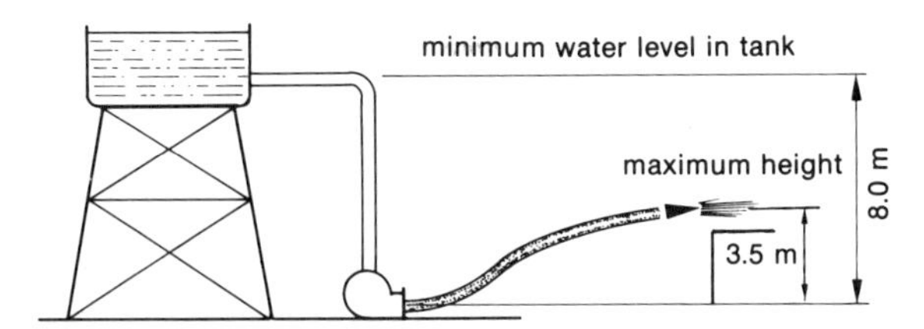

Fig. 24.4

Determine the pumping head and the shaft power, for the same conditions as specified above, if the pump is 70 per cent efficient.

Determine also the pump inlet and pump outlet gauge pressure.

24.8 Water flows from a reservoir through a 750 mm diameter pipe to a turbine located 140 m below the water level in the reservoir and is then discharged through an outlet pipe of the same diameter 5 m below the level of the turbine. If the flow rate is 4.8 m^3/s and the efficiency is 78 per cent, determine:

(a) turbine head,
(b) turbine shaft output power,
(c) gauge pressure at turbine inlet,
(d) gauge pressure at turbine outlet.

Review questions

1. Explain what is meant by the terms:
   (a) pressure head,
   (b) velocity head,
   (c) static head.
2. State Bernoulli's equation in terms of hydraulic head.
3. What is the relationship between pumping head and pumping power?
4. How can the pump shaft input power be calculated if the pumping head is known?
5. What is the relationship between turbine head and turbine power?
6. How can the turbine shaft output power be calculated if the turbine head is known?
7. State the modified Bernoulli's equation which allows for a pump or a turbine to be included in the system analysis.
8. Briefly discuss the effect of fluid friction on pressure in a straight horizontal pipeline of uniform diameter.

PART SEVEN

Thermodynamics

Thermodynamics is a funny subject.

The first time you go through the subject, you don't understand it at all.

The second time you go through it, you think you understand it, except for one or two small points.

The third time you go through it, you know you don't understand it, but by that time you are so used to the subject, it doesn't bother you any more.

Attributed to Arnold Sommerfeld
in *Order and Chaos*
by Angrist and Helper.

25

Elementary thermodynamics

The aim of this chapter is to introduce the concepts of engineering thermodynamics as preparation for more detailed discussion of their applications which follows in the subsequent chapters of this book and for further study of the subject.

We shall begin with a brief outline of the history of the basic ideas of engineering thermodynamics. In doing so we will be using some words from the specialised vocabulary of thermodynamics, whose exact meaning will be defined and explained elsewhere in this chapter. The terms, such as work, heat, temperature, system, cycle and thermal efficiency, have very precise meanings in the context of this subject and a great deal of care must be exercised in their use.

25.1 Engineering thermodynamics

Thermodynamics* is the branch of physical science dealing with the relations between properties of a substance and the quantities "work" and "heat" which cause a change in the condition of a substance. Engineering thermodynamics, as the name suggests, is concerned with the conditions of substances used as working agents in engineering systems, such as boiler plant, air compressors, refrigerators, turbines and internal combustion engines.

In particular, thermodynamics is concerned with a form of **thermal energy transfer** known as **heat**, and especially with the **conversion of heat into work**.

As a branch of physical science, engineering thermodynamics is a relatively new science which owes its impetus in the early nineteenth century to the study of the motive power of steam. Its scope has later been extended to include many other engineering applications.

Temperature was probably the earliest concept in thermodynamics to have gained recognition as an objective measure of hotness and coldness. Early in the seventeenth century, Galileo devised a crude "thermoscope" and was soon followed by the invention at the Florentine Academy of Experiment of the sealed-stem alcohol thermometer. The idea of temperature measurement is closely associated with the concept of thermal equilibrium, i.e. that all bodies

---

* The word thermodynamics originates from the Greek words *therme* = heat and *dynamis* = force or power.

exposed to the same surroundings tend to attain the same degree of hotness or coldness. The statement that **all bodies which are in thermal equilibrium are at the same temperature** is sometimes called the "zeroth law" (law number zero) of thermodynamics.

As early as 1620, the English philosopher and scientist Francis Bacon began to make an important distinction between relative hotness or coldness of substances and an influence which causes changes in temperature to occur. Initially, this influence was attributed to an imaginary agent called "caloric", which was thought to be an all-pervading fluid, contained to some extent within all real substances and capable of "flowing" from one substance into another, thus causing temperature changes.

More than a century later, Joseph Black, a chemist at the University of Glasgow, established the science of calorimetry* on the basis of the caloric theory, the main postulates of which were for the time being sufficiently accurate to account for thermal expansion of substances upon heating, and for changes in temperature and physical state of substances, e.g. boiling and freezing.

The validity of the caloric theory was undermined towards the end of the eighteenth century by Benjamin Thompson, an expatriate American engineer and scientist, also known as Count Rumford, and finally put to rest by the experimental work of an English physicist, James Joule (1818–89), who introduced the concept of thermal energy to replace the "caloric".

On the basis of his experimental work, Joule established that thermal energy is basically the same as other forms of energy, thus forming the foundation of the law of conservation of energy. The conservation of energy principle, as it applies to thermodynamic systems is known as the First Law of Thermodynamics.

The first law of thermodynamics states that **heat and work are interconvertible**, i.e. when some work is converted into heat, or some heat is converted into work, then the amounts of heat and work so converted are equal.**

In order to prove the equivalence of work and heat, Joule used an apparatus in which paddles fixed to a spindle rotated within a volume of water in an insulated container driven by a descending mass through a simple rope and pulley arrangement. The work done by the mass was compared with the heat generated in the water by friction caused by the paddles, as evidenced by a rise in temperature. By this experiment, Joule demonstrated a constant relationship for heat produced by the expenditure of work, or the heat equivalent of work.

One must not be misled by the word "equivalence" into thinking that it is always possible to convert any given quantity of heat completely into work for,

---

* Although originating from the now discredited caloric theory, the term "calorimetry" is widely used to mean the measurement of thermal constants such as specific heat, latent heat and calorific value (Chap. 26).

** The first law can be interpreted as the confirmation of the impossibility of a perpetual motion machine, i.e. a machine that would continuously produce a net output of work without consuming heat energy.

while any quantity of mechanical energy can be changed into heat, usually by means of friction, the converse is not true. It appears that there is a definite limit to the amount of work which can be obtained from a given quantity of heat. This in no way contradicts the conservation of energy principle as expressed in the statement of the first law of thermodynamics, but simply means that some energy must always remain in the form of heat and is "unavailable" for conversion into mechanical forms.

In 1824, Sadi Carnot, a French scientist and army officer, introduced the concept of the heat engine cycle and calculated a theoretical limit of efficiency of conversion of heat into work for an ideal engine operating between two temperature limits. With subsequent refinements by Lord Kelvin, the Carnot efficiency is taken to be the limit to the amount of heat which can be converted into work and forms the basis of the Second Law of Thermodynamics, stated explicitly in 1850 by Rudolf Clausius, a German mathematician and physicist.

In essence, the second law states that **complete conversion of heat into work is impossible**, i.e. it is impossible to construct a heat engine which will operate continuously, receiving heat from a fuel or another source and doing an equal amount of work. This implies that it is impossible to build a heat engine having a thermal efficiency of 100 per cent, even if the engine were mechanically perfect and frictionless. It follows, therefore, that as a consequence of the second law work is a more valuable form of energy in action than heat.

Unlike the conservation of energy principle expressed in the first law, the second law of thermodynamics would not appear quite as self-evident to a new student of engineering science, who may be tempted to challenge its validity. It is not possible here to present the full proof and complete explanation of this law or to discuss its several alternative forms. It may be helpful, however, to recognise that this law is very closely related to the well-known fact that heat can never pass spontaneously from a colder to a hotter body, and does logically follow from it.

As a result of limitations described in the second law of thermodynamics, it is not possible to drive engines or operate power stations simply by using the natural surroundings, such as atmospheric air or sea water, as a source of heat.* This means that only by burning fuels and generating high temperatures above that of the surroundings can the necessary conversion of heat to work be achieved.

In order that existing fuel reserves be preserved as long as possible, efficient use of all fuels is essential. The story of thermodynamics since the early nineteenth century has been the never-ending quest for the most efficient use of fuels to generate mechanical energy, and hence the growth of transportation, electricity generation, etc.

---

* Hydraulic and wind power are not relevant to our discussion, as these do not involve conversion of heat into work.

25.2 Temperature

Like any other fundamental concept, temperature is not easy to define precisely. To understand the meaning of temperature is it necessary to refer to the physiological sensations of hotness and coldness. In some cases, the human sense of touch, or perhaps the "temperature sense", is a sufficient indicator of the degree of hotness or coldness of a substance, such as bath water. However, the human sense of feeling is not very accurate, varies from one person to another and has a relatively limited range. It is therefore necessary to define an objective temperature scale and to devise instruments that would make accurate temperature measurement possible.

We can start by defining **temperature as a measure of the degree of hotness or coldness* of a substance** with respect to a fixed scale.

Many attempts have been made in the past to define a convenient scale of temperature and to select a suitable unit. There are two temperature scales adopted for use within the International System (SI) of Units. These are the Kelvin and the Celsius scales. Historically, the Celsius scale was the predecessor of the Kelvin scale, having been devised by Anders Celsius in the first half of the eighteenth century. The Celsius scale is often referred to as the **practical** temperature scale, while the Kelvin scale is called the **thermodynamic** or **absolute** scale.

The unit of temperature is the **kelvin**, called after William Thomson, an engineer of Glasgow, later Lord Kelvin, who was the inventor of the absolute scale. The kelvin, given the symbol K, has been defined as one hundredth part of the fundamental interval on the temperature scale. The **fundamental interval** is the difference between the temperature of melting of pure ice and the temperature at which pure water boils under atmospheric pressure. Thus, it can be said that there are 100 kelvins between the melting point and the boiling point of pure water at atmospheric pressure. The same definition of a unit applies also to the Celsius scale where the unit is commonly referred to as the Celsius degree,** written as °C. As a unit, therefore, the Celsius degree and the kelvin are identical.

The essential difference between the Kelvin and the Celsius scales lies not in the magnitude of the unit used, but in the position of the origin or zero point. The practical or Celsius scale uses the melting point of ice as the origin designated as 0°C. Celsius temperatures can be **positive**, if they are higher, i.e. hotter, than the melting point of ice, or **negative**, if they are lower, i.e. colder, than the melting point. The boiling temperature of water is, therefore, equal to 100°C.

The basic laws of thermodynamics, particularly the second law, suggest that there must exist the lowest possible temperature corresponding to a state of complete absence of thermal energy. This hypothetical temperature has

---

* "Hotness" and "coldness" are relative terms which have an unfortunate connotation of being related to some sort of a norm, or average condition. The term "temperature", by itself, does not imply a relative judgment of levels, unless specifically described as "high", "low" or "normal" temperature.

** Previously known as the centigrade degree, a name abandoned internationally since 1948 to avoid confusion with the identically named unit of angular measure equal to 1/100 of the right angle.

been adopted as the origin for the Kelvin scale and is called the **absolute zero of temperature**. There are theoretical reasons why the absolute zero of temperature is impossible to reach in practice. However, its value has been calculated to be 273 kelvins below the melting point of ice, or −273° C.*

This establishes the relationship between the practical or Celsius scale and the absolute or Kelvin scale as follows:

$$\boxed{T = t + 273}$$

where T is temperature on the Kelvin scale,
t is temperature on the Celsius scale.

This relationship is illustrated graphically in Figure 25.1.

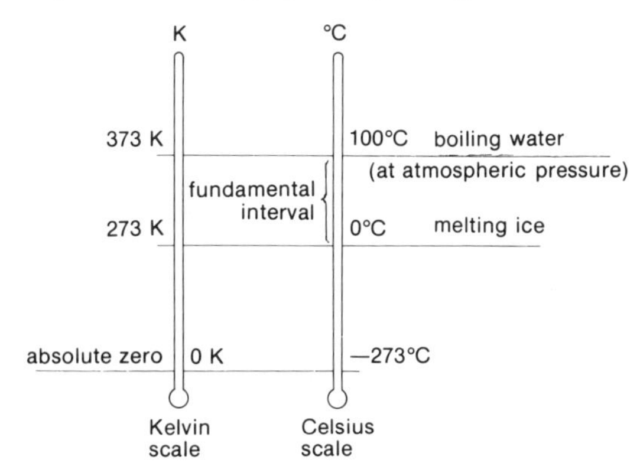

Fig. 25.1

Example 25.1

What is the atmospheric boiling point of water on the Kelvin scale?

Solution

The boiling point of water is at $t = 100°C$, therefore converting to absolute temperature

$$\begin{aligned} T &= t + 273 \\ &= 100 + 273 \\ &= 373\text{K}^{**} \end{aligned}$$

Tables 25.1 to 25.3 give some references which should assist in gaining a better appreciation of the meaning of various temperature levels.

* The exact value is −273.15°C.

** Note that by international agreement since 1968 the unit of temperature is the kelvin (K), not the degree Kelvin, and temperatures on the absolute scale do not carry the degree sign (°K). Thus the temperature 100°C is written 373K.

Table 25.1 Weather conditions

Descriptive terms		Approximate temperature °C
Heat-wave conditions	Very hot	40
	Hot	35
Normal summer	Very warm	30
	Warm	25
Most comfortable	Mild	20
Normal winter	Cool	15
	Cold	10
Cold-snap conditions	Very cold	5
	Freezing	0

Table 25.2 Some reference points on temperature scales (atmospheric pressure)

Material	t Celsius temperature °C	T Absolute temperature K
Tungsten melts at	3400	3673
Copper melts at	1083	1356
Gold melts at	1064	1337
Silver melts at	962	1235
Zinc melts at	420	693
Tin melts at	232	505
Water boils at	100	373
Ice melts at	0	273
Mercury freezes at	–39	234
Carbon dioxide solidifies at	–79	194
Oxygen becomes liquid at	–183	90
Nitrogen becomes liquid at	–196	77
Hydrogen becomes liquid at	–253	20
Helium becomes liquid at	–269	4
Absolute zero	–273	0

The measurement of temperature can be divided into two groups—direct measurement and indirect measurement. In direct measurement the sensing element comes in direct contact with the measured substance, while non-contact techniques detect temperature from emitted radiation.

The most common instrument is the **mercury-in-glass thermometer** which consists of a fine-bore glass tube with an enlargement at the bottom containing liquid mercury. Thermal expansion of mercury read against a calibrated scale, usually etched on the glass, gives the temperature reading. The range of

Table 25.3 Colour of emitted light occurring at particular temperatures

Colour	*Approximate temperature °C*
Dazzling	1500
White	1300
Yellow	1100
Orange	1000
Bright red	900
Cherry red	800
Dull cherry red	700
Dull red	600
Red just visible	500

mercury thermometers is between −35°C and +500°C. For lower temperatures other liquids, such as alcohol, sometimes dyed red, are used. Liquid-in-glass thermometers can be extremely accurate and are usually preferred where accuracy is important.

Thermal expansion of a gas is sometimes used to measure temperature in a device called the **gas thermometer**. Basically, pressure of a gas contained in the bulb of the gas thermometer is measured by means of a U-tube manometer and hence temperature calculated using the gas laws. The gas thermometer was originally used as the fundamental temperature device.

The differential expansion of two bonded metals is the basis for **bimetal thermometers**. A helical bimetal coil often forms the sensing element of temperature gauges, one end being fixed and the other attached to the dial pointer. The usual range is 0°C to 550°C with an accuracy of between 1 per cent and 5 per cent. The same bimetal strip idea is usually used in thermostats for the control of temperature.

Electrical response to temperature changes is employed in thermocouples, resistance elements and thermistors. A **thermocouple** is formed by joining two dissimilar metal wires in a closed loop. A temperature difference between the wire junctions produces an electromotive force in the circuit, which is related to the temperature difference and can be read by suitably calibrated electrical instruments. Thermocouples are the most widely used temperature-sensing elements, being simple, dependable and accurate over a very wide temperature range from extremely low negative temperatures up to about 2700°C.

Thermoresistive elements have a useful temperature range of —180°C to 650°C and consist of a very pure metal wire to which an external electric power source is connected. The change in electrical resistance of the wire serves as a very accurate measure of temperature. **Thermistors** are temperature-sensitive elements of semiconducting non-metallic materials which are used similarly to metallic resistance elements but over a narrower temperature span.

Indirect measurement devices called **pyrometers** utilise various techniques of comparing the effects of radiation emitted from hot objects. Radiation pyrometers focus the energy radiated from a hot body on a black disc, to which a number of thermocouples arranged in series are attached. Some very sophisticated equipment for selectively analysing infrared radiation is in this category. Optical pyrometers enable visual comparisons to be made between the colour of the light radiated from the hot object with an electrically heated filament.

25.3 Molecular structure of matter

Engineering thermodynamics is mainly concerned with the overall effects of mechanical and thermal energy transfer on working agents in boilers, engines, compressors, etc. However, better understanding of the principles of thermodynamics requires some degree of insight into the molecular structure of matter.

It is generally accepted that every substance consists of a very large number of separate particles called molecules. A molecule is the smallest quantity of any substance that can exist as a separate free particle and still possess the identity and all the chemical properties of that substance.

Every molecule has a mass which can be expressed on a relative scale of molecular mass units, for which a molecule of carbon has been chosen as a standard, equal to exactly 12 units.* A list of approximate molecular masses of some substances is given in Table 25.4.

Molecules themselves consist of smaller fundamental particles called atoms. Some substances, such as helium, carbon and sulphur, consist of single atom molecules. Others, including hydrogen, nitrogen and oxygen, have diatomic molecules, i.e. molecules made up of two identical atoms joined together. There are also compound substances, whose molecules are built from two or more dissimilar atoms. For example, a water molecule consists of one atom of oxygen joined with two atoms of hydrogen to form a complex molecule.

The molecular theory is quite helpful in explaining the structure of solid, liquid and gaseous phases** of matter and the transition from one phase to another. In a solid, the molecules are closely spaced and firmly fixed in their relative positions by forces of mutual attraction which give the solid its rigidity. In a liquid, the forces are not sufficient to prevent molecules from moving about relative to each other, which gives a liquid its ability to flow. On the other hand, there is adequate strength in the intermolecular attraction to maintain constant average distances between molecules of a liquid to keep it to a constant volume. In a gas, the molecules are widely spaced and the forces are negligible, hence a gas will tend to expand until it fills all the space available to it.

Likewise, the molecular theory can be used to explain the effects known as temperature and pressure. The temperature of a substance can be explained as

---

* Molecular mass units can be converted to kilograms. One molecular mass unit is approximately equal to 1.66×10^{-27} kg, an extremely small mass indeed.

** The term "phase" is used in thermodynamics to refer to the solid, liquid or gaseous forms of a substance in preference to the term "state", which can be used to describe any condition of a substance determined by its properties at any given instant of time.

Table 25.4 Molecular mass

Substance	*Molecular mass M*
Hydrogen	2
Helium	4
Carbon	12
Methane	16
Water	18
Nitrogen	28
Carbon monoxide	28
Sulphur	32
Oxygen	32
Argon	40
Carbon dioxide	44
Sulphur dioxide	64

a measure of intensity of vibration of the molecules, which can be related mathematically to the kinetic energy of vibration on the molecular scale. The pressure of a fluid can similarly be explained as the summation of all the elementary forces caused by collisions of the molecules of the fluid with the walls of the vessel containing the fluid.

When heat is added to a solid, the energy content of its molecules increases, i.e. the molecules vibrate with an increased degree of agitation while being held in their fixed relative positions, and the temperature of the solid increases. With further addition of heat, the energy of vibrating molecules enables some molecules to break away from the rest, i.e. melting begins. Further input of heat does not increase the energy level of the already moving molecules until all molecules are freed, i.e. until the melting process is complete. Only then will the temperature continue to rise.

Similarly, the transition from liquid to gaseous state can be explained in terms of the increase in the energy level of molecules, the weakening of the intermolecular forces and consequent increase in the mobility of the molecules.

25.4 Terminology

As was mentioned previously, thermodynamics uses a vocabulary of specialised terms which need precise and careful definition. We will now explain some of the more important of these terms.

System, surroundings, boundary

The idea of a system is very important in thermodynamics. By the word **system** we mean a region in space containing a quantity of matter. In the engineering context, a gas expanding within the cylinder of an internal combustion engine, or steam generated from water within a boiler, are examples of thermodynamic systems. A system is always separated from its **surroundings** by a **boundary**, which may be a physical boundary, such as the walls of a cylinder and the face of a piston, or by some imaginary planes, such as cross-sections through the inlet and outlet pipes of a boiler.

When a system has fixed contents, even for a short duration of time, such as the gas expanding in a cylinder against a moving piston, the system is called a **closed system** and only energy in the form of heat and work can enter the system or leave it.

An **open system**, on the other hand, is a system in which fluid is continually flowing into and out of the system, across its boundary, e.g. water flowing into a boiler and steam leaving the boiler. Thus in an open system, in addition to heat and work transfer, mass flow is also taking place, and the flow energy is a relevant factor.

Thermodynamic properties, equilibrium

At any given instant of time the state or condition of a thermodynamic system can be described in terms of a number of quantities known as its **thermodynamic properties**.

Three familiar thermodynamic properties are volume, pressure and temperature. For example, we can describe the state of air in a balloon by stating that its volume is 10 m^3, its pressure is 101.3 kPa and its temperature is 20°C.

Later in this chapter we shall introduce two additional thermodynamic properties called internal energy and enthalpy.

The mass contained within a closed system and the mass flow rate through an open system are necessary for a full description of the system itself but are not regarded as thermodynamic properties. The essential difference is that while volume, pressure and temperature may change from one state to another, the mass involved is subject to the conservation law and remains constant at all times within closed systems and in open systems with steady flow.

When changes of state of a closed system, as measured by variation in pressure, volume and temperature, have ceased to occur, a system is said to be in **thermodynamic equilibrium**. The properties, such as pressure and temperature, must be uniform throughout the system when it is in equilibrium.

In an open system in steady flow, changes in pressure, temperature and volume occur between different points along the flow path. However, at any one point in the system these properties do not change in time.

Process, cycle

The changes in the state of the fluid within a system are referred to as **thermodynamic processes**.

The processes undergone by the fluid in a closed system are described as **non-flow processes**. Such processes are characterised by a constant mass contained within the system and by two different states of the fluid separated by an interval of time, as for example, when 0.75 kg of water is heated in an

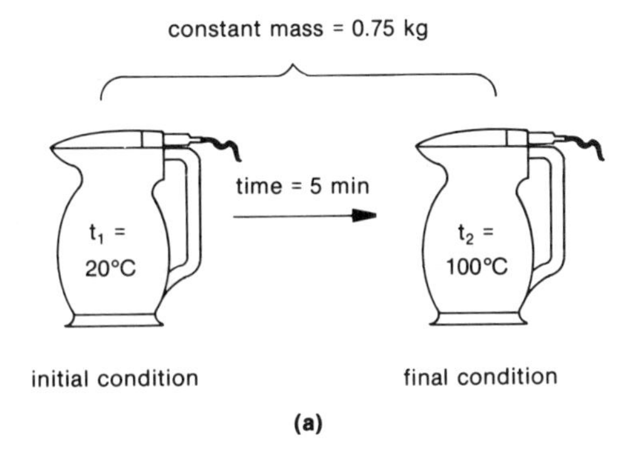

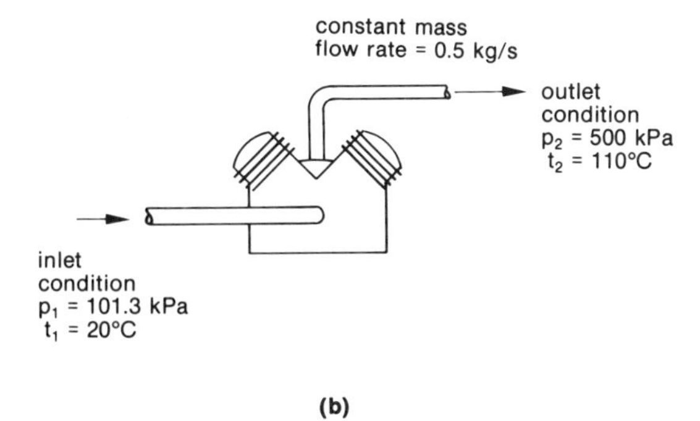

Fig. 25.2(a) *Closed system* **(b)** *Open system*

electric jug from 20°C to 100°C in 5 minutes. We can therefore refer to the *initial* and *final* states of the system and the process which occurs between them.

On the other hand, the processes undergone by the fluid in an open system are described as **flow processes**. These are characterised by a constant mass flow rate through the system and by different states of the fluid at the inlet and outlet of the system, as when 0.5 kg of air per second is compressed from 101.3 kPa and 20°C to 500 kPa and 110°C (see Fig. 25.2). We can, therefore, refer to the *inlet* condition and the *outlet* condition of the fluid flowing through the system.

Many mechanical devices, such as reciprocating engines and compressors, operate in repeating cycles. A system is said to undergo a **cycle** when it passes through a series of connected processes in such a way that its final state is equal in all respects to its initial state. The four-stroke engine cycle, consisting of suction, compression, expansion and exhaust, is a good example of a repeatable thermodynamic cycle (Fig. 25.3).

Cycles, as well as separate processes, can usually be conveniently represented in the form of pressure-volume diagrams. It is very useful for students to

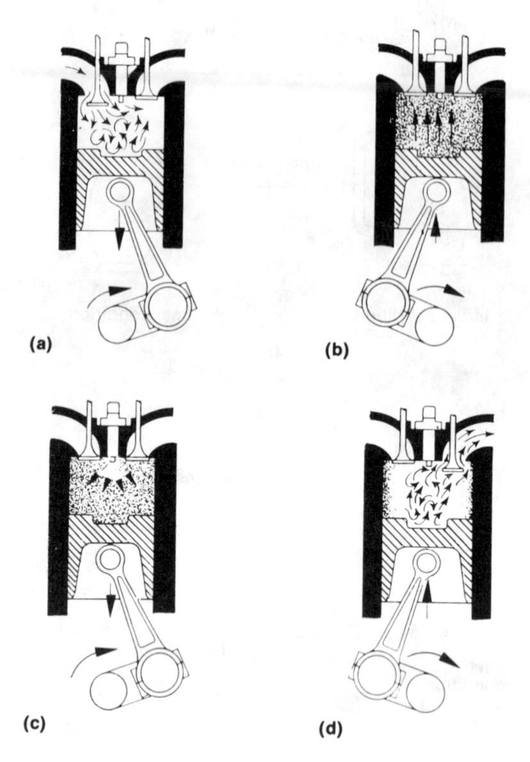

Fig. 25.3(a) *Suction stroke* **(b)** *Compression stroke* **(c)** *Power stroke* **(d)** *Exhaust stroke*

at least sketch the p-V lines for each problem involving expansion, compression and pumping of fluids, in order to gain a better appreciation of the path of the process or cycle involved.

25.5 Internal energy and enthalpy

In addition to temperature, pressure and volume, two other thermodynamic properties, called **internal energy** and **enthalpy**, are often necessary in order to conveniently describe various thermodynamic states and processes, especially in their relationships to heat and work.

Internal energy

The molecular theory of matter suggests that all substances, including liquids and gases used in engineering systems, consist of separate particles, called molecules. Each molecule contains a certain amount of kinetic energy associated with its vibration, and in the case of fluids, with its random linear motion. Furthermore, following the molecular theory, the temperature of a substance is mathematically related to the kinetic energy of its molecules.

However, temperature itself is not a direct measure of the energy stored in a substance by virtue of its molecular structure. It may be said that temperature and internal energy are two very closely related thermodynamic properties but they are not equivalent. The mathematical relationships between temperature and internal energy can be very complex, especially when changes of phase are involved. It is, therefore, helpful to consider the concept of internal energy as a separate thermodynamic property, alongside temperature, pressure and volume.

Let us define internal energy as **the energy stored within a substance, due to the kinetic energy of its molecules**. As a measure of energy, internal energy should be expressed in joules, or their multiples, such as kilojoules. The usual symbol for internal energy is U. Internal energy can also be expressed in units of energy per unit of mass, i.e. in joules per kilogram or kilojoules per kilogram, with lower-case u as the symbol. Internal energy per unit mass is called specific internal energy of a substance.

It follows that:

$$u = \frac{U}{m}$$

where u is the specific internal energy (J/kg)
U is the total internal energy stored in a substance in J
m is the mass of the substance in kg

It is important to understand that internal energy is a thermodynamic property which can be used to describe a thermodynamic state of a particular fluid, in much the same way as temperature, pressure and volume are often used to do so. For example, we can describe the state of a gas in a particular system by saying that it is at 101.3 kPa pressure and has specific internal energy of 14.4 kJ/kg.*

The advantage in using internal energy as a thermodynamic property lies in the fact that internal energy can readily be related to heat and work transfer between a closed system and its surroundings. However, it is impossible to measure internal energy directly, in a way that temperature or pressure can be measured. It is, therefore, usual practice to measure and describe conditions of working fluids in terms of measurable quantities, such as temperature, and hence determine internal energy by calculation or from tables.

For any given substance, such as water or air, it is possible to compile tables or charts relating specific internal energy to other properties, such as temperature and pressure. The computed magnitude of specific internal

---

* According to the theory of thermodynamics, any state of a system containing only one phase of matter is completely defined when any two of its independent properties are known. However, in most cases it is not possible to use internal energy in conjunction with temperature for this purpose, because both temperature and internal energy are related to the kinetic energy of molecules and are, therefore, not completely independent of each other.

Table 25.4 Specific internal energy of air and liquid water

Temperature °C	*Specific internal energy—u* kJ/kg	
	Air[a]	*Water*[b]
0	0	0
25	17.9	104.7
50	35.8	209.2
75	53.8	313.8
100	71.8	419.0
125	89.9	525.0[b]
150	108.1	632.0
175	126.3	740.5
200	144.7	851.2
225	163.2	964.2
250	181.7	1081.0
275	200.5	1203.0
300	219.3	1333.0
325	238.3	1475.0
350	257.5	1643.0

a The values of specific internal energy for gases and liquids are practically independent of pressure, i.e. they are a function of temperature only.
b It is possible to have liquid water at temperatures above 100°C provided the pressure is sufficient to prevent boiling, e.g. water is liquid at 150°C if the pressure is 500 kPa absolute.

energy depends on the choice of an arbitrary datum level to which internal energy values are referred, which is often taken to be the condition of the substance at 0°C. Table 25.4 gives typical values of specific internal energy for two of the most common substances used in engineering systems, i.e. air and water (liquid). The values in the table are relative to 0°C datum at which specific internal energy is assumed to be zero. This, of course, does not mean that there is no energy contained in the substance when it is below 0°C, but simply that only the excess internal energy over its value at 0°C has been tabulated.*

The actual numerical value of internal energy is not particularly important by itself, so long as changes in internal energy can be calculated for a system undergoing a process. An increase in internal energy is always taken to be positive, while a decrease is negative. The change in internal energy of a substance can be calculated from

$$\Delta U = U_2 - U_1 = m.u_2 - m.u_1 = m(u_2 - u_1)$$

where ΔU is the change which occurs between the initial and final states.

* This method is somewhat similar to expressing elevations of a particular point, relative to ground level, sea level or some other convenient reference at which elevation is assumed to be zero. Doing this does not mean that there are no other points located below the reference level.

Example 25.2

Three kilograms of air initially at 25°C undergo a process as a result of which the temperature increases by 150 degrees. Determine the total change in the internal energy of the air.

Solution

The final temperature is 25° + 150° = 175°C.

Specific internal energy values from Table 25.4 are:

at 25°C $\quad u_1 = 17.9$ kJ/kg

at 175°C $\quad u_2 = 126.3$ kJ/kg

Therefore, the change in internal energy between the two states is:

$$\Delta U = m(u_2 - u_1) = 3 \text{ kg} \times (126.3 \text{ kJ/kg} - 17.9 \text{ kJ/kg})$$
$$= 325.2 \text{ kJ}$$

The answer is positive, which signifies an increase in the internal energy of the air.

Enthalpy

In many engineering applications, particularly those involving continuous flow through open systems, such as centrifugal compressors, refrigeration cycles and steam generation, another thermodynamic property, called **enthalpy**, is very useful.

Enthalpy is defined as **the sum of the internal energy and the product of the pressure and volume** of a substance. Total enthalpy is given the symbol, H, and specific enthalpy, i.e. enthalpy per unit mass of substance, is given the symbol, h. Therefore, by definition:

$$\boxed{H = U + pV}$$

and
$$h = u + pv$$

where V is the total volume of a substance
v is the specific volume of substance
p is the absolute pressure
U and u are the total and specific internal energy, respectively.

It follows that

$$\boxed{h = \frac{H}{m}}$$

where m is the mass of the substance.

The units of enthalpy are the same as those of internal energy, i.e. joules, kilojoules, etc. Obviously, in order to add the pressure-volume product and internal energy, the two must be in the same units. This can easily be demonstrated by substitution of unit symbols for the pressure and volume as follows:

$$Pv = \text{Pa} \times \text{m}^3 = \frac{\text{N}}{\text{m}^2} \times \text{m}^3 = \text{N.m} = \text{J}$$

One must not be misled by the apparent identity of units into believing that enthalpy is a form of energy, a mistake often made even in technical literature.* It will be shown later that enthalpy plays an important part in the relationships between heat and work transfer taking place in open systems. Furthermore, in the particular case of a fluid entering or leaving such a system, enthalpy of the fluid does indeed represent its energy content. However, it must be strongly emphasised that, in general, enthalpy as such is not equivalent to energy. For example, if a quantity of air is contained in a closed pressure vessel, it has certain values of pressure, volume and internal energy, which can be used to calculate enthalpy of the air. However, in this case the only true energy content of the air in the vessel is represented by its internal energy, and not by its enthalpy. An analogy from mechanics may also help to illustrate this point. You may recall that torque and work are expressed in the same combination of units, i.e. newton metres, and in the case of rotational motion they are closely related quantities. However, like enthalpy and energy, torque and work are not identical concepts.

The main reason for introducing the property enthalpy is to effect a simplification in certain types of engineering problems, in which $(U + pV)$ consistently appears as a group of variables, by substituting a single variable H for the group. Since pressure, volume and internal energy are thermodynamic properties, enthalpy itself is a thermodynamic property, i.e. it can be used to describe a thermodynamic state of a substance.

Tables of thermodynamic properties of substances, such as steam usually contain enthalpy as one of the most useful properties. Table 25.5 gives some typical values of specific enthalpy for water and steam at atmospheric pressure and for air at any pressure, as a function of temperature. The table follows on from Table 25.4, in that at the datum condition internal energy of the substances is taken to be zero and enthalpy for all temperatures has been calculated from $h = u + pv$.

Changes in enthalpy of a substance undergoing a process can be calculated in a similar manner to changes of internal energy. The corresponding equation is as follows:

$$\boxed{\Delta H = H_2 - H_1 = mh_2 - mh_1 = m(h_2 - h_1)}$$

---

* In some dictionaries, as well as in some early textbooks on the subject, enthalpy has been described as "the heat content of a substance". Such a description is extremely misleading, since heat is a form of energy transfer, while enthalpy is not.

Table 25.5 Specific enthalpy of air, water and steam

	Specific enthalpy—h kJ/kg		
Temperature °C	*Air at any pressure*[a]	*Water*	*Steam*
		at atmospheric pressure[b]	
0	78.4	0.1	—
25	103.4	104.8	—
50	128.5	209.3	—
75	153.7	313.9	—
100	178.9	419.1	2676
125	204.1	—	2727
150	229.5	—	2777
175	254.9	—	2827
200	280.5	—	2876
225	306.1	—	2926
250	331.8	—	2975
275	357.8	—	3025
300	383.8	—	3075
325	409.9	—	3126
350	436.3	—	3177

a The values of specific enthalpy for air are practically independent of pressure, i.e. they are a function of temperature only.
b The boiling point of water and the properties of steam are a function of pressure. Therefore, the values of specific enthalpy of steam tabulated above are applicable only to steam at atmospheric pressure.

Example 25.3

Thirty-five kilograms of steam are generated at atmospheric pressure from feed-water, which is initially at 25°C. If the total enthalpy of the substance increases by 97 MJ during the process, what is the final temperature of steam?

Solution

The mass of water/steam, $m = 35$ kg

The total change of enthalpy, $\Delta H = 97\,000$ kJ

The initial specific enthalpy of water at 25°C:

$$h_1 = 104.8 \text{ kJ/kg (from Table 25.5)}$$

Substitute into: $\Delta H = m(h_2 - h_1)$

$$97\,000 \text{ kJ} = 35 \text{ kg}(h_2 - 104.8 \text{ kJ/kg})$$

Hence, the final specific enthalpy of steam is

$$h_2 = \frac{97\,000 \text{ kJ}}{35 \text{ kg}} + 104.8 \text{ kJ/kg}$$

$$= 2876 \text{ kJ/kg}$$

From the table of specific enthalpies, the temperature of steam, corresponding to $h = 2876$, is 200°C.

Therefore, the temperature of steam generated is 200°C.

Fig. 25.4 *Steam generator and associated equipment*

Problems

Use Tables 25.1 to 25.5 where required.

25.1 Convert the following temperatures from the Celsius to Kelvin scale:
(a) 150°C,
(b) 0°C,
(c) −150°C.

25.2 Convert the following temperatures from the Kelvin to Celsius scale:
(a) 500K,
(b) 273K,
(c) 150K.

25.3 What is the difference between the melting points of zinc and silver, in kelvins and in degrees Celsius?

25.4 Approximately what colour of light would you expect to be emitted from melting copper?

25.5 Calculate the atomic masses of hydrogen and oxygen and hence prove that the molecular mass of water is 18.

25.6 Determine specific internal energy of:
(a) air at 250°C,
(b) water at 125°C.

25.7 Determine specific enthalpy of:
(a) air at 300°C,
(b) water at 50°C,
(c) steam at 275°C and 101.3 kPa.

25.8 One kilogram of air occupies 0.57 m^3 at 150°C and 213 kPa absolute. Given its specific internal energy, equal to 108.1 kJ/kg, calculate its specific enthalpy and then compare your answer with the corresponding value from Table 25.5.

25.9 150 kg of water, initially at 25°C are converted into steam with a final temperature of 250°C, at atmospheric pressure. Determine the change in total enthalpy which occurs during each of the following stages:
(a) Heating of water from 25°C to the boiling point at 100°C.
(b) Conversion of water into steam at 100°C.
(c) Heating of steam from 100°C to 250°C.

25.10 A cylinder of 5 L volume contains 0.041 kg of argon at 500 kPa absolute and 20°C. Specific internal energy of argon at 20°C is 6.34 kJ/kg. Calculate the total internal energy and total enthalpy of argon in the cylinder.

25.6 Work and heat transfer

Thermodynamics is the science of energy transfer and energy conversion. In particular, it is concerned with the quantities "work" and "heat". These quantities have a well-defined meaning and a sign convention which must be adhered to at all times when thermodynamic systems are discussed.

Work

Work, W, may be defined as a transfer of mechanical energy between a thermodynamic system and its surroundings.

It may be recalled that in mechanics work was defined as the product of force and the distance moved by the force, $W = F \times S$. This means that for any work to be done there must be a force and there must be motion, without which there can be no work done.

In thermodynamics we are dealing with fluids, i.e. liquids and gases, contained within or flowing through systems like compressor cylinders and boilers. It is, therefore, more convenient to consider distributed forces, called pressure, as causing work to be done, and the changes in volume associated with movement or expansion of the fluid.

You will recall* that work done by a constant pressure is given by:

$$W = p(V_2 - V_1)$$

where W is work done
p is constant pressure or mean effective pressure
V_1 is initial volume
V_2 is final volume

This also corresponds to the area under the pressure-volume graph.

If we consider a system comprising a fluid in a cylinder subjected to a change in volume, two alternative directions may be possible, namely expansion and compression. It is, therefore, necessary to adopt a sign convention to indicate the direction of work transfer.

The work is regarded as *positive* if it is done *by* the system, i.e. by the fluid *on* the surroundings, e.g. when the fluid expands in a cylinder pushing a piston outwards.

The work is regarded as *negative* if it is done *on* the system, i.e. on the fluid, *by* the surroundings, e.g. when an external force is applied to a piston to compress a fluid.

It follows from the formula above that if a fluid is restricted within a closed container of fixed dimensions, without any possibility for expansion or compression, i.e. without any change in volume ($V_1 = V_2$), there can be no work done ($W = 0$) by, or on, the fluid. This fact is well worth remembering, i.e. **if there is no change in volume of a closed system, the work done is zero**, regardless of any changes in pressure and temperature.

Heat

Heat, Q, may be defined as a transfer of thermal energy between a thermodynamic system and its surroundings. As a form of energy transfer, heat is measured in joules.

---

* See Chapter 22, Section 22.5.

In the case of heat flow, a temperature difference is a necessary condition for the transfer of energy to take place. Heat has a natural tendency to flow from hotter objects to colder objects. In engineering systems thermal energy is usually obtained from fuels and is then transferred to various fluids, such as steam in a boiler, in the form of heat. Steam itself can be used as a heating medium or as a working agent, e.g. to drive a turbine.

Like work, heat transfer can occur in one of two possible directions. Heat transfer is taken to be *positive* if thermal energy flows *into* the system *from* its surroundings, i.e. a gain of heat by the system is positive. Conversely, if heat flows *from* the system *to* the surroundings it is said to be *negative*, i.e. a loss of heat from the system is negative.

Thermal insulation placed between a system and its surroundings tends to reduce the amount, or the rate of heat transfer. If a system is completely insulated from the surroundings, there will be no heat transfer at all. Systems which are impervious to heat, and processes which take place without loss or gain of heat, i.e. $Q = 0$, are known as **adiabatic*** systems or processes. In practice, while there is no such thing as perfect insulation, many thermodynamic systems closely approach the adiabatic characteristics. In addition, it must be remembered, that since heat transfer depends on temperature difference, there will be no heat transfer if the system and its surroundings are at the same temperature.

25.7 Energy equations for thermodynamic systems

When energy exchange takes place between a thermodynamic system and its surroundings, the relationships between heat transfer (Q), work transfer (W) and the change of thermodynamic state of the system are known as the energy equations.

There are two basic forms of energy equations, corresponding to the two types of systems, namely open and closed systems. These are discussed separately below.

Closed systems

Let us consider a typical closed system comprising a gas enclosed by a cylinder and piston as shown in Figure 25.5. There are two possible ways in which energy can cross the boundary of such a system. First, heat, Q, can flow from an external source, e.g. fuel, into the system. The opposite direction of heat flow would simply be regarded as negative. Secondly, if the gas expands, external work, W, will be done by the system on the surroundings. Again, the reverse, i.e. work done by the surroundings to compress the gas, would be negative.

---

* The term adiabatic comes from Greek *adiabatos* (*a* = not, *dia* = through, *batos* = passable).

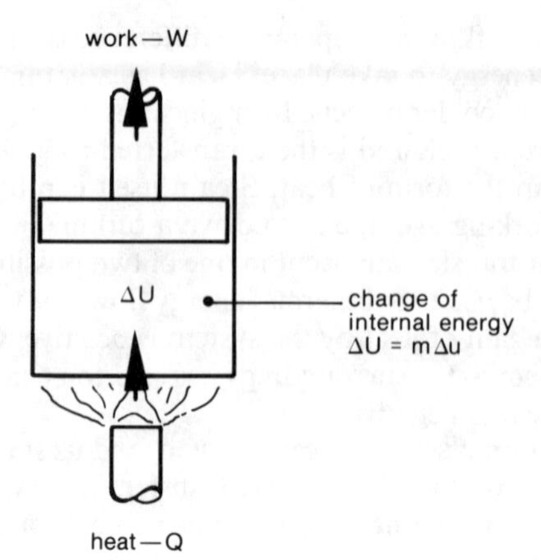

Fig. 25.5

Many times, in systems like this, when a substance is heated or cooled, expanded or compressed, the amounts of thermal and mechanical energy entering the system are not exactly balanced by the amounts of energy leaving the system. The resultant imbalance between the incoming and outgoing energy quantities cannot disappear, but becomes a gain or deficit in the amount of energy stored internally in the substance contained within the system. We have previously learned to recognise stored energy as internal energy, U.

We can express this in the form of an equation, which states that the net change in the internal energy, ΔU, of a closed system undergoing a process is exactly equal to the net energy transfer into the system in the form of heat, Q, and work, W.

$$\boxed{\Delta U = Q - W}$$

where $\Delta U = U_2 - U_1 = m(u_2 - u_1)$ is the change of internal energy of the system,
Q is net heat input into the system,
W is net work output from the system.

This equation applies to all closed systems in which, as we know, only non-flow processes take place. The equation is, therefore, called the **non-flow energy equation**.

Example 25.4

0.0229 kg of air, initially at 50°C, is heated in a cylinder by the addition of 1.55 kJ of heat. The air expands, doing 0.31 kJ of work against a piston. Determine the total change in the internal energy of the system and the final temperature of the air.

Solution

Given, $Q = 1.55$ kJ and $W = 0.31$ kJ, substitute into $\Delta U = Q - W$, to find total change in internal energy:

$$\Delta U = 1.55 \text{ kJ} - 0.31 \text{ kJ}$$
$$= 1.24 \text{ kJ}$$

We can now write:

$$\Delta U = m(u_2 - u_1)$$

where $m = 0.0229$ kg and $u_1 = 35.8$ kJ/kg, from Table 25.4.

Substitute, $1.24 \text{ kJ} = 0.0229 \text{ kg} \times (u_2 - 35.8 \text{ kJ/kg})$

Hence, $u_2 = 89.9 \text{ kJ/kg}$

The final temperature is found from the table to correspond with specific internal energy of 89.9 kJ/kg.

The final temperature is 125°C.

Open systems

Let us now turn to a typical open system, which may consist of a steam turbine plant comprising a boiler followed by the turbine itself. For our purposes, at this stage, it is convenient to define the system by its boundary, as shown in Figure 25.6 in dotted line.

It can readily be seen that heat, Q, and work, W, can flow into or out of the system, across the system boundary. As before, with the closed system, heat and work transfer are seldom balanced, thus producing a net gain or loss of energy by the system, equal to $Q - W$.

However, in this case, the system is open for continuous flow of fluid, e.g. steam, into and out of the system. Under these conditions, the energy content of the fluid entering and leaving the system is not limited to its internal energy, U, but contains an additional component, equal to the product of pressure and volume, pV, associated with the fact that a pumping process must occur somewhere within or outside the system, to cause the fluid to flow through the system. The energy content of the fluid is the sum of these components which,

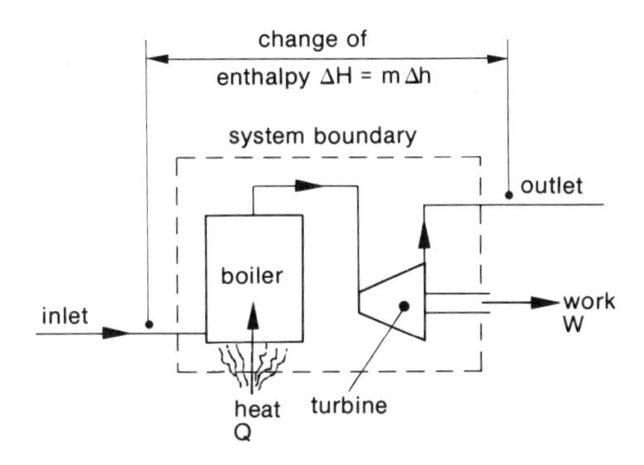

Fig. 25.6

by definition, equals the enthalpy of the fluid. Thus, the energy content at the inlet and outlet of the system is given by

$$U_1 + p_1V_1 = m(u_1 + p_1v_1) = H_1$$

and

$$U_2 + p_2V_2 = m(u_2 + p_2v_2) = H_2$$

The change of enthalpy between inlet and outlet of an open system, equal to $\Delta H = H_2 - H_1 = m(h_2 - h_1)$, represents the gain, or loss, of energy by the fluid. We can, therefore, write the following equation:

$$\boxed{\Delta H = Q - W}$$

This equation applies to open systems, in which flow processes occur. The equation is, therefore, known as the **steady-flow energy equation.***

Example 25.5

Determine the work output from an open system in which a fluid enters at the rate of 3 kg/s with an initial enthalpy of 3200 kJ/kg and leaves with an enthalpy of 2750 kJ/kg. The heat loss from the system is at the rate of 350 kJ/s.

Solution

The change of enthalpy of the system is:

$$\Delta H = H_2 - H_1 = m(h_2 - h_1) = 3 \text{ kg/s} \times (2750 \text{ kJ/kg} - 3200 \text{ kJ/kg})$$
$$= -1350 \text{ kJ/s}$$

Note that heat transfer is also negative, i.e. a loss of heat, $Q = -350$ kJ/s.

Substitute into $\qquad \Delta H = Q - W$

$$-1350 \text{ kJ/s} = -350 \text{ kJ/s} - W$$

Hence, $\qquad W = 1000$ kJ/s

This is a positive result, meaning that work is being done by the system on the surroundings at the rate of 1000 kJ per second.

The rate of doing work is called power. Therefore, the result can be described as power output of 1000 kW.

25.8 Internal energy, enthalpy and heat transfer

There are special circumstances under which heat transferred to, or from, a substance can be related directly to changes in its internal energy or enthalpy. These are examined below.

---

* In its complete form, the steady-flow energy equation should also include kinetic and potential energy terms from Bernoulli's equation. However, these terms are usually of negligible magnitude, except in some specialised applications, e.g. nozzles, and have, therefore, been omitted from the equation used in this book.

Constant volume heating

Imagine a closed rigid container in which a substance is heated, as shown in Figure 25.7. Due to the inability of the substance to expand, there is no work being done.

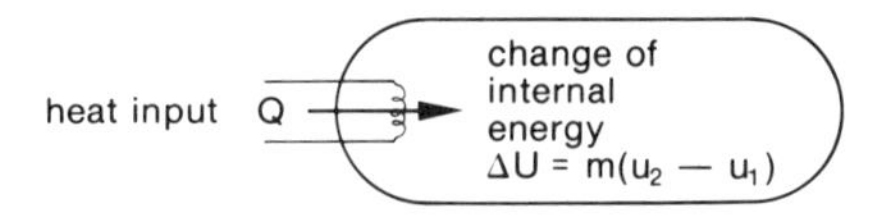

Fig. 25.7

The system is a closed system, for which the non-flow energy equation, $\Delta U = Q - W$, applies.

Substituting $W = 0$, results in:

$$\Delta U = Q$$

This means that during a constant volume heating of a substance, the change of internal energy is equal to the heat added.

Example 25.6

Determine the amount of heat required to increase the temperature of 1.2 kg of air from 0°C to 250°C, if the air is contained within a closed cylinder.

Solution

Using Table 25.4, $u_1 = 0$ and $u_2 = 181.7$ kJ/kg.

$$\begin{aligned}\Delta U &= m(u_2 - u_1)\\ &= 1.2 \text{ kg } (181.7 - 0) \text{ kJ/kg}\\ &= 218.0 \text{ kJ}\end{aligned}$$

Hence,
$$\begin{aligned}Q &= \Delta U\\ &= 218 \text{ kJ}\end{aligned}$$

Constant pressure heating: Flow process

Consider now a fluid flowing inside a pipe, a duct or a heat exchanger at constant pressure, receiving heat through the walls of the pipe, as in Figure 25.8.

In this case there is no mechanical movement other than the flow of the fluid. Therefore, there is no external work done, $W = 0$.

The process is a flow process for which $\Delta H = Q - W$.

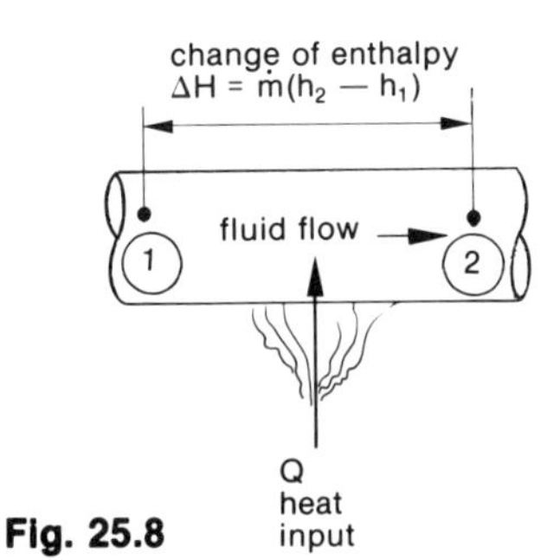

Fig. 25.8

Hence,

$$\boxed{\Delta H = Q}$$

In words, during a constant pressure heating of a substance flowing in a pipe, duct or heat exchanger, the change of enthalpy is equal to the heat added.

Example 25.7

Determine the amount of heat required to increase the temperature of 10 kg of steam from 100°C to 350°C at atmospheric pressure.

Solution

From Table 25.5, $h_1 = 2676$ kJ/kg and $h_2 = 3177$ kJ/kg.

$$Q = \Delta H = m(h_2 - h_1) = 10 \text{ kg } (3177 \text{ kJ/kg} - 2676 \text{ kJ/kg})$$
$$= 5010 \text{ kJ}$$

Constant pressure heating: Non-flow process

Let us consider a quantity of gas contained in a cylinder with a piston loaded by a constant weight, as shown in Figure 25.9. In a system like this, the gas remains at constant pressure at all times.

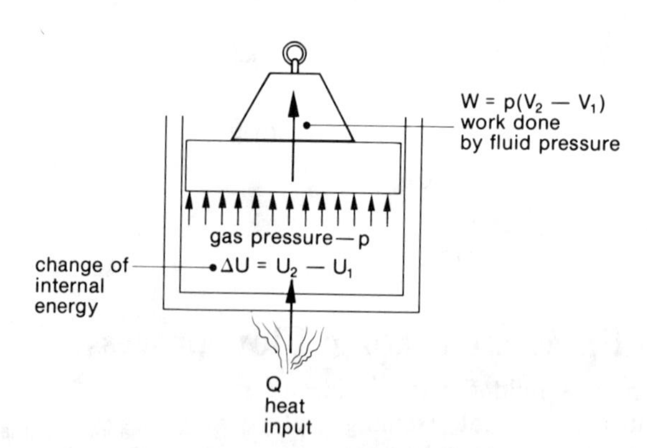

Fig. 25.9

When the gas is heated by the addition of Q units of heat, it will expand and lift the piston, doing an amount of work given by $W = p(V_2 - V_1)$.

For a closed system, $\Delta U = Q - W$, where $\Delta U = U_2 - U_1$.

Substitute and rearrange as follows:

$$U_2 - U_1 = Q - p(V_2 - V_1)$$
$$U_2 - U_1 + p(V_2 - V_1) = Q$$
$$(U_2 + pV_2) - (U_1 + pV_1) = Q$$
$$H_2 - H_1 = Q$$

Therefore,

$$\boxed{\Delta H = Q}$$

Once again, during a constant pressure heating of a gas, this time within a closed system, the change of enthalpy is equal to the heat added.

Example 25.8

0.025 kg of air initially at 25°C is heated within a cylinder under constant pressure to a final temperature of 150°C.

Determine the amount of heat used and the work done by the air in lifting the load on the piston.

Solution

Using Tables 25.4 and 25.5, the following data are obtained:

$t_1 = 25°C \qquad u_1 = 17.9 \text{ kJ/kg} \qquad h_1 = 103.4 \text{ kJ/kg}$

$t_2 = 150°C \qquad u_2 = 108.1 \text{ kJ/kg} \qquad h_2 = 229.5 \text{ kJ/kg}$

Heat transfer under constant pressure, $Q = \Delta H$,

$$Q = \Delta H = m(h_2 - h_1) = 0.025 \text{ kg } (229.5 \text{ kJ/kg} - 103.4 \text{ kJ/kg})$$
$$= 3.153 \text{ kJ}$$

The non-flow energy equation also applies:

$$Q - W = \Delta U$$

where

$$\Delta U = m(u_2 - u_1)$$
$$= 0.025 \text{ kg}(108.1 \text{ kJ/kg} - 17.9 \text{ kJ/kg})$$
$$= 2.255 \text{ kJ}$$

Substitute, $3.153 \text{ kJ} - W = 2.255 \text{ kJ}$

work done, $W = 0.898 \text{ kJ}$

25.9 Summary

It may be useful to summarise the energy exchange and related equations from this chapter for future reference.

1. *Change of internal energy of a substance.*
   $\Delta U = U_2 - U_1 = m(u_2 - u_1)$
   Refer to Table 25.4 for u_1 and u_2.
2. *Change of enthalpy of a substance.*
   $\Delta H = H_2 - H_1 = m(h_2 - h_1)$
   Refer to Table 25.5 for h_1 and h_2.
3. *Energy exchange in any closed system—non-flow energy equation.*
   $\Delta U = Q - W$

4. *Energy exchange in any open system—steady-flow energy equation.*
   $\Delta H = Q - W$
5. *Constant volume heating (special case)*
   $\Delta U = Q$
6. *Constant pressure heating (special case)*
   $\Delta H = Q$

Problems

Use Tables 25.4 and 25.5.

25.11 Determine the change of internal energy of a closed system which has 25 kJ of work done on it and loses 10 kJ of heat during a process.

25.12 If the increase of enthalpy of a gas flowing through an open system is 35 kJ, while the system is doing 65 kJ of work on the surroundings, determine the amount and direction of heat transfer.

25.13 A closed system contains 0.05 kg of gas. During a process its specific internal energy changes from 45 kJ/kg to 105 kJ/kg, while 5 kJ of heat are transferred to the system. Determine the amount and direction of work transfer.

25.14 An open system with a flow rate of gas equal to 0.7 kg/s undergoes a process during which the specific enthalpy of the gas increases from 150 kJ/kg to 250 kJ/kg. If heat is transferred to the gas at the rate of 100 kJ/s, determine the power output from the system.

25.15 A closed system contains 0.03 kg of air. During a process the temperature of the air increases from 25°C to 125°C, while 4 kJ of heat are transferred to the system. Determine the amount and direction of work transfer.

25.16 An open system with a flow rate of air equal to 1.5 kg/s undergoes a process during which the temperature of the air decreases from 300°C to 75°C. If the air is doing work at the rate of 200 kJ/s, determine the rate of heat loss from the system.

25.17 A closed pressure vessel contains 3.44 kg of air at 25°C. As a result of fire in the vicinity of the vessel, 500 kJ of heat are gained by the air. What is the final temperature of air?

25.18 Steam at atmospheric pressure is used as a heating medium in a heat exchanger, entering at 200°C and leaving as condensate, i.e. liquid water, at 50°C. Determine the flow rate of steam required per hour if the rate of heat transfer required is 80 kJ/s.

25.19 0.018 kg of air initially at 25°C expands in a cylinder at constant pressure when 2.27 kJ of heat are added. Determine the final temperature of the air and the amount of work done by the air.

25.20 Determine the cost of heating 300 litres of water from 25°C to 75°C, if the efficiency* of the heating elements is 80 per cent and the cost of electricity is 8 cents per kilowatt-hour. (Note: 1 kW.h = 3.6 MJ.)

Review questions

1. Define engineering thermodynamics.
2. Briefly explain the main points of the zeroth, first and second laws of thermodynamics.
3. What is temperature? Is temperature a synonym of heat? Explain.
4. Explain the two temperature scales.
5. State the formula for converting Celsius temperature to Kelvin temperature.
6. Describe some temperature measuring devices.
7. Briefly explain the solid, liquid and gaseous phases of matter in terms of the molecular theory.
8. Define the terms molecular mass and atomic mass.
9. Define the terms:
(a) thermodynamic system,
(b) surroundings,
(c) boundary,
(d) closed system,
(e) open system.
10. What is meant by thermodynamic properties? Give some examples.
11. What is meant by thermodynamic processes? Give some examples of non-flow and of steady-flow processes.
12. Explain internal energy.
13. Explain enthalpy.
14. State the sign convention for work transfer.
15. State the sign convention for heat transfer.
16. State the sign convention for changes in internal energy or in enthalpy.
17. State and explain the non-flow energy equation.
18. State and explain the steady-flow energy equation.
19. State the relationships between heat transfer and changes in internal energy or enthalpy during:
(a) constant volume heating,
(b) constant pressure heating.

* Efficiency is the ratio of useful output to the initial input, in this case the ratio of heat actually transferred to the water to the input of electrical energy.

26

Calorimetry of heat

Calorimetry has been defined as the measurement of thermal constants such as specific heat, latent heat and calorific value. Such measurements usually involve the use of a calorimeter, a device which consists typically of an insulated enclosure within which some form of heat exchange takes place which is measurable in terms of temperature increase produced and/or in terms of change of phase observed.

In a broad sense, the word calorimetry means a careful mathematical account of heat quantities involved when a substance undergoes a change in temperature, a change of phase, or when heat is liberated from a fuel during combustion.

When heat is transferred to a substance, one of two possible results may be observed, i.e. the temperature may rise or a change of phase may occur. Under any given pressure, these two effects never occur simultaneously and we shall, therefore, discuss them separately.

Later in this chapter we will also discuss the heat liberated from the chemical energy of fuels.

26.1 Sensible heat

In general, we recognise heat transfer by the effect it has on the fluids to which it flows. When a single phase substance, such as water, receives heat, its temperature tends to rise in proportion to the amount of heat transferred. The temperature rise is observable by our sense of touch, or temperature sense. Therefore, **the heat transfer which is accompanied by a temperature rise is called sensible heat**.

The change of temperature produced by the addition of sensible heat depends on the amount of heat added to each unit mass of substance and on the ability of the substance to absorb heat, known as its specific heat capacity.

The **specific heat capacity of a substance** is defined as the amount of heat required to produce one degree of temperature rise in a unit mass of the substance. The unit of specific heat capacity is the joule per kilogram kelvin (J/kg.K). It is a small unit and its multiple, the kilojoule per kilogram kelvin (kJ/kg.K), is usually employed in practice. Different substances have different specific heat capacities as shown in Table 26.1.

Fig. 26.1 *Water calorimeter*

Table 26.1 Specific heat capacities of some solids and liquids

Substance	*Specific heat capacity kJ/kg.K*
Lead	0.13
Copper	0.39
Mild steel	0.48
Cast iron	0.50
Aluminium	0.91
Oil	1.80
Cork	2.03
Ice	2.04
Kerosene	2.10
Glycerine	2.43
Paraffin	2.89
Water	4.19

If we let,

c = specific heat capacity of substance, kJ/kg.K,
m = mass of substance, kg,
Δt = change of temperature °C (or K),
Q = heat transferred to produce temperature change, kJ,

then it follows from the definition of specific heat that

$$Q = m.c.\Delta t$$

Example 26.1

Determine the amount of heat required to increase the temperature of 300 L of water from 25°C to 75°C.

Solution

Specific heat capacity of water is 4.19 kJ/kg.K.

Substitute m = 300 kg, t_1 = 25°C and t_2 = 75°C into $Q = m.c.\Delta t$:

$$Q = 300 \text{ kg} \times 4.19 \text{ kJ/kg.K} (75 - 25)\text{K}$$
$$= 62\,850 \text{ kJ*}$$

Example 26.2

A steel drum of 1.4 kg mass contains 12 kg of kerosene. If the initial temperature was 40°C and the final temperature is 15°C, determine the total amount of heat lost by the drum and by the kerosene.

* There is a small discrepancy between this answer and the corresponding value for Problem 25.20 due to the fact that specific heat capacity is not constant but varies slightly with temperature. For example, the specific heat capacity of water falls from 4.217 kJ/kg.K at 0°C to a minimum of 4.178 kJ/kg.K at about 35°C and then begins to rise again reaching 4.19 kJ/kg.K at 100°C. However, the error introduced by assuming constant specific heat is quite insignificant.

Solution

Heat lost by the steel drum:

$$\begin{aligned} Q_s &= m.c.\Delta t \\ &= 1.4 \text{ kg} \times 0.48 \text{ kJ/kg.K } (40 - 15)\text{K} \\ &= 16.8 \text{ kJ} \end{aligned}$$

Heat lost by the kerosene:

$$\begin{aligned} Q_k &= m.c.\Delta t \\ &= 12 \text{ kg} \times 2.1 \text{ kJ/kg.K } (40 - 15)\text{K} \\ &= 630 \text{ kJ} \end{aligned}$$

$$\text{Total heat loss} = 16.8 + 630 = 646.8 \text{ kJ}$$

26.2 Heat balance principle

When heat exchange takes place between two substances, or between two quantities of the same substance, at different initial temperatures, the conservation of energy law finds its expression in the **heat balance principle**.

The heat balance principle simply states that when heat exchange takes place between substances A and B, heat lost by substance A is equal to the heat gained by substance B.

heat lost by A	=	heat gained by B

Two assumptions underlie this principle. First, it is assumed that no heat is lost to the surroundings or gained from the surroundings. Secondly, it is assumed that after the heat exchange ceases a common final temperature will have been reached by both substances. Substituting $m.c.\Delta t$ on each side of the heat balance equation results in the following expression

$$m_A.c_A(t_A - t) = m_B.c_B(t - t_B)$$

where m_A and m_B are the masses of substances A and B
c_A and c_B are the specific heat capacities of A and B
t_A and t_B are the initial temperatures of A and B
t is the final common temperature

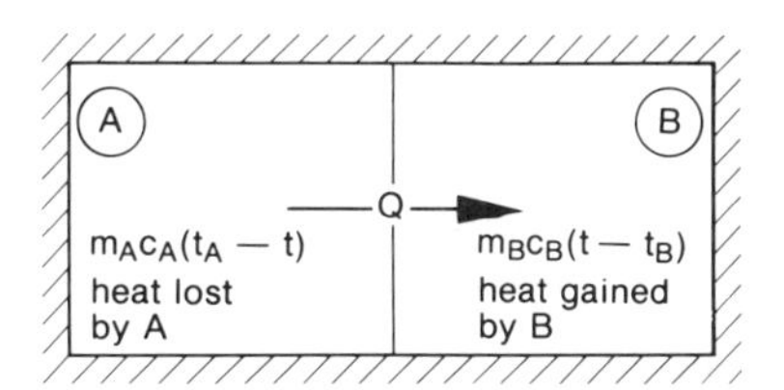

Fig. 26.2

Example 26.3

A steel component of mass 8 kg initially at 230°C is quenched in a bath of oil of mass 20 kg, which is initially at 25°C. Neglecting external heat gain or loss, determine the final common temperature.

Solution

Heat lost by steel, $m.c.\Delta t = 8 \text{ kg} \times 0.48 \text{ kJ/kg.K } (230 - t)\text{K}$

Heat gained by oil, $m.c.\Delta t = 20 \text{ kg} \times 1.8 \text{ kJ/kg.K } (t - 25)\text{K}$

heat lost = heat gained,

$$\therefore 8 \times 0.48\,(230 - t) = 20 \times 1.8\,(t - 25)$$

Solving for t gives the final common temperature:

$$t = 44.8°\text{C}$$

It can be shown that by making the common final temperature the subject of the heat balance equation, the following expression can be applied directly

$$t = \frac{\Sigma(m.c.t)}{\Sigma(m.c)}$$

Problems can be solved very quickly by substitution of all given data into this expression. For example, the solution to the previous example will appear as a single line:

$$t = \frac{\Sigma(m.c.t)}{\Sigma(m.c)} = \frac{8 \times 0.48 \times 230 + 20 \times 1.8 \times 25}{8 \times 0.48 + 20 \times 1.8} = 44.8°\text{C}$$

This method applies to any number of substances, so long as they all reach a common final temperature and any external heat loss or gain is negligible.

Problems

Use values from Table 26.1 if required.

26.1 Determine the amount of heat required to raise the temperature of 3 kg of aluminium from 20°C to 150°C.

26.2 Determine the final temperature of a 5 kg aluminium casting cooled from 200°C by removing 728 kJ.

26.3 Determine the mass of ice that can be brought to the melting point at 0°C from an initial temperature of -30°C by 306 kJ of heat.

26.4 A copper vessel of mass 2.5 kg contains 7.5 kg of water. If the initial temperature of the vessel and water is 15°C, how much heat is required to increase the temperature to 75°C?

26.5 Determine the time taken to bring 1 L of water from 20°C to the boiling point at 100°C if the heater is rated at 1.5 kW and has an efficiency of 85 per cent.

26.6 In an industrial process, hot water is used to heat paraffin. Given the following information, determine the mass of water required per tonne of paraffin.
Temperature of water: initial 95°C
final 55°C
Temperature of paraffin: initial 15°C
final 50°C

26.7 Determine the final temperature of bath water made by mixing 120 L of hot water at 80°C and 140 L of cold water at 15°C.

26.8 One kilogram of lead shot at a temperature of 90°C is poured into 0.5 L of water contained in a copper vessel of mass 0.25 kg. If the initial temperature of the water and the container is 22°C and external heat losses are negligible, determine the final common temperature of the lead, water and container.

26.9 During an experiment, a piece of solid of mass 0.52 kg was heated to a temperature of 95°C and then immersed in 0.87 kg of water contained in a calorimeter, initially at 22°C. The final common temperature was observed to be 26.4°C.
Neglecting the mass of the calorimeter and any external heat loss, calculate the specific heat capacity of the solid.

26.10 A journal bearing supports a shaft of 120 mm diameter running at 620 revolutions per minute with a normal load of 25 kN and the coefficient of friction between the shaft and bearing surface of 0.05.
If the heat generated by friction is removed by a continuous flow of lubricating oil at the rate of 6.5 kg per minute, determine the temperature increase of the oil.

26.3 Latent heat

So far we have discussed only sensible heat, i.e. heat associated with a temperature change of a substance. However, after a sufficient amount of sensible heat has been added to a substance, saturation condition will be reached, i.e. a temperature at which change of phase may occur. At saturation temperature, the substance is ready to melt, if it is a solid, or to boil, if it is a liquid.

Saturation condition can also be reached from the other side of the temperature scale, i.e. by cooling or removal of sensible heat, until a gas is ready to condense or a liquid is about to freeze.

It is important to understand that, once saturation temperature is reached, phase change does not occur spontaneously, but requires a considerable amount of heat to be transferred into, or out of, the substance, and a sufficient time for it to do so. For example, it may take 5 minutes to bring a kettle of water to the boiling point, but it would take much longer to boil all the water away.

It is an observable fact that, if pressure remains constant, change of phase, such as boiling or condensation, occurs at constant temperature, i.e. the temperature of the substance does not change during the phase change. For example, while water is actually boiling at atmospheric pressure its temperature remains at 100°C. Likewise, the average temperature* in a bucketful of crushed ice, while melting, will remain at 0°C.

Therefore, unlike sensible heat, heat added or removed during a phase change is not accessible to our sense of touch and is not observable by reading a thermometer. For this reason, **heat associated with a phase change is called latent** heat**.**

It must be emphasised that the comparison between sensible and latent heat does not imply a heat energy of a different type at the source. The distinction lies purely in the response produced in the substance which receives, or loses, the heat. For example, consider an electric jug before and after the water starts boiling and ponder about the heat source and the effect produced on the water. Is the heat originating from the element any different before and after boiling begins?

Different substances require different amounts of latent heat per kilogram of mass. Furthermore, the same chemical substance requires different amounts for melting and for boiling. Latent heat required for melting is called **latent heat of fusion** and applies equally to melting and freezing. Latent heat required for boiling is called **latent heat of evaporation**, and is also applicable to the reverse process, i.e. condensation.

The total amount of latent heat, Q, involved in effecting a complete phase change of a given mass, m, is given by

$$Q = m.l$$

where l is the specific latent heat of the substance, which is the quantity of heat required to change a unit mass of the substance from one phase to another without change of temperature.

In this book we are only concerned with the water-substance at atmospheric pressure, for which:

specific latent heat of fusion = 335 kJ/kg

specific latent heat of evaporation† = 2257 kJ/kg

Example 26.4

Determine the total amount of heat required to convert 14 kg of water at 100°C to steam at 100°C under atmospheric pressure.

---

* A large lump of solid ice may have different, i.e. lower, temperatures within, while melting on the surface, due to the inability of its particles to move relative to each other and thus achieve uniform temperature. Liquid or gaseous substances are much more likely to be at a uniform temperature throughout.

**From Latin *latere* = to lie concealed, not apparent.

† Compare this value with the difference in enthalpy between steam and liquid water at 100°C in Table 25.5. What conclusion do you draw from this comparison?

Solution

$$Q = m.l = 14 \text{ kg} \times 2257 \text{ kJ/kg} = 31\ 598 \text{ kJ} = 31.6 \text{ MJ}$$

Example 26.5

How many 5 kg blocks of ice at 0°C can be made from water at 0°C by extracting 67 MJ of heat?

Solution

The mass of ice produced is:

$$m = \frac{Q}{l} = \frac{67\ 000 \text{ kJ}}{335 \text{ kJ/kg}} = 200 \text{ kg}$$

Therefore the number of 5 kg blocks is $\dfrac{200 \text{ kg}}{5 \text{ kg}} = 40.$

26.4 Heat account: Sensible and latent heat

In many engineering problems, such as generation of steam or ice making, total heat transfer associated with a sequence of sensible heating or cooling and phase change must be considered. The solution of such problems usually involves separate calculations for each stage in the process followed by the summation of all heat quantities. A heat-temperature graph is very helpful in recognising where one stage ends and another begins.

Example 26.6

Thirty-five kilograms of steam are generated at atmospheric pressure from feed-water, which is initially at 25°C. If the final temperature of steam is 200°C, what is the total amount of heat required? (Average specific heat of steam at atmospheric pressure may be taken as 2 kJ/kg.K.)

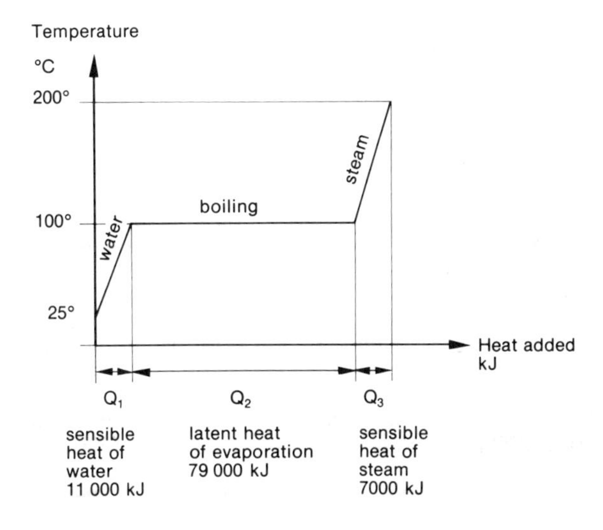

Fig. 26.3

Solution

Figure 26.3 is a diagram showing the three stages of the process. These are:

1. Heating of the water from 25°C to the boiling point at 100°C, during which sensible heat is added.

   $Q_1 = m.c.\Delta t = 35\ \text{kg} \times 4.19\ \text{kJ/kg.K}\ (100 - 25)\text{K} = 11\,000\ \text{kJ}$

2. Conversion of water into steam at 100°C, involving latent heat.

   $Q_2 = m.l = 35\ \text{kg} \times 2257\ \text{kJ/kg} = 79\ 000\ \text{kJ}$

3. Heating the steam from 100°C to its final temperature of 200°C, for which a further amount of sensible heat is necessary.

   $Q_3 = m.c.\Delta t = 35\ \text{kg} \times 2\ \text{kJ/kg.K}\ (200 - 100)\text{K} = 7000\ \text{kJ}$

The total amount of heat required is the sum of heat transferred during each of the three processes.

$Q = Q_1 + Q_2 + Q_3 = 11\,000 + 79\,000 + 7000 = 97\,000\ \text{kJ} = 97\ \text{MJ}$

Problems

Where required use specific heat of ice, water and steam as 2.04, 4.19 and 2.0 kJ/kg.K respectively.

26.11 Determine the amount of heat required to convert 5 kg of water at 100°C to steam at 100°C.

26.12 Determine the amount of heat required to melt one tonne of ice at 0°C.

26.13 An electric jug contains 1.2 kg of water and the net heat transfer into the water is at the rate of 1.3 kJ/s. If the thermostat fails to switch the element off, calculate the percentage of the initial quantity of water that will be boiled away in 15 minutes after the beginning of the boiling process.

26.14 Steam at atmospheric pressure is used in a calorifier (heat exchanger) to heat 200 L of water from 15°C to 85°C. If only the latent heat of steam is used, how much steam must condense to provide the necessary amount of heat?

26.15 Determine the amount of heat required to convert 5 kg of ice, initially at −15°C, to water at 20°C.

26.16 Steam is generated at atmospheric pressure at the rate of 5400 kg/h. If the temperature of feed-water is 40°C and the temperature of steam is 120°C, determine the rate of heat transfer in the boiler in kilojoules per second.

26.17 Ice at −20°C is manufactured from water at 25°C at the rate of 500 blocks of 10 kg each per hour. Determine the required refrigeration capacity, i.e. heat-removing capacity, of the ice-making plant in kilowatts.

26.18 Determine the final temperature of the mixture after 5 kg of ice at 0°C are mixed with 14 kg of water at 53°C.

26.5 Fuels and combustion

Fuels are combustible substances from which most of the heat energy used in industry is obtained. Combustion of a fuel is a rapid chemical reaction during which the component substances of the fuel combine with oxygen from the air. Combustion is usually accompanied by the generation of light in the form of flame and liberation of a considerable amount of energy in the form of heat.

The mechanism of combustion

The chemical and physical aspects of combustion were not properly understood until some pioneering discoveries in 1772 by the French chemist Antoine-Laurent Lavoisier, who was the first to recognise that combustion was a reaction between the burning substance and a gas present in a limited quantity in the atmosphere. Lavoisier called this gas "oxygen". Later, English scientist Sir Humphry Davy performed a number of important experiments, including measurements of flame temperatures. However, it was Benjamin Thompson's experiments with heat and the evolution of the molecular theory of matter that finally explained the energy aspects of combustion.

The mechanism of combustion is a complex process, involving a number of steps. It is usually initiated by external factors such as heat, friction or sparks, in the presence of flammable material or vapours given off by the fuel under certain conditions of temperature and access to air. The ignition temperature, or flash point, depends on the volatility of a particular substance and on the pressure of the combustible mixture. The structure and propagation of the flame depends on the specific condition of movement of the fuel and air mass and of the energy. For example, when a wax candle burns, the heat produced by the combustion maintains of reservoir of molten wax, which is then drawn by capillary action up the wick, where sufficient surface area accessible to air is available for continuous combustion. Formation of soot is a feature of most flames, which makes the flame luminous and non-transparent.

The main engineering uses of combustion fall into two categories, i.e. heating devices and internal combustion engines. The former include steam generating boilers, metallurgical furnaces and a large variety of industrial and commercial heating processes, air conditioning, etc. The latter comprise diesel and petrol engines, gas turbines and jet engines. In addition to the main uses, combustion is also employed in rocket propulsion and in explosives.

Fuels

The vast majority of fuels are substances found to consist of some combination of carbon and hydrogen, called hydrocarbons, from which most of the fuel's energy originates. Sulphur and a number of other elements are often found in natural fuels in relatively small quantities. Their contribution to the energy of the fuel is usually small, but some side effects, such as air pollution, can be quite significant. The majority of the fuel sources, such as coal and petroleum, are natural fossil deposits, from which a variety of manufactured fuel products are developed.

The desirable characteristics of a fuel would include high release of heat energy during combustion, low cost, low content of constituents which produce harmful combustion products, ease and safety in storage, handling and transportation. The combustion should be readily initiated and easy to control.

Fuels can be divided into three classes: solid, liquid and gaseous.

Solid fuels include wood, coal, coke and charcoal. Wood has ceased to be a commercially important fuel. An increase in the carbon content characterises the change from wood (50 per cent carbon) to coal (80 per cent). Anthracite or stone coal, is about 95 per cent carbon. It burns with intense heat, without smoke, and with little flame. Coal is widely used in electric-power-generating stations. Coke and charcoal are artificial fuels, made by a process known as carbonisation from coal and wood respectively. Coke is used most extensively in the metallurgical industries.

The majority of **liquid fuels** are produced from petroleum. These include petrol, kerosene and various fuel oils. Other liquid fuels can also be produced by a distillation process from shale, coal and vegetable matter containing cellulose. Liquid fuels are essential in many types of engines and are also used for heating purposes and for steam generation, particularly in small packaged boilers.

Gaseous fuels are either natural or manufactured. The natural gas is a mixture of hydrocarbons, of which the chief is methane. The other gaseous fuels, such as town gas, are produced artificially from solid fuels, e.g. coal. Combustion of gaseous fuels is relatively clean, efficient and permits easy control of the flame.

Combustion control

The main problem of combustion control consists of air and fuel supply to the furnace or combustion chamber, at a proper rate and with satisfactory mixing of fuel and air quantities. The air-fuel ratio can be defined as the ratio of the quantity of air supplied per unit of fuel used. For solid and liquid fuels this ratio is usually expressed in units of mass, i.e. in kilograms of air per kilogram of fuel. For gaseous fuels the volume ratio at normal atmospheric pressure and temperature is preferred.

It is possible in each case to calculate the exact theoretical air-fuel ratio required for complete combustion. An average solid or liquid fuel contains about 85 per cent carbon and 15 per cent hydrogen by mass, and simple chemistry determines the correct theoretical air-fuel ratio to be about 15:1 by mass. For gaseous fuels, the ratio depends on the varying composition of the fuel. However, in practice it is very difficult to ensure perfect mixing between the air and the fuel, so that every particle of fuel finds a molecule of oxygen to combine with. It is usual to introduce some excess air quantity to ensure complete combustion. If the air-fuel ratio is insufficient, energy potential is wasted because of unburnt fuel, and harmful combustion products, such as carbon monoxide, may be produced. On the other hand, if too much excess air is supplied, energy is wasted in heating excess air which performs no useful function, thus reducing useful energy release per unit of fuel. In either case, combustion of incorrect air-fuel mixtures is often difficult to initiate and to maintain.

For good combustion of any fuel, the combustible elements must be surrounded with oxygen necessary for complete combustion. This is relatively easy to achieve for gaseous fuels. In the case of liquid and solid fuels, it is desirable to break the fuel up into small droplets or particles, thus increasing the surface contact area between the fuel and the oxygen required for combustion. The process of splitting up a liquid fuel into a fine spray, in order to improve its rate of vaporisation and its distribution within the air stream is called atomisation. Many types of fuel injectors are fitted to internal combustion engines and atomising oil burners form part of the equipment of oil-fired boilers and furnaces.

It is also possible to burn coal and low-grade solid fuels efficiently if they are pulverised. During the process of pulverisation, the fuel is subjected first to crushing, and then reduced to a very fine powder by pulverisers, which are usually air-swept, the velocity of the air being so regulated that fine particles of fuel when sufficiently reduced are carried away from the mill. By forcing the fuel into the combustion chamber in powdered form, more thorough mixing with the air is achieved, giving better efficiency of combustion. It is claimed that pulverised coal firing is as flexible as oil firing and that the same burners can be used for both.

The mixing of air with petrol, to form an explosive mixture in the correct proportion for use in the cylinders of internal combustion (petrol) engines, is called carburation, and the device for achieving this is called a carburettor. It consists essentially of a jet, or jets, discharging the fuel into the air stream under the pressure difference created by the velocity of the air as it flows through a nozzle-shaped constriction. As petrol is drawn through the jet, it mixes with the air stream, evaporating into it and producing an explosive mixture which is fed to the engine. Various methods adopted to achieve and control the required mixture strength to correspond to changing loads and engine speeds make the carburettor a delicate and sensitive instrument.*

26.6 Calorific value

Combustion is an exothermic** chemical reaction, i.e. a process during which energy is liberated in the form of heat. The heating value of fuels is an item of great interest to the engineer.

The amount of thermal energy, i.e. heat, released per unit mass of fuel is called the **calorific value** of that fuel.

* Fuel technology, carburation, furnace and burning equipment is a broad area involving chemistry, metallurgy and mechanical engineering. The scope and nature of this book do not permit more than a sketchy outline as presented above. Students seeking further knowledge in this field should refer to other texts.

** From the Greek *exo* = outward and *therme* = heat.

Table 26.2 Typical calorific values of some fuels

Fuel	*Approximate calorific value* kJ/kg	kJ/m^3
Wood	16 000	—
Coke	30 800	—
Coal	34 000	—
Fuel oil	45 400	—
Petrol	46 800	—
Kerosene	46 400	—
Methane	—	41 500
Natural gas	—	37 500
Town gas	—	20 000

For a given mass of fuel, m_f, the amount of heat released, Q, is given by

$$Q = m_f . C_f$$

where C_f is calorific value.

Calorific values of gaseous fuels are often expressed per unit volume, instead of unit mass, at standard atmospheric conditions, in which case:

$$Q = V_f . C_f$$

where V_f is volume of fuel burned.

Gross calorific values of some common fuels are listed in Table 26.2.

Example 26.7

It was calculated in Example 26.6 that 97 MJ of heat are required to convert 35 kg of water at 25°C into steam at 200°C at atmospheric pressure.

If fuel oil is used and the efficiency of heat transfer is 75 per cent, how much fuel is necessary?

Solution

Amount of heat required from the fuel, to allow for heat losses is:

$$Q = \frac{97 \text{ MJ}}{0.75} = 129.3 \text{ MJ}$$

Now using $Q = m_f . C_f$, where $C_f = 45\ 400$ kJ/kg from Table 26.2,

$$129\ 300 \text{ kJ} = m_f \times 45\ 400 \text{ kJ/kg}$$

The mass of fuel oil required is 2.85 kg.

If the carbon-hydrogen composition of a hydrocarbon fuel is known, its calorific value can be estimated. The method is illustrated by the following example, given that carbon has a calorific value of 33.7 MJ/kg and an atomic mass of 12 units, and hydrogen has a calorific value of 144 MJ/kg and an atomic mass of 1 unit.

Example 26.8

Determine the calorific value of octane (chemical formula C_8H_{18}).

Solution

First convert the chemical formula to composition by mass,

8 atoms of carbon at 12 units each	=	96 units
18 atoms of hydrogen at 1 unit each	=	18 units
Total mass of octane	=	114 units

As part of each kilogram,

carbon accounts for $\frac{96}{114} = 0.842$ kg

and hydrogen accounts for $\frac{18}{114} = 0.158$ kg

Therefore, heat liberated from each constituent is:

carbon, 33.7 MJ/kg × 0.842 kg = 28.38 MJ

hydrogen, 144 MJ/kg × 0.158 kg = 22.75 MJ

Hence, calorific value of octane is the sum of the two parts, or 28.38 MJ + 22.75 MJ = 51.13 MJ per kg.

Problems

Use values from Table 26.2 if required.

26.19 Determine the amount of heat liberated by burning 50 kg of coal.

26.20 Determine the rate in kW at which heat is liberated if town gas is used at the rate of 4 litres per minute.

26.21 In a test to measure calorific value of a sample of fuel, 1.2 g of fuel was burned and the heat transferred to 3.36 kg of water in a calorimeter. The temperature rise of the water was observed to be 3.2 degrees Celsius. What was the calorific value of the fuel?

26.22 Natural gas heaters are used to heat an assembly hall 20 m long by 10 m wide. It is estimated that 0.15 kW is required per square metre of floor area. Determine the cost of heating per hour, if the gas costs 40 cents per cubic metre.

26.23 A hot water boiler requires 1500 MJ of heat per hour. If an oil of relative density 0.85 is used and the heat transfer efficiency is 70 per cent, determine the rate in litres per hour at which the boiler burns fuel.

26.24 The chemical formula of benzene is C_6H_6. Convert to percentage carbon-hydrogen composition by mass and hence determine its calorific value per kilogram.

26.25 Hexahydroxylene, a hydrocarbon present in petrol, consists of 85.7 per cent carbon and 14.3 per cent hydrogen. Determine its calorific value.

26.26 A sample of coal is found to consist of 76.5 per cent carbon, 13.5 per cent hydrogen and 10 per cent incombustible elements. Determine the calorific value per kilogram of coal.

26.27 Methane gas has density of 0.677 kg/m^3 and its chemical formula is CH_4. Determine its calorific value per cubic metre and compare with the value in Table 26.2.

Review questions

1. What is meant by the term calorimetry?
2. What is sensible heat?
3. Define specific heat capacity of a substance.
4. State the formula for calculating sensible heat.
5. Explain the heat balance principle.
6. What is latent heat?
7. Define:
   (a) specific latent heat of fusion,
   (b) specific latent heat of evaporation.
8. Briefly describe the mechanism of combustion.
9. List some common fuels and discuss their desirable characteristics.
10. From other textbooks or industrial catalogues find suitable diagrams and descriptive material relating to the operation of:
    (a) gas and liquid fuel burners,
    (b) stokers and pulverisers for solid fuels,
    (c) carburettors for petrol engines.
11. Define calorific value.

27

Laws of gases

In the field of engineering, wide use is made of gaseous substances such as air, oxygen, nitrogen and combustible mixtures of natural or manufactured gases. The compressibility and expansion properties of air are utilised extensively in pneumatic tools and control of machinery. Explosive mixtures of air and fuel are used in internal-combustion engines, guns and turbines. Gases are also employed in industrial processes, such as welding, for deep-sea diving and as raw materials in the chemical industry.

It is, therefore, necessary to describe and understand the behaviour of gases when they are heated, cooled, expanded or compressed, in terms of their thermodynamic properties.

Fig. 27.1 *Air compressor*

27.1 Air and other gases

Air

The Earth's atmosphere consists of a mixture of gases, which is called air. Air consists of 75.5 per cent nitrogen, 23.2 per cent oxygen and just under 1.3 per cent argon by mass. The rest, which is a small fraction of a per cent, includes carbon dioxide, neon, helium, methane, sulphur dioxide, hydrogen and minor constituents such as krypton and ozone.

The apparent molecular mass of air, which is the average molecular mass of all components, is 28.96 molecular units, based on the carbon-12 scale. For most practical purposes a figure of 29 units is sufficiently accurate.

The Earth's weather is largely the result of changes in the air temperature, which lead to pressure changes and to the amount of water vapour the air can absorb. Cloud formation, winds and rain can be explained in terms of the thermodynamic properties of air. Some of the engineering applications of air will form examples and problems in this and the following chapter.

Diatomic gases

Diatomic gases have molecules consisting of two identical atoms joined together. These include hydrogen, H_2, nitrogen, N_2, and oxygen, O_2. Their molecular masses are 2, 28 and 32 units respectively and each atom has a mass equal to one-half of the molecular mass.

Oxygen and nitrogen are produced by liquefaction of air followed by fractional distillation into its two major components at extremely low temperatures. Nitrogen, with the lower boiling point, evaporates out, leaving oxygen in the liquid form. Hydrogen is separated from natural gas, coke or coal, by chemical reactions with steam or oxygen at high temperatures.

Oxygen, the most important of industrial gases, is used in large quantities in the steel industry and in other primary metal processing. The second largest demand for oxygen is for the manufacture of plastics and other industrial chemicals and for oxygen-acetylene welding and cutting.

The principal use of nitrogen is in chemical processing and the petroleum industry. Large amounts of pure nitrogen are used where an inert gas is required to prevent oxidation, fire or explosion. Nitrogen is also used for food processing and storage, e.g. controlled atmosphere storage of apples.

Hydrogen is the lightest element in nature and an industrial gas with one of the fastest growing demands in the world. The greatest quantities of hydrogen are used in the production of ammonia and various organic compounds and in petroleum refining processes. Hydrogen is also used as a propellant in space rocket flight.

Compound gases

Many gaseous substances are chemical compounds, i.e. their molecules consist of two or more dissimilar atoms joined together. The more common of these include methane, CH_4, carbon monoxide, CO, carbon dioxide, CO_2, and sulphur dioxide, SO_2. Their molecular masses can be determined by simple summation of the atomic masses of their constituent parts. (See also Table 25.4, Chap. 25.)

Methane is the major component of the combustible mixture of gases known as natural gas, which occurs in reservoirs beneath the surface of the Earth. It is estimated that the largest reserves of natural gas exist in the Soviet Union, which supplies eastern Europe, and the next largest in the United States. Important quantities are also found in the North Sea off the United Kingdom and in the Middle East. The deposits found in Australia are presently being developed at a rapidly expanding rate. The major use of natural gas is as fuel for industrial and domestic consumption and as raw material for the manufacture of methanol, acetylene and hydrogen. In the Soviet Union, a considerable quantity of natural gas is used for electric power generation.

Carbon monoxide and carbon dioxide are compounds of carbon and oxygen. The former is used as a raw material in chemical plant and the latter for the carbonation of beverages, which requires a gas of high degree of purity so that their taste or odour will not be affected. Dry ice, or frozen carbon dioxide, is widely used as a refrigerant for ice cream and other frozen foods, especially during their transport.

Sulphur dioxide is produced by the combustion of sulphur-bearing minerals in specially designed burners. It is used as an intermediate product in the manufacture of sulphuric acid and for pulping and bleaching wood in the paper-making industry. The combustion of fuels containing sulphur produces sulphur dioxide and related products, a major source of air pollution.

The noble gases

The rare, or noble, gases, such as argon, neon and helium, are all found in the atmosphere. All of the rare gases are colourless, odourless and tasteless and do not readily react chemically with other materials. They are produced as a by-product of the liquefaction of air. Helium is also extracted from natural gas.

Helium has the lowest boiling point of all elements, only 5 kelvins above absolute zero. For this reason it finds many applications as a cryogenic* fluid where the lowest possible temperatures are required. Argon is very stable and inert and is in demand for shielded arc welding of materials, such as stainless steel, in which the welds must be protected from atmospheric oxygen while being formed. Neon is still used in the familiar neon signs, but also finds application in complex scientific and electronic apparatus.

27.2 The simple gas laws

Let us consider a gas to be contained in the cylinder of an engine or a compressor. Given the identity of the gas and its mass enclosed within the

* From the Greek *kruos* = frost and *gen* (root of *genesis*) = becoming.

cylinder, there are three variable quantities, namely volume of the gas and its temperature and pressure, which can change from one condition to another. The relationships between these three measurable thermodynamic properties can be expressed in the form of statements called the gas laws.

Boyle's Law

The first gas law, discovered by the English physicist Robert Boyle (1627–91), and independently of him by a Frenchman, Edme Mariotte (1620–84),* states that **during a change of state of any gas in which the mass and the temperature remain constant, the product of the absolute pressure and the volume remains constant**. In symbols, $pV =$ constant, if m and T are kept constant. That is, if a gas is compressed to half its volume the pressure will be twice as much, or, if the pressure of the gas is doubled, the volume will be reduced by one-half, provided that the temperature does not change, for if it does, the law does not apply.

Charles' Law

The second gas law was worked out a century after Boyle's discovery of the first law** and is attributed to the French physicist Jacques Charles (1746–1823) and named after him.

Charles' law states that **during a change of state of any gas in which the mass and the pressure remain constant, the volume is proportional to the absolute temperature**. In symbols, $\frac{V}{T} =$ constant, if m and p are kept constant. That is, if the absolute temperature of a gas is increased, the volume will increase proportionately, as long as the pressure does not change.

Combined Boyle-Charles' Law

So far we know that, for a given mass of gas under certain conditions, volume is inversely proportional to absolute pressure (Boyle's Law) and directly proportional to absolute temperature (Charles' Law).

These two laws can be combined into a single law which states that **during a change of state of any gas in which the mass remains constant, the volume is directly proportional to the absolute temperature and inversely proportional to the absolute pressure**. In symbols, $\frac{pV}{T} =$ constant, if m is constant.

This is a very useful relationship for calculating the unknown value of one of the three variables after a change of state, if the other two are known. We should think in terms of the following scheme:

* In France the law is named after Mariotte, and in some other countries is referred to as the Boyle-Mariotte Law.

** The reason it took so long to discover Charles' Law after Boyle's Law was known was the need to introduce a difficult concept of absolute temperature, coupled with the practical difficulties of measuring temperature accurately over a sufficiently wide range.

State 1	PROCESS	State 2
initial condition of a gas	$\longrightarrow$	final condition of a gas

$$\frac{p_1 V_1}{T_1} = \text{constant} = \frac{p_2 V_2}{T_2}$$

where p_1 and p_2 are initial and final pressures
V_1 and V_2 are initial and final volumes
T_1 and T_2 are initial and final temperatures.

Example 27.1

Air initially at 101.3 kPa abs. and 15°C occupies a volume of 0.02 m^3. The air is compressed to a new volume of 0.005 m^3 and a pressure of 550 kPa abs. What is the final temperature of the air?

Solution

Compare the initial and final states:

$$\left.\begin{aligned} p_1 &= 101.3 \text{ kPa} \\ V_1 &= 0.02 \text{ m}^3 \\ T_1 &= 15 + 273 = 288\text{K} \end{aligned}\right\} \xrightarrow{\textbf{Compression process}} \left\{\begin{aligned} p_2 &= 550 \text{ kPa} \\ V_2 &= 0.005 \text{ m}^3 \\ T_2 &= ? \end{aligned}\right.$$

$$\frac{p_1 V_1}{T_1} = \text{const.} = \frac{p_2 V_2}{T_2}$$

Substitute:

$$\frac{101.3 \times 0.02}{288} = \frac{550 \times 0.005}{T_2}$$

Solving for T_2:

$$T_2 = 288 \times \frac{550 \times 0.005}{101.3 \times 0.02}$$

$$= 390.9\text{K}$$

Final temperature, $t = 390.9 - 273 = 117.9°\text{C}$.

The simple gas laws are useful for solving a large variety of practical problems in industry and science which involve gases. However, it should be noted that these laws are not equally applicable to all gases under all conditions. It is difficult to specify exactly under what conditions the amount of deviation in the behaviour of a particular real gas, from that predicted by the laws,* becomes unacceptable. In general, the deviation increases gradually as the temperature decreases or as the pressure increases, temperature being the more significant factor. As a simple rule, it can be said that the laws are sufficiently accurate for most practical purposes if the temperature is above 0°C. However, if sub-zero temperatures are encountered, at a level approaching liquefaction

* An imaginary ideal substance which obeys Boyle's and Charles' laws exactly is called a perfect gas. Needless to say, a perfect gas does not exist, except as a mathematical model.

of a gas, considerable error might result if the laws were applied indiscriminately.

27.3 The equation of state

We have now seen how changes of state for any fixed mass of gas are related by the combined Boyle-Charles' law, i.e. how the volume of a gas is connected to its pressure and temperature by the equation:

$$\frac{p_1 V_1}{T_1} = \text{constant} = \frac{p_2 V_2}{T_2}$$

Sooner or later it becomes necessary to relate these thermodynamic properties to the mass of a particular gas in a particular thermodynamic state. For example, we may be interested in the actual mass of air involved in Example 27.1 above.

The answer to this question lies in examining the nature of the constant in the equation:

$$\frac{pV}{T} = \text{constant}$$

It may be suspected that this constant depends on the two factors which, so far, have been mentioned, but not included as variables in the equation, namely the mass and the identity of the particular gas.

The results of numerous experiments involving a large variety of different gases at different conditions of temperature and pressure have shown that the constant is always equal to $8.314 \frac{m}{M}$ kilojoules per kelvin,

where m is the actual mass of gas in kilograms,
M is the molecular mass in molecular units.

It is common practice to combine 8.314 with the molecular mass of a particular gas into what is known as the characteristic gas constant R.

$$R = \frac{8.314}{M}$$

Using the values of molecular masses for various gases from Table 25.4, the corresponding values of the characteristic gas constants can be calculated as shown in Table 27.1.

The unit used here for the characteristic gas constants is kJ/kg.K, which is consistent with the use of kilopascals for pressure and kelvins for temperature.

If we now substitute the new constant into the combined Boyle-Charles' Law, we obtain the following expression:

$$\frac{pV}{T} = \text{constant} = 8.314 \frac{m}{M} = mR$$

Table 27.1 Characteristic gas constants R

Gas	Molecular mass M	Characteristic gas constant $R = 8.314 \div M$
Hydrogen	2	4.157
Helium	4	2.079
Methane	16	0.520
Nitrogen	28	0.297
Carbon monoxide	28	0.297
Air	29*	0.287
Oxygen	32	0.260
Argon	40	0.208
Carbon dioxide	44	0.189
Sulphur dioxide	64	0.130

* *Average*

This relationship is known as the **equation of state** and is usually written in the form:

$$pV = mRT$$

where, for any given state of a particular gas,

p is absolute pressure in kilopascals,
V is volume in cubic metres,
m is mass in kilograms,
R is characteristic gas constant in kJ/kg.K,
T is absolute temperature in kelvins.

Unlike the gas laws, this equation does not refer to a change in state of a gas, but rather to each of its separate states, i.e. the initial and the final state. It is, therefore, helpful to visualise the following scheme, when solving problems concerning a fixed mass of a particular gas undergoing a process.

State 1 initial condition	Process	*State 2* final condition
p_1, V_1, T_1	$\longrightarrow$	p_2, V_2, T_2
$p_1V_1 = mRT_1$	$\dfrac{p_1V_1}{T_1} = \dfrac{p_2V_2}{T_2}$	$p_2V_2 = mRT_2$

Example 27.2

A cylinder of 200 L capacity contains compressed oxygen at 2.5 MPa and 15°C.

Determine the mass of oxygen in the cylinder and the new pressure when the temperature increases to 40°C.

Solution

At the initial conditions:

$$p_1 = 2500 \text{ kPa}$$
$$V_1 = 0.2 \text{ m}^3$$
$$T_1 = 15 + 273 = 288\text{K}$$

Therefore, using $pV = mRT$, where $R = 0.26$ kJ/kg.K for oxygen:

$$m = \frac{pV}{RT} = \frac{2500 \times 0.2}{0.26 \times 288} = 6.677 \text{ kg}$$

At the final conditions:

$$V_2 = 0.2 \text{ m}^3$$
$$T_2 = 40 + 273 = 313\text{K}$$

Therefore, from $\dfrac{p_1 V_1}{T_1} = \dfrac{p_2 V_2}{T_2}$

$$\frac{2500 \times 0.2}{288} = \frac{p_2 \times 0.2}{313}$$

Hence, $p_2 = 2717$ kPa or 2.72 MPa.

Alternatively, once the mass of the gas is known, $p_2 V_2 = mRT_2$ could be used to calculate p_2.

The equation of state is particularly useful when mass does not remain constant, e.g. when gas is taken from, or added into, a container.

Example 27.3

If 5 kg of oxygen have been used from the cylinder in Example 27.2 and the final temperature is 20°C, what is the final pressure?

Solution

The mass left in the cylinder is equal to the initial mass minus the mass removed:

$$m = 6.677 - 5 = 1.677 \text{ kg}$$

The final conditions are

$$V = 0.2 \text{ m}^3 \text{ and } T = 20 + 273 = 293\text{K}.$$

Therefore, using $pV = mRT$, final pressure is:

$$p = \frac{mRT}{V} = \frac{1.677 \times 0.26 \times 293}{0.2}$$
$$= 638.8 \text{ kPa}.$$

Problems

Use values from Table 27.1 if required.

27.1 A quantity of carbon dioxide has an initial pressure of 170 kPa absolute and occupies 120 L. It is then

compressed to a pressure of 850 kPa absolute while the temperature remains constant.
What is the final volume of the gas?

27.2 A quantity of nitrogen has an initial volume of 0.05 m^3 and a temperature of 20°C. It is heated and expanded at constant pressure to a volume of 0.09 m^3. Calculate the final temperature of the nitrogen.

27.3 A quantity of air in a closed cylinder has an initial pressure of 101.3 kPa absolute and a temperature of 15°C. The air is heated within the cylinder, at constant volume, until its temperature is 300°C. Determine the final absolute pressure of the air.

27.4 A litre of air has an initial pressure of 1.5 MPa absolute and a temperature of 180°C. It is expanded to a final pressure of 200 kPa absolute and a final temperature of 50°C. Calculate the final volume.

27.5 Neon and krypton are noble gases with molecular mass of 20 and 84 units respectively. Determine their characteristic gas constants.

27.6 Calculate the volume occupied by one kilogram of argon at 15°C and 101.3 kPa absolute.

27.7 A certain gas has a density of 1.86 kg/m^3 at 101.3 kPa absolute and 15°C. If the gas is one of those listed in Table 27.1, determine which gas it is.

27.8 Compressed air is stored in a receiver of 500 L capacity at a pressure of 900 kPa absolute and temperature 20°C. Determine:
(a) mass of air in the receiver,
(b) the gauge pressure reading if the temperature rises to 40°C.

27.9 A storage tank of 12 m^3 capacity contains methane at 2 MPa absolute and 20°C. The tank is used to refill containers each of which takes 0.4 kg of gas from the tank. Determine the number of refills that can be made before the storage tank pressure drops to 0.5 MPa absolute, while the temperature remains constant.

27.10 Two containers of 50 and 100 litre capacity are filled with air to a pressure of 500 kPa and 900 kPa absolute, respectively. After a valve connecting the two containers is opened, the pressure equalises. Assuming that the temperature remains at 20°C throughout, determine the final pressure.

27.4 Specific heat capacity of air

Let us now consider two of the many possible ways in which a gas can be heated, i.e. heating at constant volume and heating at constant pressure. To simplify our discussion at this stage we will take the gas to be air. However, the general conclusions reached in this section are applicable to other gases, provided that appropriate numerical values of specific heat capacities are used for each particular gas.

Constant volume heating of air

Let us now consider a closed rigid container filled with air to which an amount of heat, Q, is added.

Due to the inability of the air to expand, there will be no work done and the non-flow energy equation, $\Delta U = Q - W$, reduces to $\Delta U = Q$, as has been discussed in Section 25.8.

The observable effect of the heat transfer is an increase in the temperature of the air in the container. We know that the heat transfer which is accompanied by a temperature rise is called sensible heat, and is given by $Q = mc\Delta t$, where c is the specific heat capacity of the substance.

Combining the two equations shows that for air heated under constant volume conditions, i.e. inside a closed rigid container, the change of internal energy is equal to the sensible heat, as follows:

$$\boxed{\Delta U = m.c_v.\Delta t}$$

where c_v is the specific heat capacity of air at constant volume.

Example 27.4

Using the specific internal energy table for air, Table 25.4, determine the amount of heat required to heat 0.5 kg of air from 25°C to 125°C inside a closed rigid container and, hence, the specific heat capacity of air at constant volume.

Solution

For a closed system at constant volume, $\Delta U = Q$. Therefore,

$$Q = \Delta U = m(u_2 - u_1) = 0.5 \text{ kg } (89.9 - 17.9) \text{ kJ/kg} = 36 \text{ kJ}$$

This is sensible heat represented by $Q = m.c_v.\Delta t$.

Substitute:

$$36 \text{ kJ} = 0.5 \text{ kg} \times c_v \times (125 - 25)\text{K}$$

Hence, specific heat capacity of air at constant volume is:

$$c_v = \frac{36 \text{ kJ}}{0.5 \text{ kg} \times (125 - 25)\text{K}} = 0.72 \text{ kJ/kg.K}$$

By repeating this example, using different values of mass and temperature, it can be demonstrated that the value of specific heat capacity of air, under constant volume conditions, is not exactly constant. It varies between 0.716 kJ/kg.K at normal atmospheric temperatures and 0.768 kJ/kg.K at 350°C, increasing further at higher temperatures.

For many practical purposes, a standard value of 0.718 kJ/kg.K is often used and is sufficiently accurate when taken to be constant. However, if temperatures in excess of 200°C are encountered, the error involved may become significant for some applications.

In this chapter we will use:

$$c_v = 0.718 \text{ kJ/kg.K}$$

where c_v is specific heat capacity of air at constant volume.

Constant pressure heating of air

In Section 25.8, constant pressure heating of a substance was discussed. It may occur as a flow process inside a duct or heat exchanger, or as a non-flow process when expansion of a gas is allowed within a cylinder with a piston loaded by a constant weight.

In either case, it has been shown that when heat is added to a substance under constant pressure conditions, the amount of heat added is equal to the change in enthalpy produced, i.e. $\Delta H = Q$.

At the same time, the corresponding rise in temperature can also be related to the amount of heat added by $Q = m.c.\Delta t$.

If we designate the specific heat capacity of air at constant pressure by c_p and combine the two equations, it can be seen that, for air heated under constant pressure conditions, the change in enthalpy of air is equal to the sensible heat, as shown by the following equation:

$$\Delta H = m.c_p.\Delta t$$

Example 27.5

Using the specific enthalpy table for air, Table 25.5, determine the amount of heat required to heat 0.5 kg of air from 25°C to 125°C in a heat exchanger or in a cylinder under constant load and, hence, the specific heat capacity of air at constant pressure.

Solution

For air at constant pressure, $\Delta H = Q$.

Therefore:

$$Q = \Delta H = m(h_2 - h_1) = 0.5 \text{ kg } (204.1 - 103.4) \text{ kJ/kg}$$
$$= 50.35 \text{ kJ}$$

This is sensible heat represented by $Q = m.c_p.\Delta t$.

Substitute:

$$50.35 \text{ kJ} = 0.5 \text{ kg} \times c_p \times (125 - 25)\text{K}$$

Hence, specific heat capacity of air at constant pressure is:

$$c_p = \frac{50.35 \text{ kJ}}{0.5 \text{ kg } (125 - 25)\text{K}}$$
$$= 1.007 \text{ kJ/kg.K}$$

This example can also be repeated to show that the value of specific heat capacity of air, under constant pressure conditions, is not constant, but varies between 1.0 kJ/kg.K at normal atmospheric conditions and 1.056 kJ/kg.K at 350°C, increasing further at higher temperatures.

It is common practice to use a standard constant value of 1.005 kJ/kg.K for most practical purposes at moderate temperatures. However, if the temperatures exceed 200°C, an error, which increases with the increase in temperature, can become significant.

For this chapter we will use:

$$c_p = 1.005 \text{ kJ/kg.K}$$

where c_p is the specific heat capacity of air at constant pressure.

Joule's Law

Since changes in internal energy and enthalpy of gases are closely related to energy exchange in closed and open systems, it is often desirable to be able to estimate the internal energy term ΔU, or the enthalpy term ΔH, without reference to the tables of specific internal energy or specific enthalpy.

An experiment carried out by Joule showed that the internal energy of a gas is a function of temperature only and is independent of changes in pressure and volume.* Translated into the language of systems and processes, Joule's law means that the change of internal energy of a gas in a system is *always* equal to $\Delta U = m.c_v.\Delta t$, irrespective of whether the process occurs at constant volume, or under any other conditions, so long as there is a temperature change, Δt, suffered by the gas.

If we now turn our attention to enthalpy, which by definition is $H = U + pV$, and remember that for a gas $pV = mRT$, substitution yields $H = U + mRT$. Each of the two terms can be seen as a function of temperature only: U according to Joule's law and mRT because T is the only variable for a given quantity of a given gas. We can conclude, therefore, that, for a gas, the enthalpy is a function of temperature only, independent of changes in pressure and volume. Again, in terms of systems and processes, it means that the change of enthalpy of a gas in a system is *always* equal to $\Delta H = m.c_p.\Delta t$, irrespective of the type of process which occurs within the system, so long as a change of temperature, Δt, has taken place.

The foregoing conclusions should not be confused with the formula for heat received by the gas during a process, $Q = m.c.\Delta t$, in which the value of specific heat must correspond to the conditions under which heat is being transferred, i.e. c_v for constant volume heating and c_p for constant pressure heating.

---

* Later experiments by Joule and Kelvin showed that Joule's original conclusion was not completely accurate. However, like Boyle's and Charles' laws it is sufficiently accurate for most practical purposes at moderate temperatures and pressures.

Consider the following examples.

Example 27.6

0.03 kg of air is compressed in a cylinder from an initial condition of 100 kPa absolute and 25°C to a final condition of 600 kPa absolute and 175°C. The work done on the air during the process is 4.38 kJ while 1.13 kJ of heat are lost by the air.

Calculate the change in internal energy of the air and the change of its enthalpy. How do these quantities compare to heat and work transfer?

Solution

(a) The change of internal energy:

$$\Delta U = m \times c_v \times \Delta t$$
$$= 0.03 \text{ kg} \times 0.718 \text{ kJ/kg.K} \times (175 - 25)\text{K}$$
$$= 3.23 \text{ kJ}$$

(b) The change of enthalpy:

$$\Delta H = m \times c_p \times \Delta t$$
$$= 0.03 \text{ kg} \times 1.005 \text{ kJ/kg.K} \times (175 - 25)\text{K}$$
$$= 4.52 \text{ kJ}$$

The energy exchange in a closed system is given by $Q - W = \Delta U$.

Substitute $Q = -1.13$ kJ

$W = -4.38$ kJ

$\Delta U = +3.23$ kJ

$$(-1.13 \text{ kJ}) - (-4.38 \text{ kJ}) = (+3.23 \text{ kJ})$$

Hence, $+3.25 \text{ kJ} = +3.23 \text{ kJ}$

That is, the energy equation is satisfied, except for a slight discrepancy* which is due to the fact that specific heats of gases are not exactly constant.

In the case of a closed system, such as in this example, the change of enthalpy, while it can be calculated as shown above, does not represent energy, and is not relevant to the energy equation.

Example 27.7

Determine the amount of heat and work transfer if air in the previous example was heated from 25°C to 175°C,

(a) at constant volume,
(b) at constant pressure.

---

* In this example the error is less than 1 per cent and is, therefore, quite acceptable. However, if higher temperatures are encountered, the error will increase and may become significant, in which case exact tabulated values of internal energy and/or enthalpy can be used.

Solution

The change of internal energy of the air is

$$\Delta U = m.c_v.\Delta t = 3.23 \text{ kJ}$$

as before, regardless of the type of process involved, so long as the temperature rise is $175 - 25 = 150$ degrees.

On the other hand, heat and work transfer do depend on the conditions under which the process takes place.

(a) For constant volume heating:

$$Q = m \times c_v \times \Delta t$$
$$= 0.03 \times 0.718 (175 - 25)$$
$$= 3.23 \text{ kJ}$$

and since there is no expansion, $W = 0$,

which also follows from $Q - W = \Delta U$,

where after substitution $3.23 - W = 3.23$,

or $W = 0$.

(b) For constant pressure heating:

$$Q = m \times c_p \times \Delta t = 0.03 \times 1.005 (175 - 25)$$
$$= 4.52 \text{ kJ}$$

The work done is found from

$$Q - W = \Delta U,$$

by substitution,

$$4.52 \text{ kJ} - W = 3.23 \text{ kJ}$$

Hence, work done by the expanding air is

$$W = 1.29 \text{ kJ}$$

27.5 Specific heat capacities of gases

It is obvious from the previous discussion that for any given gas, such as air, there is no single value of specific heat capacity, the actual value being dependent on the conditions under which the heating process takes place.

There are at least two significant values, that at constant volume, which applies to heating of a gas confined within a rigid vessel, and that at constant pressure, which applies to heat exchangers, ducts and cylinders with pistons under constant load. The difference is largely due to the compressibility of gases, as compared with solids and liquids, for which no distinction between constant volume and constant pressure specific heat capacities is necessary (see Table 26.1).

Table 27.2 gives typical average values of specific heat capacities for some gases.

Table 27.2 Specific heat capacities of some gases

Gas	*Specific heat capacity kJ/kg.K* Constant pressure c_p	Constant volume c_v
Hydrogen	14.36	10.22
Helium	5.23	3.15
Methane	2.23	1.69
Carbon monoxide	1.04	0.75
Nitrogen	1.04	0.74
Air	1.01	0.72
Oxygen	0.92	0.66
Carbon dioxide	0.85	0.66
Sulphur dioxide	0.80	0.67
Argon	0.52	0.32

Problems

Use specific heat capacities from Table 27.2.

27.11 Determine the amount of heat required to heat 5 kg of nitrogen enclosed within a rigid container from 20°C to 250°C.

27.12 Determine the amount of heat required to heat 3 kg of oxygen flowing inside a duct from 25°C to 50°C.

27.13 Determine the final temperature and the change of specific internal energy of 10 kg of hydrogen which is heated at constant volume from 20°C by the addition of 2.76 MJ of heat.

27.14 Air is cooled from 200°C by removing 152 kJ of heat per kilogram of air at constant pressure. Determine the final temperature and the change in specific enthalpy of air.

27.15 0.02 kg of air is expanding within a cylinder at constant pressure. If the temperature of the air changes from 25°C to 75°C during the process, calculate:
(a) change of internal energy of the air,
(b) heat transferred to the air,
(c) work done by the air.

27.16 Air is heated by passing it across a heating coil at the rate of 0.5 kg/s, where the temperature of the air increases from 15°C to 40°C.

If water is used as a heating medium at the rate of 15 L/min, what is the temperature drop of the water?

27.17 Determine the amount of heat which can be recovered from 1 kg of flue gas, initially at 300°C, by cooling it at constant pressure to 150°C. Assume that one kilogram of the gas consists of 0.7 kg of nitrogen, 0.25 kg of carbon dioxide and 0.05 kg of sulphur dioxide.

27.18 An oxygen bottle of 50 L capacity is at 2.5 MPa absolute and 25°C. Calculate the amount of heat which, when transferred to the oxygen in the bottle, will increase its temperature to 65°C.

27.19 Air is heated at constant pressure from an initial condition at 101.3 kPa absolute and 15°C until its volume is doubled.
Determine the amount of heat required per kilogram of air and the final temperature.

27.20 Determine the amount of heat that must be removed to cool 0.3 kg of air at constant volume from an initial condition of 200 kPa absolute and 300°C until its pressure drops to 100 kPa absolute. What will the final temperature be?

27.21 Five cubic metres of air, initially at atmospheric pressure and 20°C, is heated to 65°C in a rigid container, i.e. at constant volume. Calculate:
(a) heat required,
(b) change in internal energy,
(c) change in enthalpy.

27.22 Ten kilograms of air at atmospheric pressure are heated from 20°C to 65°C by a heating coil at constant pressure. Calculate:
(a) heat required,
(b) change in internal energy,
(c) change in enthalpy.

Review questions

1. In what ways is a gas different from a liquid?
2. List some engineering applications, that you are familiar with, in which gases are used.
3. State and explain Boyle's Law.
4. State and explain Charles' Law.
5. State the equation representing combined Boyle-Charles' Law.
6. What is the equation of state?
7. Define characteristic gas constant and show how it can be determined for a particular gas.
8. Discuss specific heat capacities of air at constant volume and constant pressure.
9. State two useful approximate formulas which are based on Joule's Law.

28

Thermal systems

In this chapter a brief discussion is given of typical applications of concepts of thermodynamics to some practical systems and equipment, with particular reference to energy exchange equations. The chapter is divided into three parts covering heat and work transfer relationships for closed systems, open systems and heat engines.

Essentially, no new material is introduced here, but already familiar concepts and equations are being summarised, as they apply to each particular system.

In this chapter, "special case" equations have been developed for each type of system or apparatus, followed by some problems illustrating their use. At this stage, it is being left to the students to apply the equations* to the solution of these problems, without any solved examples in the text of the chapter.

In general, it is usually necessary to recognise a system as either a closed or an open system and to apply the non-flow or the steady-flow energy equation accordingly. Any special conditions, such as the absence of heat transfer to or from a thermally insulated, i.e. adiabatic system, or the absence of work done by a gas enclosed within a rigid vessel, must be noted. These special conditions, when substituted into the appropriate energy equation, help to reduce that equation to a form suitable for the solution of problems specifically related to the particular system.

Heat engines, discussed very briefly in the last section of this chapter, require an additional concept of thermal efficiency, as a measure of the effectiveness of conversion of heat into work.

28.1 Non-flow processes

Non-flow processes occur within closed systems, when heat and/or work transfer occurs between the system and its surroundings.

A closed system is a system which has fixed mass contents, while undergoing a process. During the process only energy in the form of heat (Q) or work (W) can enter the system or leave it, but there is no transfer of mass across the system boundary.

* It is not suggested that students should learn these equations by rote. On the contrary, when solving problems, one must always try to understand the basic principles involved and not merely substitute information into the formulae to obtain numerical answers.

The difference, if any, between the net amounts of heat and work transfer results in the change of internal energy of the system, according to the non-flow energy equation:

$$Q - W = \Delta U$$

The total change of internal energy of the system (ΔU), can be related to the mass (m) of the substance, e.g. air, water or steam, contained within the system and to the change in its specific internal energy ($u_2 - u_1$).

$$\Delta U = m(u_2 - u_1)$$

This equation also makes it possible to relate changes in internal energy to the corresponding changes in temperature of the substance by reference to Table 25.4.

Autoclave, pressure vessel

An autoclave is a vessel, usually constructed of thick-walled steel, able to withstand high temperatures and pressures, used in the chemical industry, for vulcanisation of rubber and in the curing of certain timber and plastic materials. A device operating on a similar principle used in cooking is known as a pressure cooker. Our discussion also applies to any closed rigid container, e.g. pressure vessel, to which heat is transferred, while the volume of the material contained within remains constant.

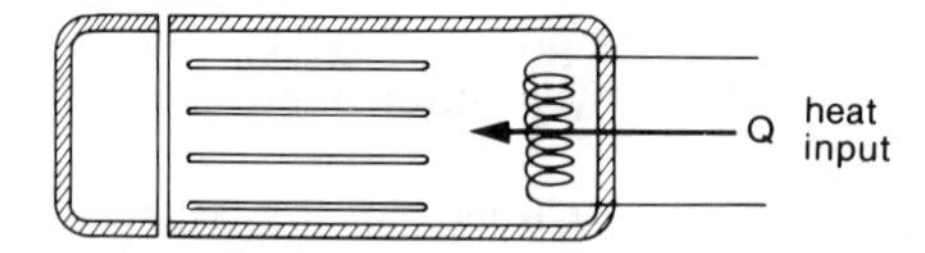

Fig. 28.1

Heat transferred under constant volume conditions is given by $Q = m.c_v(t_2 - t_1)$, provided that there is no change of phase taking place.

There is no external work done, due to the inability of the substance to expand, i.e. $W = 0$.

Substitution into the non-flow energy equation, $Q - W = \Delta U$, yields:

$$m.c_v(t_2 - t_1) - 0 = m(u_2 - u_1),$$

which reduces to

$$c_v(t_2 - t_1) = (u_2 - u_1)$$

This equation can be regarded as a special case of the non-flow energy equation applicable to the constant volume heating of a substance in an autoclave or a closed pressure vessel.

Problems

28.1 The total mass of the material contained in an autoclave is 350 kg and its average specific heat capacity at constant volume is 2.23 kJ/kg.K. If the temperature must rise from 20°C to 165°C, determine:
(a) the amount of heat required,
(b) the total change in internal energy,
(c) the change in specific internal energy.

28.2 A radiator of a heating system has a volume of 20 litres. When the radiator is filled with steam, all valves to the radiator are closed. Initially, the steam has specific internal energy of 2506 kJ/kg and density of 0.598 kg/m^3. After some heat has been transferred to the room, a mixture of steam and condensate remains in the radiator. The final specific internal energy of the mixture is 160 kJ/kg.
Determine how much heat has been transferred to the room.

28.3 A pressure vessel of 750 L capacity contains nitrogen at 300 kPa absolute and 20°C. If, as a result of fire in the vicinity of the vessel, the nitrogen absorbs 400 kJ of heat, determine its final temperature and pressure. Determine also the increase in its specific internal energy.

Bomb calorimeter

A calorimeter is a device for measuring heat energy liberated during a mechanical, electrical or chemical reaction. Calorimeters have been designed in a great variety of forms to suit different purposes.

A form of apparatus used for determining the calorific values of liquid or solid fuels is known as the bomb calorimeter. The bomb consists of a thick-walled stainless steel vessel in which a carefully measured quantity of the fuel is ignited in an atmosphere of compressed oxygen. The bomb is immersed in a known volume of water that absorbs the heat of combustion and thus increases in temperature.

Measurement of this temperature rise and the knowledge of the mass and specific heat capacity of water permit the total amount of heat generated to be calculated.

In order to account for the heat-absorbing capacity of the container and all other components of the apparatus, their combined effect is often expressed as an additional mass of water requiring the same quantity of heat to raise its temperature by the same amount as the container, etc. This term is called water equivalent of the calorimeter and is a fixed quantity for any given assembly of parts.

Fig. 28.2(a) *Bomb calorimeter parts*

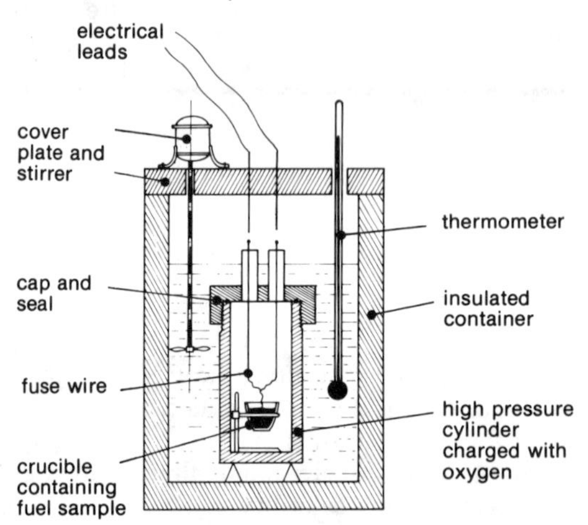

Fig. 28.2(b) *Bomb calorimeter assembly*

Considering the combustion process, the amount of heat liberated from the fuel is given by:

$$Q = m_f . C_f$$

where m_f is the mass of fuel burned
C_f is the calorific value of the fuel

Precision calorimeters are very carefully designed and constructed to minimise heat losses and to account for any losses that do occur. The effect of the stirrer used to achieve uniform temperature distribution is also allowed for.

It is, therefore, possible to relate the rise of temperature of the water directly to the amount of heat liberated from the bomb, provided that by the total mass we mean the actual mass of water within the calorimeter plus the water equivalent of the calorimeter parts.

Therefore, if we let

m_w = actual mass of water
m_e = water equivalent of the calorimeter
c_w = specific heat capacity of water
t_1 = initial temperature of water
t_2 = final temperature of water

then heat absorbed by the water and calorimeter is equal to

$$Q = (m_w + m_e).c_w.(t_2 - t_1)$$

Equating these two expressions for heat transfer between the bomb and its surroundings, i.e. water and calorimeter, gives

$$m_f.C_f = (m_w + m_e).c_w.(t_2 - t_1)$$

The work done is zero, as there are no mechanical moving parts in the device, $W = 0$. (Work input from the stirrer is negligible.)

Furthermore, using the non-flow energy equation, the change of internal energy of the system can be seen to be equal to the amount of heat received from the fuel, $Q = \Delta U$. However, this step does not contribute much further to our understanding of the process, or to the accuracy of the device in practical use.*

Problems

28.4 The following readings were taken during a bomb calorimeter experiment:

Mass of fuel burned = 0.75 g
Mass of water in the calorimeter = 1.364 kg
Water equivalent of the calorimeter = 0.682 kg
Initial temperature = 21.7°C
Final temperature = 25.2°C

Calculate the calorific value of the fuel.

28.5 In order to determine the water equivalent of a calorimeter, a complete assembly, containing 1.248 kg of water, was heated by an immersion heating element with a known output of 0.5 kW. The temperature rise after 3 minutes was 10 degrees Celsius.

Calculate the water equivalent.

* The temperature of the water does not usually increase by more than a few degrees, in which case a constant value of specific heat of water equal to 4.19 kJ/kg.K is very accurate, making the use of specific internal energy tables quite unnecessary.

Stirrer, marine propeller

A marine propeller is a device with a central hub carrying two, or more, radiating blades, placed so that each forms part of a spiral surface. By its rotation in water, a propeller produces the thrust to drive a water craft forward. Propeller-like devices are also used as stirrers in all kinds of processes to produce mixing or agitation of liquid substances in tanks and other vessels.

When a propeller or a stirrer rotates in water, it is doing work which can be calculated and included in the overall energy balance of a particular system. It is not our purpose here to discuss the hydrodynamics of propeller operation, but simply to account for the mechanical energy transfer, where a stirrer is involved.

One typical application involves an insulated container, filled with water in which a propeller, such as that of an outboard marine motor, may be tested (Fig. 28.3).

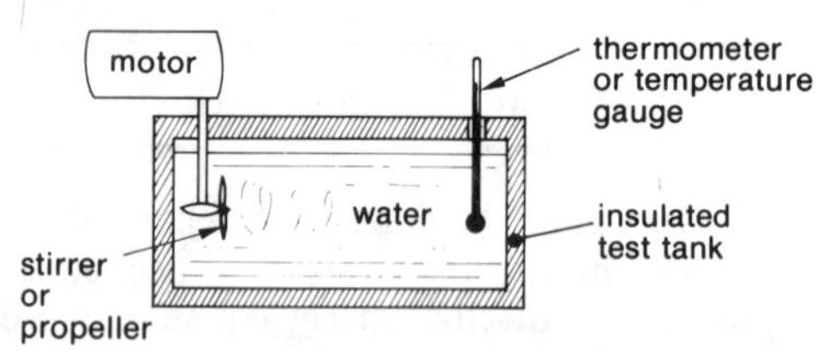

Fig. 28.3

If we let W be the work input from the propeller and recognise that, in a well-insulated test tank, the loss of heat, if any, to the surroundings is minimal, i.e. $Q = 0$, the non-flow energy equation reduces to:

$$-W = m(u_2 - u_1)$$

Since at moderate temperatures the volume of water in the tank does not change significantly and the specific heat capacity is nearly constant at $c =$ 4.19 kJ/kg.K, we can write $c(t_2 - t_1) = (u_2 - u_1)$ and substitute, resulting in:

$$\boxed{-W = m.c(t_2 - t_1)}$$

It is obvious that the input of mechanical energy in the form of work done by the stirrer, or propeller, on the water in the tank, results in the temperature rise, from which the work done can be calculated.

Problems

28.6 A stirrer is used to mix a slurry produced from some solid materials in water in a large vessel, which can be regarded as thermally insulated from the surroundings. It is found that, when the work done by the stirrer is 35.7 MJ per tonne of the slurry, the temperature of the mixture increases from 22°C to 45°C.

Determine the average specific heat of the mixture.

28.7 An insulated test-tank for an outboard motor contains 1.5 m^3 of water initially at 15°C. If, after a 12-minute test, the water temperature increases to 24.2°C, calculate the power delivered by the propeller into the water.

Engine or compressor cylinder

When a gas expands in a cylinder fitted with a movable piston, work is done by the gas on the piston. Conversely, when a gas is compressed in a cylinder of a reciprocating compressor, work is done by the piston on the gas.

The relationships between the amount of work done and the pressure and volume of the gas can be rather complex and are generally outside the scope of this book.* However, when the pressure remains constant during a process, or when an average, or mean, effective pressure is used, the work done is given by

$$W = p(V_2 - V_1)$$

where W is work done by the gas at constant pressure
p is actual constant pressure, or mean effective pressure
V_1 is initial volume of the gas
V_2 is final volume of the gas.

(A quick revision of Section 22.5 "Work done by fluid pressure" may be useful here.)

Expansion or compression of gases is usually accompanied by some heat transfer. In the case of reciprocating engines, the heat results from combustion within the cylinder of a combustible mixture. In reciprocating compressors, heat is taken away by a cooling agent, such as water circulating through a cooling jacket surrounding the compressor cylinder.

The difference between heat added or removed and the work done results in the change in internal energy of the gas, $Q - W = \Delta U$.

Hence,

$$\boxed{Q - p(V_2 - V_1) = m(u_2 - u_1)}$$

* It can be shown that pressure and volume are usually related by the law pV^n = constant, where n is the compression or expansion index, with a numerical value which depends on the nature of a particular gas and on the conditions under which the process takes place. Furthermore, the work done is given by the area under the p-V curve describing the expansion or compression process.

Problems

28.8 Air is compressed in a cylinder from an initial volume of 0.9 L to a final volume of 0.225 L, with a mean effective pressure of 228.2 kPa, while 39.9 J of heat are lost to the cylinder cooling water. Calculate:
(a) the amount of work done on the air by the piston, and
(b) the total change in internal energy of the air.

28.9 If the initial condition of the air in the previous problem is 25°C and 100 kPa absolute, determine:
(a) the mass of the air in the cylinder,
(b) the initial and final specific internal energy,
(c) the final temperature,
(d) the final pressure.

28.10 In the theoretical analysis of the working stroke of an engine it is assumed that, while the fuel is burning within the cylinder of the engine, constant pressure expansion of the gas occurs against the moving piston. After that, the working stroke is completed by adiabatic expansion, i.e. expansion without any loss or gain of heat.

Assuming the following information, calculate the total amount of work done by the gas on the piston during one working stroke.

Mass of the gas in the cylinder is 0.016 kg.
Specific internal energy:
at commencement of the working stroke is 585 kJ/kg
when combustion ends and adiabatic expansion begins, is 1279 kJ/kg
at the end of the working stroke is 429 kJ/kg
Mass of fuel used per working stroke is 0.335 g
Calorific value of the fuel is 43.3 MJ/kg.

28.2 Steady-flow processes

Steady-flow processes occur within open systems, when heat and/or work transfer occurs between the fluid flowing through the system and the surroundings.

An open system is a system through which a fluid is flowing at a steady* mass flow rate, while undergoing a process. A continuous transfer of energy occurs, to or from the fluid, in the form of heat (Q) or work (W), or both. Since the flow of mass is often expressed per unit time, it is usually convenient to think of heat and work transfer also as a rate.

* Non-steady flows are much more difficult to deal with mathematically and are not covered in this book.

The difference, if any, between the net amounts of heat and work transfer results in the change of enthalpy of the fluid between the inlet and outlet of the system, according to the steady-flow energy equation:

$$Q - W = \Delta H$$

The total change of enthalpy of the fluid (ΔH), can be related to the mass (m) of the substance, e.g. air, water or steam, flowing through the system and to the change in its specific enthalpy ($h_2 - h_1$):

$$\Delta H = m(h_2 - h_1)$$

This equation makes it possible to relate changes in enthalpy to the corresponding changes in temperature and phase of the substance by reference to Table 25.5.*

Heat exchangers

A heat exchanger is a device that transfers heat from a hot to a cold fluid, decreasing the temperature of the hot fluid and increasing that of the cold.

Heat exchangers are constructed in many different designs and are used extensively in power plants, air conditioning, refrigeration and in the chemical industry.

In the absence of moving parts, no external mechanical work is done in a heat exchanger, i.e. $W = 0$.

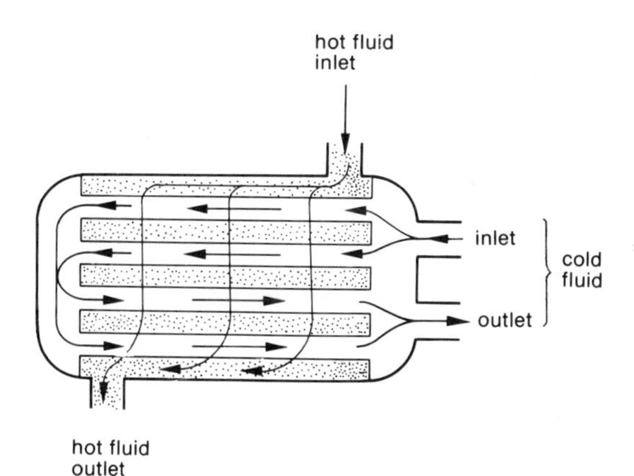

Fig. 28.4

* One should remember, that tables presented in this book are only a small extract from more detailed tables available elsewhere.

Therefore, only heat exchange between the two substances needs to be considered. In each case, heat transfer results in the change of enthalpy of the fluid. Thus:

$$Q = m(h_2 - h_1)$$

It is usually safe to assume that external heat losses, or gains, are minimal and the heat exchange equation becomes:

$$\underbrace{m(h_2 - h_1)}_{\text{cold fluid}} = \underbrace{m(h_1 - h_2)}_{\text{hot fluid}}$$

If a particular fluid does not undergo a phase change when it flows through a heat exchanger, the amount of heat received, or lost, by the fluid can also be expressed as $Q = m.c_p.(t_2 - t_1)$, since heat exchange occurs essentially at constant pressure.

When neither fluid undergoes a phase change, the heat exchange equation becomes:

$$\underbrace{m.c_p.(t_2 - t_1)}_{\text{cold fluid}} = \underbrace{m.c_p.(t_1 - t_2)}_{\text{hot fluid}}$$

Problems

28.11 Air is cooled from 40°C to 17°C at the rate of 0.5 m³ per second by means of chilled water initially at 10°C, circulating in the cooling coil at the rate of 23.5 litres per minute.

Assuming air density of 1.13 kg/m³ and specific heat capacities of air and water of 1.01 and 4.19 kJ/kg.K respectively, determine the final temperature of the water.

28.12 Steam at atmospheric pressure is used in a calorifier to heat 400 litres of water from 25°C to 75°C. If steam enters the calorifier coil at 125°C and leaves as condensate, i.e. liquid water, at 100°C at the rate of 2.42 kg/min, determine the time taken to accomplish the heating required.

Gas calorimeter

A gas calorimeter is a continuous flow device used for determining the calorific values of gaseous fuels. The calorimeter consists of a well-insulated heat exchanger, in which heat liberated by the combustion of a carefully metered volume of a gaseous fuel is absorbed by cooling water, whose mass flow rate and temperature rise are also measured accurately (Fig. 28.5).

Fig. 28.5(a) *Gas calorimeter*

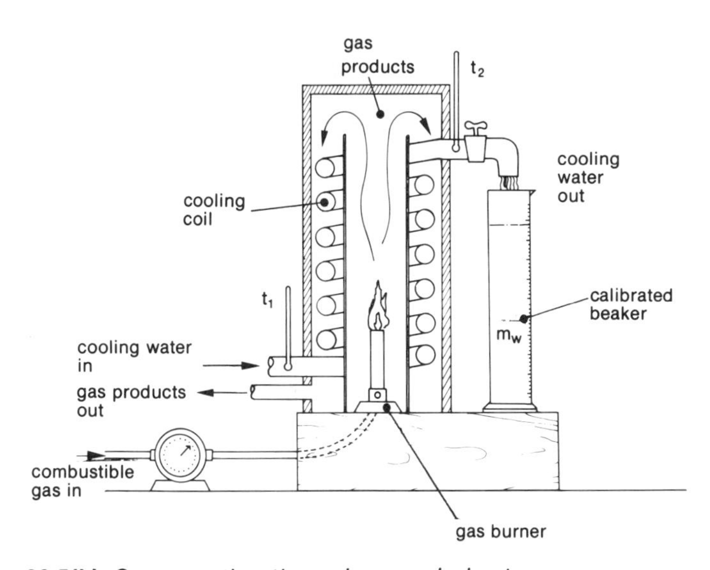

Fig. 28.5(b) *Cross-section through gas calorimeter*

In a calorimeter, constructed and used according to specifications, the heat transfer is assumed to be totally between the combustion products and the cooling water.

Heat liberated from the fuel is given by $Q = V_f \times C_f$, while the heat absorbed by the water is $Q = m_w.c_w.(t_2 - t_1)$. Equating these two expressions for heat transfer yields:

$$V_f \times C_f = m_w . C_w . (t_2 - t_1)$$

where V_f is volume of the fuel gas,
m_w is mass of cooling water used over the same time interval,
C_f is calorific value of the fuel,
c_w is specific heat capacity of water,
t_1 is inlet water temperature,
t_2 is outlet water temperature.

Problems

28.13 A gas calorimeter was used over a period of 5 minutes, during which 6.1 litres of gas were consumed and 1.95 litres of water flowed through the calorimeter. The temperatures of water in and out of the calorimeter were 18.8°C and 33.5°C. Calculate the calorific value of the gas in MJ/m³.

Boilers and condensers

Boilers and condensers are essentially heat exchangers designed and used for the purpose of producing phase changes of water by supplying heat to it, or by removing heat from steam. Condensers are also used in refrigeration and chemical plants.

A boiler, or steam-generator, consists of a furnace, a surface to transmit heat from the furnace to water and a space where steam can accumulate. Most conventional boilers are classified as either fire-tube or water-tube types. In the fire-tube type, the hot gases flow inside steel tubes running through the water, while in the water-tube boiler, the boiling water is inside tubes with the hot furnace gases circulating outside the tubes. The largest boilers usually of water-tube type are found in electricity generating stations, while a large variety of smaller steam generators, both water-tube and fire-tube, are used in hospitals, factories, air conditioning and for many other applications.

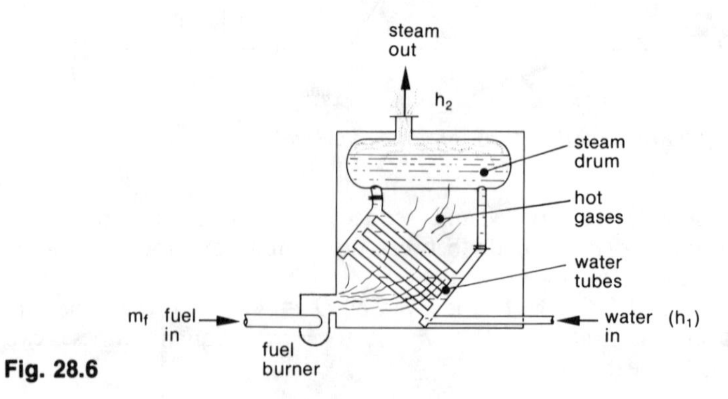

Fig. 28.6

Since there is no work done, $W = 0$, the heat transferred to the water is equal to the change of enthalpy produced, i.e. $Q = \Delta H$. Thus for a boiler:

$$Q = m(h_2 - h_1)$$

The heat (Q) supplied to the water in a boiler comes from fuel burned and would be equal to $Q = m_f C_f$ if there were no losses of heat to the surroundings. ($Q = m_f V_f$, if the fuel is gaseous.) However, if such losses exist, they can be allowed for by introducing an efficiency factor (η), which is the ratio of the heat actually supplied by a boiler in heating and evaporating the water to the heat supplied to the boiler from the fuel. Boiler efficiencies usually vary between 60 per cent and 90 per cent. The formula incorporating boiler efficiency would be:

$$\eta . m_f . C_f = m(h_2 - h_1)$$

A condenser is a device for reducing steam, or some other vapour, to a liquid. In power plants, condensers are used for the condensation of exhaust steam from turbines and in refrigeration plants for the condensation of refrigerants, such as ammonia.

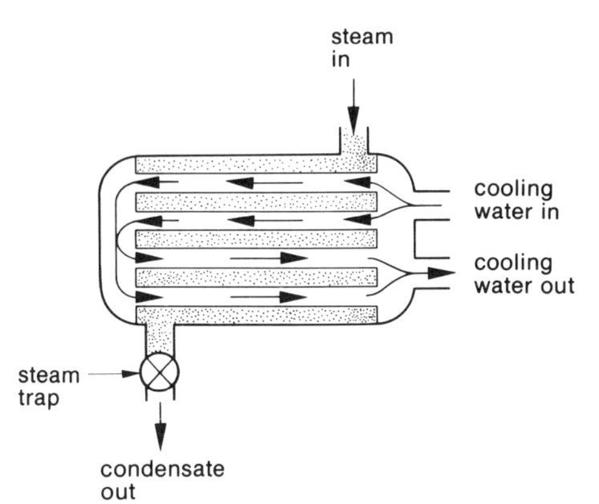

Fig. 28.7

Here again, there is no external work done, $W = 0$, and the amount of heat transfer is equal to the change of enthalpy of the substance, i.e. $Q = \Delta H$. Thus for a condenser:

$$Q = m(h_2 - h_1)$$

The heat removed from the condensing substance passes to the cooling agent, which is usually water circulating through the condenser, thus producing a temperature rise in the water according to $Q = m . c . \Delta t$.

After taking into account the direction of heat transfer, substitution yields

$$\underbrace{m . c_p . (t_2 - t_1)}_{\text{cooling water}} = \underbrace{m(h_1 - h_2)}_{\text{condensing substance}}$$

Problems

28.14 Determine the amount of heat required to produce 50 kg of steam at atmospheric pressure and 275° C from feed-water at 50°C.

28.15 A boiler is generating steam at high pressure and 450°C from feed water at 40°C at the rate of 3000 kg per hour. Fuel oil of calorific value 45.4 MJ/kg is used at the rate of 281 kg per hour.

If the specific enthalpies of water and steam, corresponding to the boiler pressure, are 167.5 kJ/kg and 3357 kJ/kg, calculate the boiler efficiency.

28.16 Determine the rate at which heat must be removed from steam at atmospheric pressure and 175°C, flowing to a condenser at the rate of 0.35 m³/s and leaving as condensate at 75°C, if the density of steam at inlet conditions is 0.493 kg/m³.

28.17 A water-cooled refrigerant condenser uses 25 litres of water per minute with a temperature rise of 12 degrees Celsius. If the change in the specific enthalpy of the refrigerant is 122 kJ/kg, what is the mass flow rate of the refrigerant through the condenser?

Turbines and centrifugal compressors

A turbine is a machine in which steam, or another gaseous substance, is made to do work by acting on a set of curved blades attached to a rotor. Since the average velocity of flow through a turbine is usually very high, the time available for heat transfer to take place is quite negligible and the process is assumed to be adiabatic, i.e. $Q = 0$.

It follows from the steady-flow energy equation that $-W = \Delta H$, or

$$-W = m(h_2 - h_1)$$

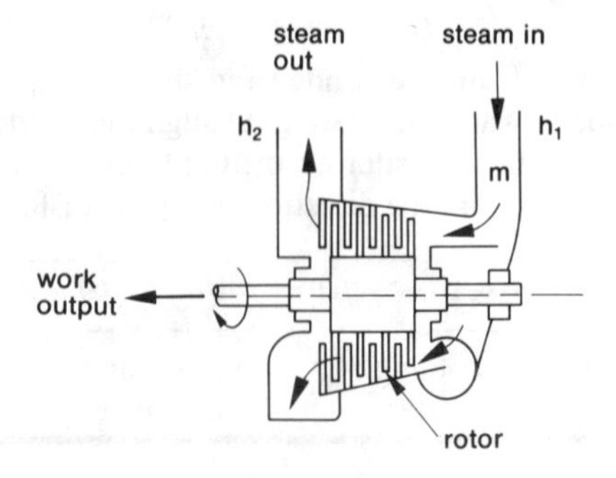

Fig. 28.8

The centrifugal compressor can be regarded as a reversed turbine, work being done on the fluid by the rotor to raise its enthalpy and pressure. Therefore, with due regard to the sign convention for work done, the same formula is applicable to both turbines and centrifugal compressors.

It should be remembered that the work calculated by the abovementioned formula represents work transfer between the rotor and the fluid and not the shaft input or output. Mechanical efficiency factor must be included if shaft work must be known.

In the case of turbine work, shaft output is given by $W \times \eta$ and in the case of a compressor by $\frac{W}{\eta}$. Can you see why?

Problems

28.18 A steam turbine uses steam at the rate of 50 kg per minute, with an initial specific enthalpy of 3357 kJ/kg and a final specific enthalpy of 2772 kJ/kg. The mechanical efficiency is 80 per cent.

Determine:
(a) the work done by the steam on the rotor per second,
(b) shaft output power, i.e. output work per second.

28.19 A centrifugal air compressor, which requires 69 kW of shaft power input, takes in air at the rate of 0.3 m^3/s at 101.3 kPa absolute and 25°C. The outlet conditions are 175°C and 422 kPa absolute. Use Table 25.5 to determine specific enthalpy of air at inlet and outlet and, hence, calculate:
(a) the work done by the rotor on the air per second,
(b) the mechanical efficiency of the compressor.

28.3 Heat engines

This concluding section is not an explanation of how particular engines work. Instead, it is a brief summary of the basic thermodynamic principle underlying the operation of all devices, called heat engines, in which energy is supplied in the form of heat and some of this energy is transformed into work.

Essentially, an engine is a device which operates continuously, receiving heat from a heat source, such as hot gases resulting from combustion of a fuel, converting some of that energy into work. Unfortunately, complete transformation of heat into work cannot possibly occur, even in a mechanically perfect engine. This limitation, which is not possible for us to discuss in detail here, is a consequence of the Second Law of Thermodynamics, introduced briefly in Chapter 25. It means that some of the energy supplied must remain in the form of heat and is rejected from the engine into its surroundings.

If we consider a heat engine simply in terms of heat received (Q) and work done (W), as shown in Figure 28.9, the rejected heat must be equal to the difference between the heat input and the work output.

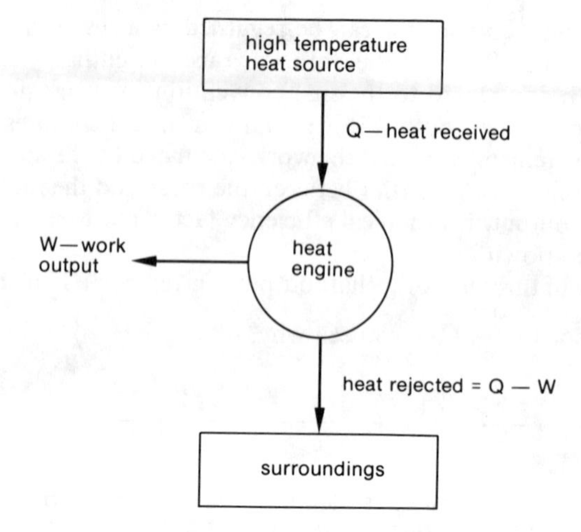

Fig. 28.9

The ratio of work output (W) and heat received (Q) is called the **thermal efficiency** and determines what fraction of the heat input has been converted into work.

$$\eta_t = \frac{W}{Q}$$

This ratio is always less than one hundred per cent and is limited to a theoretical maximum value* obtainable between any two temperature limits. Most engines and power generating cycles have thermal efficiencies much lower than the theoretical maximum, due to friction and undesirable losses of heat to the surroundings.

Problems

28.20 An engine receives heat at the rate of 95 kW and develops 26.6 kW of mechanical power.
Calculate the amount of heat rejected and the thermal efficiency.

28.21 During a test on a diesel engine, the following data were collected:
Engine power output—120 kW
Fuel consumption rate—0.76 kg/min
Calorific value of the fuel—41.2 MJ/kg
Determine the thermal efficiency of the engine.

---

* The theoretical maximum efficiency is known as Carnot efficiency and is given by $\eta_c = \frac{T_H - T_L}{T_H}$, where T_H and T_L are the highest and lowest absolute temperatures between which a particular engine cycles operates.

28.22 Estimate the fuel consumption rate in litres per hour for an engine with power output of 75 kW and thermal efficiency of 25 per cent, if the fuel density is 0.718 kg/L and calorific value is 43 MJ/kg.

28.23 A coal-fired power station has an output of 500 MW and its thermal efficiency is 27 per cent. The calorific value of the coal used is 31.3 MJ/kg. Calculate the fuel consumption rate in tonnes per hour.

28.24 An electric power generating system consists of a boiler, as described in problem 28.15, supplying steam to a turbine, as in problem 28.18.

The turbine is used to drive an electric generator, which has a mechanical efficiency of 84.5 per cent.

Calculate the electric power generated and the overall thermal efficiency of the plant and draw a diagram of the plant, showing all mass and energy transfer terms. (Ignore possible heat recovery from condensed steam.)

Review questions

1. From other books, industrial catalogues or other sources, find and summarise additional information relating to the construction and operation of the following equipment:
   (a) autoclave
   (b) bomb calorimeter
   (c) reciprocating air compressor
   (d) heat exchanger
   (e) gas calorimeter
   (f) boiler
   (g) steam turbine
   (h) petrol engine
2. Define thermal efficiency.
3. Discuss the concept of thermal efficiency with reference to:
   (a) automotive engine,
   (b) power station.

PART EIGHT

Appendices

Science is organised common sense.

Norman Feather
in *Mass, Length and Time*

APPENDIX A

Centroids and moments of inertia

There are many problems in engineering which involve the evaluation of geometrical properties of areas and rigid bodies.

In particular, the evaluation of bending stresses in beams requires the knowledge of **centroids** and **moments of inertia** of cross-sectional areas of beams (see Chap. 19).

Similarly, consideration of rotational motion requires the knowledge of **mass moments of inertia** and **radii of gyration** of rotating rigid bodies (see Chap. 9).

The methods summarised in this appendix can be applied to a large number of engineering and mathematical examples. Some of these are illustrated here, but in a book of this size and purpose only a few of the essential examples can be given.

A.1 Centroids of plane areas

The centroid of a plane area is the unique point which is the geometrical centre of the area distribution.

The concept of the centroid is often explained in terms of the centre of gravity of a thin homogeneous plate of uniform thickness. Although this analogy is helpful, one should keep in mind that the centroid is a geometrical concept and not a mass-related concept.

Centroids of most of the common shapes have been determined by integration and are available in tables and handbooks.

The positions of centroids of symmetrical shapes such as circles, squares and rectangles are easily identifiable. In a right-angled triangle, the centroid is located one-third of the side length from the 90° angle in each direction. This is the point of intersection of the medians (see Table A.1).

Centroids of composite shapes can be calculated by dividing the area into its simple component parts and then applying the principle of first moments of area, which states that the **moment of an area about any reference line equals the algebraic sum of the moments of its component areas about the same line.**

Mathematically this principle can be expressed as

$$\bar{x} = \frac{\Sigma(A.x)}{\Sigma(A)}$$

where $\Sigma(A)$ = sum of all component areas, $A_1 + A_2 + A_3$, etc., and $\Sigma(A.x)$ = sum of all area moments, $A_1x_1 + A_2x_2 + A_3x_3$, etc.

Similarly in the y direction:

$$\bar{y} = \frac{\Sigma(A.y)}{\Sigma(A)}$$

$\bar{x}$ and $\bar{y}$ are the locating coordinates relative to the chosen axes.

Example A.1

Locate the centroid of the composite area shown in Figure A.1(a) with respect to the x-x and y-y axes.

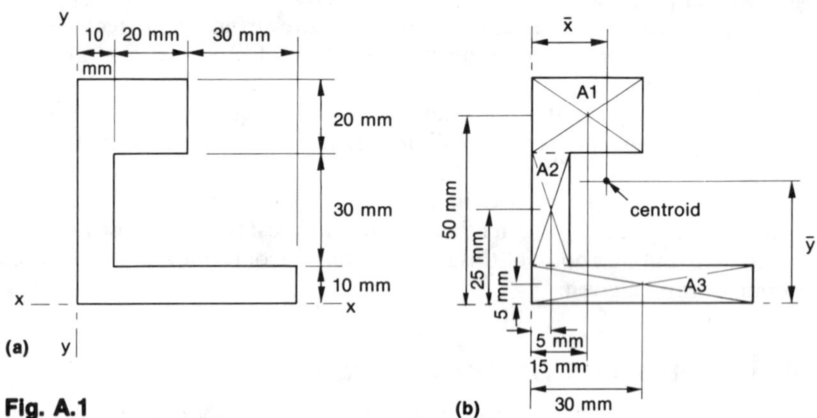

Fig. A.1

Solution

Divide the area into three rectangular elements, $A1$, $A2$ and $A3$ and locate the centroid of each element as in Figure A.1(b).

Calculate each elementary area:

$$A_1 = 30 \times 20 = 600 \text{ mm}^2$$
$$A_2 = 30 \times 10 = 300 \text{ mm}^2$$
$$A_3 = 60 \times 10 = 600 \text{ mm}^2$$
$$\text{Total area } \Sigma(A) = 1500 \text{ mm}^2$$

Calculate the area moments in x-x direction:

$$A_1x_1 = 600 \times 15 = 9000 \text{ mm}^3$$
$$A_2x_2 = 300 \times 5 = 1500 \text{ mm}^3$$
$$A_3x_3 = 600 \times 30 = 18\,000 \text{ mm}^3$$
$$\text{Total area-moment } \Sigma(Ax) = 28\,500 \text{ mm}^3$$

The position of the centroid with respect to y-y axis, i.e. located in the x-x direction is:

$$\bar{x} = \frac{\Sigma(A.x)}{\Sigma(A)} = \frac{28\,500}{1500} = 19 \text{ mm}$$

Similarly in the *y-y* direction:

$$\bar{y} = \frac{\Sigma(A.y)}{\Sigma(A)} = \frac{600 \times 50 + 300 \times 25 + 600 \times 5}{600 + 300 + 600} = 27 \text{ mm}$$

It is obvious from this example that the centroid of a composite area can lie outside the outline of the area itself.

It is often convenient to enter all intermediate values in a simple table as illustrated below:

	Area	Distance		Area moment	
Element	A	x	y	Ax	Ay
1 2 3	600 300 600	15 5 30	50 25 5	9 000 1 500 18 000	30 000 7 500 3 000
Σ =	1500	—	—	28 500	40 500

Hence

$$\bar{x} = \frac{28\,500}{1500} = 19 \text{ mm}$$

and

$$\bar{y} = \frac{40\,500}{1500} = 27 \text{ mm}$$

Example A.2

Locate the centroid of the shape shown in Figure A.2.

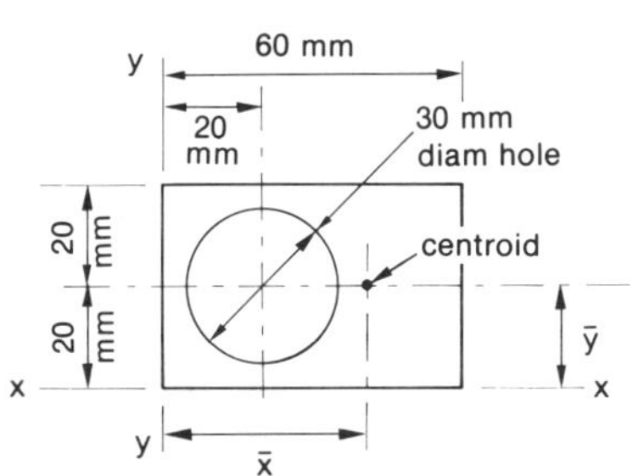

Fig. A.2

Solution

Areas removed, e.g. holes, can be regarded as negative areas. In all other respects the method of solution remains as explained before:

$$\bar{x} = \frac{\Sigma(A.x)}{\Sigma(A)} = \frac{(60 \times 40) \times 30 + \left(-\frac{\pi . 30^2}{4}\right) \times 20}{(60 \times 40) + \left(-\frac{\pi . 30^2}{4}\right)} = 34.2 \text{ mm}$$

From symmetry $\bar{y}$ = 20 mm.

A.2 Moments of inertia of plane areas

The **moment of inertia (I)** of a plane area, more appropriately called the second moment of area, is the sum of all elementary products of area elements and the square of each respective distance from the centroidal axis. As such, it is a mathematical concept which is best illustrated by examples such as Examples A.3 to A.5 below.

For our purposes, moment of inertia of a cross-section of a beam, combined with another dimension of that cross-section, is a measure of the resistance offered by the cross-section to bending moment and can be related to the bending stresses induced in the beam by bending (see Chap. 19).

Table A.1 Centroids and moments of inertia of elementary plane areas

Shape	*Area* *A*	*Position of centroid*	*Centroidal moment of inertia* I_c
Circle (D)	$\frac{\pi.D^2}{4}$	At centre	$\frac{\pi.D^4}{64}$
Square (a, a)	a^2	At intersection of diagonals	$\frac{a^4}{12}$
Rectangle (h, b)	$b.h$	At intersection of diagonals	$\frac{b.h^3}{12}$
Right-angle triangle (h, $\frac{h}{3}$, b)	$\frac{b.h}{2}$	At intersection of medians (1/3 of altitude)	$\frac{b.h^3}{36}$

Moments of inertia of most common shapes have been calculated by means of integral calculus and can be found in tables and handbooks. The moments of inertia of simple geometrical shapes are given in Table A.1.

Example A.3

Determine the moment of inertia of a rectangular area, 30 mm base and 20 mm height, about its horizontal centroidal axis.

Solution

From Table A.1, the moment of inertia of a rectangle is given by

$$I_c = \frac{bh^3}{12}$$

Substitute and solve:

$$I_c = \frac{30 \times 20^3}{12} = 20\ 000\ \text{mm}^4$$

For a given shape, the moment of inertia is not unique but depends on the particular reference axis. Table A.1 gives the moments of inertia for each shape about its horizontal centroidal axis.

The moment of inertia about any other axis parallel to the centroidal axis is given by the relationship known as the parallel axis theorem:

$$I = I_c + Ad^2$$

where I is moment of inertia about any axis
I_c is moment of inertia about centroidal axis
A is area
d is distance between the axes.

Example A.4

Determine the moment of inertia of the rectangular area shown in Figure A.3 about a parallel axis 23 mm away from its centroidal axis.

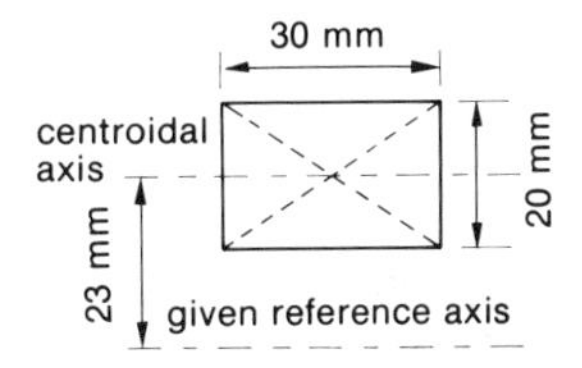

Fig. A.3

Solution

For the rectangular area, the moment of inertia about its centroidal axis is:

$$I_c = \frac{bh^3}{12} = \frac{30 \times 20^3}{12} = 20\ 000\ \text{mm}^4$$

The transfer term is:

$$Ad^2 = (30 \times 20) \times 23^2 = 317\ 400\ \text{mm}^4$$

Therefore, the required moment of inertia about the given reference axis is:

$$I = I_c + Ad^2 = 337\ 400\ \text{mm}^4$$

Example A.5

Determine the moment of inertia of the composite area given in Example A.1 about its horizontal centroidal axis.

Solution

The moment of inertia of a composite area about its centroidal axis is equal to the sum of the individual moments of inertia of its component areas relative to the common centroidal axis.

$$I = I_1 + I_2 + I_3$$

A tabulated solution is again helpful:

Element	*Area* A	*Distance from centroid* d	*Centroidal moment of inertia* I_c	*Transfer term* Ad^2	*Transferred moment of inertia* $I = I_c + Ad^2$
1	30 × 20 = 600	23	$\frac{30 \times 20^3}{12} = 20\,000$	600 × 23^2 = 317 400	337 400
2	30 × 10 = 300	2	$\frac{10 \times 30^3}{12} = 22\,500$	300 × 2^2 = 1200	23 700
3	60 × 10 = 600	22	$\frac{60 \times 10^3}{12} = 5000$	600 × 22^2 = 290 400	295 400
Total moment of inertia of the area about its centroidal axis $\Sigma(I)$ = 656 500 mm^4					

Problems

A.1 For each of the areas shown below locate the centroid relative to x-x and y-y axes.

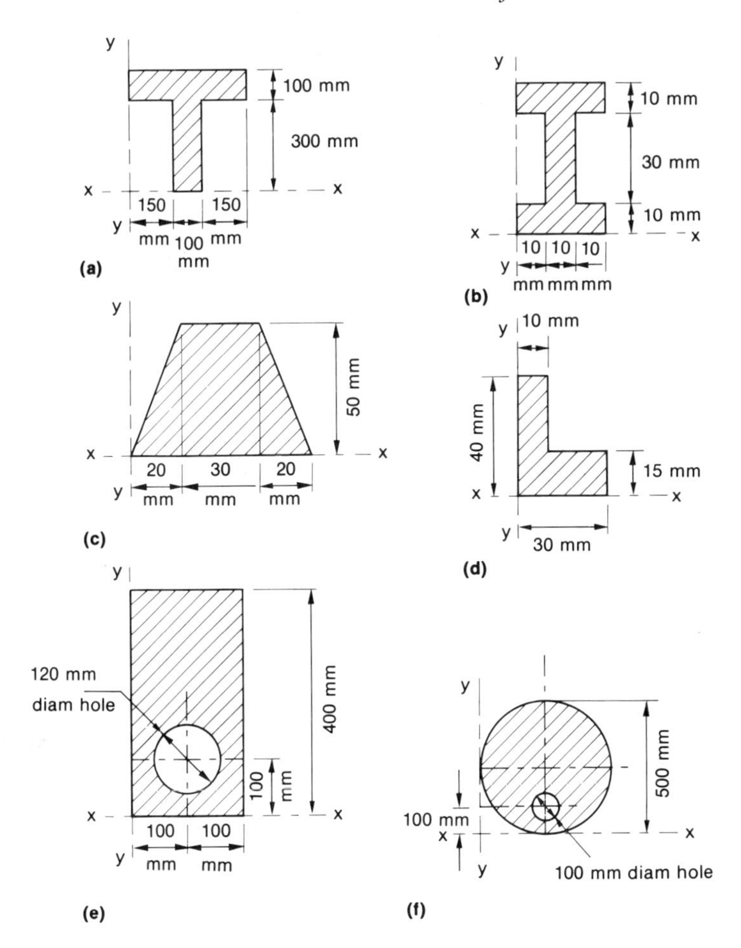

Fig. A.4

A.2 Determine the moments of inertia of the following areas about their horizontal centroidal axes:

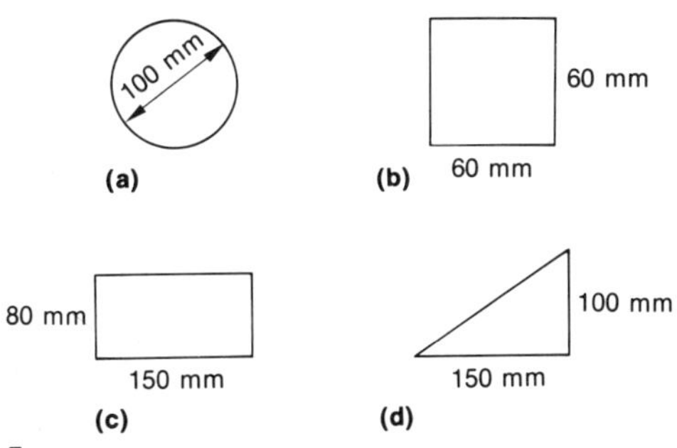

Fig. A.5

A.3 For each of the areas in problem A.1, determine the moment of inertia about the horizontal centroidal axis.

A.3 Mass moment of inertia of rigid bodies

The **mass moment of inertia** (I)* of a rigid body is mathematically similar to moments of inertia of plane areas, except that elements of mass distribution are used instead of area elements in its definition.

In rotational dynamics, mass moment of inertia is a measure of the rotational inertia of a body, i.e. the resistance that the body offers to having its rotational speed changed by the application of a torque.

Mathematically, mass moments of inertia can be calculated with respect to any given line. However, within the limitations of this book the mass moment of inertia of a rigid body will always be referred to a line which is, or can be, its axis of rotation.

Expressions for the mass moments of inertia of most common rotating solids such as a disc (e.g. flywheel), solid cylinder (e.g. shaft) and hollow cylinder (e.g. hollow shaft), as well as a mass concentrated at a fixed distance from the centre of rotation, have been derived by integration and are given in Table A.2.

It can be seen from these expressions that the units of mass moment of inertia are those of mass multiplied by those of linear measure squared, i.e. kilogram times square metre ($kg.m^2$).

Example A.6

Determine the mass moment of inertia of a flywheel in the form of a disc 50 mm wide × 300 mm diameter, if the material is steel (density** of steel = 7800 kg/m^3).

Solution

Volume of material $V = l \times \frac{\pi D^2}{4} = 0.05 \times \frac{\pi 0.3^2}{4} = 0.00353\ m^3$

Mass of flywheel, $V \times \rho = 0.00353 \times 7800 = 27.57\ kg$

Moment of inertia (disc) $I = m\frac{r^2}{2} = 27.57 \times \frac{0.15^2}{2} = 0.3101\ kg.m^2$

Answer $I = 0.310\ kg.m^2$

* It is common practice to use the same symbol, I, for both the moment of inertia of plane areas and the mass moment of inertia of rigid bodies. If any likelihood of confusion is to be avoided, J is recommended as an alternative symbol for the mass moment of inertia.

** Density of a substance is the mass of unit volume of that substance, expressed in such units as kilograms per cubic metre (see Chap. 15).

Table A.2 Mass moment of inertia and radii of gyration of rigid bodies about their axis of rotation (m = total mass, r = radius)

Rotating body	I *Mass moment of inertia* $kg.m^2$	k *Radius of gyration* m
Point mass at radius r 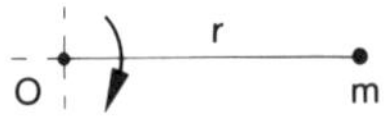	$m.r^2$	r
Thin shell or hoop about its axis 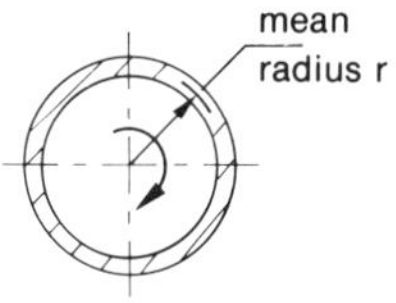 (mean radius r)	$m.r^2$	r
Solid cylinder about its axis 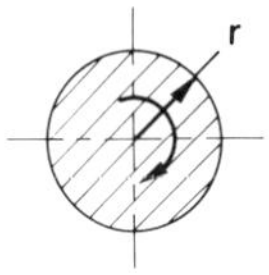	$\frac{m.r^2}{2}$	$\frac{r}{\sqrt{2}}$
Hollow cylinder about its axis 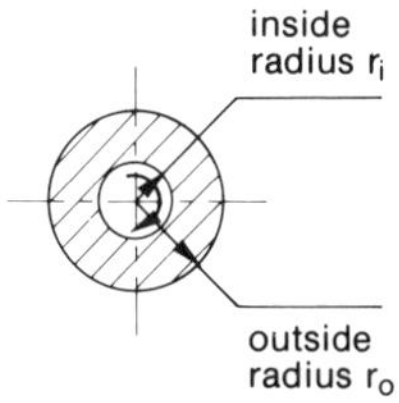 (inside radius r_i, outside radius r_o)	$\frac{m}{2}(r_o^2 + r_i^2)$	$\frac{\sqrt{r_o^2+r_i^2}}{\sqrt{2}}$

The mass moment of inertia of a composite rotating component can be calculated in a manner similar to calculating moments of inertia of composite areas, bearing in mind that the transfer term on the parallel axis formula uses mass instead of area, i.e.

$$I = I_c + m.d^2$$

Any mass removed, e.g. a hole, can be regarded as negative mass and its mass moment of inertia as negative. The following example should help to illustrate this.

Example A.7

Determine the mass moment of inertia of the flywheel shown in Figure A.6, if the material is steel.

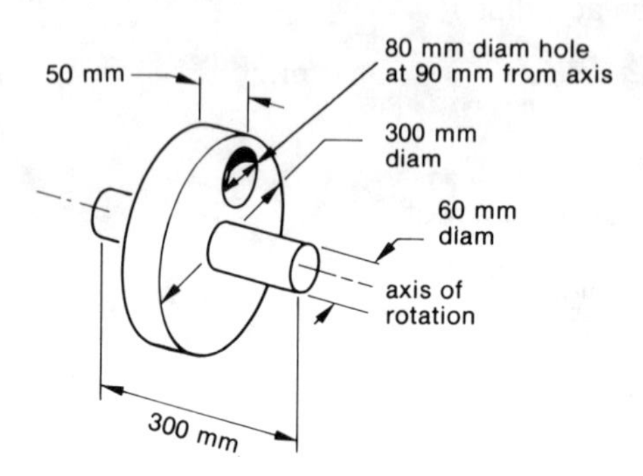

Fig. A.6

Solution

There are three components to be considered: the disc, shaft and hole.

(a) Disc:

The mass moment of inertia has already been calculated (see Example A.6 above).

$$I_{disc} = 0.3101 \text{ kg.m}^2$$

(b) Shaft:

The length of the shaft not included as part of the disc is 250 mm.

Volume, $V = l\dfrac{\pi D^2}{4} = 0.25 \times \dfrac{\pi 0.06^2}{4} = 0.000707 \text{ m}^3$

Mass, $V \times \rho = 0.000707 \times 7800 = 5.51 \text{ kg}$

Mass moment of inertia, $I = m.\dfrac{r^2}{2} = 5.51 \times \dfrac{0.03^2}{2} = 0.0025 \text{ kg.m}^2$

(c) Hole:

This can be regarded as a solid disc removed, i.e. all values below are regarded negative.

Volume, $V = l\dfrac{\pi D^2}{4} = 0.05 \times \dfrac{\pi 0.08^2}{4} = 0.000251 \text{ m}^3$ (negative)

Mass, $V \times \rho = 0.000251 \times 7800 = 1.96$ kg (negative)

Centroidal mass moment of inertia,

$$I = m\frac{r^2}{2} = 1.96 \times \frac{0.04^2}{2} = 0.00157 \text{ (negative)}$$

Transfer term, $m.d^2 = 1.96 \times 0.09^2 = 0.0159$ (negative)

Mass moment of inertia about common axis of rotation,

$$I = I_c + md^2 = 0.00157 + 0.0159 = 0.0175 \text{ kg.m}^2 \text{ (negative)}$$

The total mass moment of inertia of the flywheel is

$$I = \Sigma(I) = 0.3101 + 0.0025 - 0.0175 = 0.2951 \text{ kg.m}^2$$

Answer $I = 0.295 \ kg.m^2$

For convenience this solution can also be tabulated as follows:

Element	m	d	I_c	md^2	I
Disc	27.57	—	0.3101	—	0.3101
Shaft	5.51	—	0.0025	—	0.0025
Hole	-1.96	0.09	-0.0016	-0.0159	-0.0175
					$\Sigma(I) = 0.2951$ kg.m^2

A.4 Radius of gyration

The **radius of gyration** (k) of any body capable of rotation about an axis is the distance out from that axis where all the mass of the body would have to be concentrated for it to have the same moment of inertia as the body actually possesses about that axis.

From this definition of k it follows that $I = m.k^2$, or that

$$k = \sqrt{\frac{I}{m}}$$

For example, for a disc of uniform cross-section, the moment of inertia is $I = m\frac{r^2}{2}$.

Therefore, for a disc,

$$k = \sqrt{\frac{I}{m}} = \sqrt{\frac{m\frac{r^2}{2}}{m}} = \frac{r}{\sqrt{2}} = 0.707\, r$$

The relationships for calculating radii of gyration are also given in Table A.2.

Example A.8

What is the radius of gyration of the flywheels in Examples A.6 and A.7?

A.6 $k = \dfrac{r}{\sqrt{2}} = \dfrac{150}{\sqrt{2}} = 106 \text{ mm}$

A.7 $k = \sqrt{\dfrac{I}{m}} = \sqrt{\dfrac{0.2951 \text{ kg.m}^2}{(27.57 + 5.51 - 1.96) \text{ kg}}} = 0.0974 \text{ m} = 97.4 \text{ mm}$

Problems

A.4 For each of the following steel shafts (density of steel = 7800 kg/m³) determine the mass moment of inertia:
(a) 750 mm long, 60 mm diameter, solid shaft
(b) 750 mm long, 75 mm outside diameter, 45 mm inside diameter, hollow shaft.

A.5 What is the radius of gyration for each of the shafts in problem A.4?

A.6 Determine the mass moment of inertia and radius of gyration of an aluminium pulley shown in Figure A.7. Density of aluminium is 2560 kg/m³.

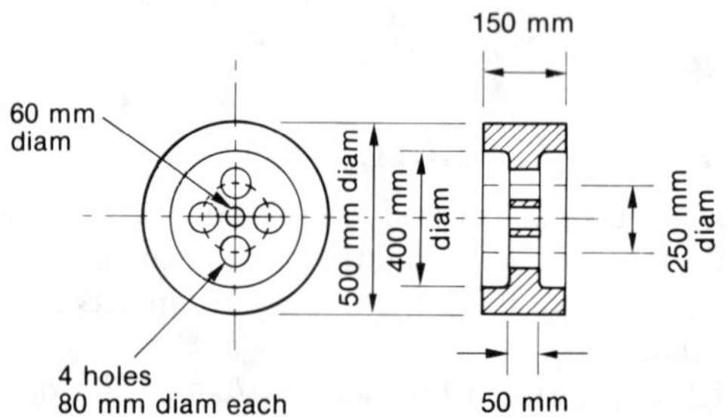

Fig. A.7

APPENDIX B

Summary of metric system of units

This appendix is an abridged summary of the SI units* used in engineering and of the rules governing their use based on the Metric Conversion Board publication called *Australia's Metric System*.

The metric system of measurement for Australia has been defined as measurement in terms of:

1. the units comprised in the International System of Units,
2. units decimally related to those units, and
3. other units, declared from time to time to be within the metric system.

Parts (1) and (2) of the definition mean that the Australian metric system includes all International System (SI) units and all the decimal multiples and submultiples of each such unit that can be formed by the attachment of SI prefixes to it, so long as such units are within the approval by the International Conference on Weights and Measures.

Part (3) provides for the extension of the system by introducing other units, which may owe their derivation to units outside the SI (e.g. hour, degree Celsius) or are specially named multiples or submultiples of SI units (e.g. litre, tonne).

The rules governing the formation and use of units and symbols and the correct spellings of unit names and abbreviations must always be followed if ambiguity and incorrect statements of measurements are to be avoided.

B.1 The International System of Units (SI)

The International System of Units is founded on seven base units and two supplementary units, of which only the units tabulated below are relevant to mechanical engineering at the level treated in this book.

All physical quantities may be measured in terms of SI base and/or supplementary units taken singly or in mathematical combinations (by multiplication and/or division). Where a unit of a physical quantity is defined in terms of such a combination of SI units without the use of a numerical factor or arbitrary constant, the unit is called an SI derived unit.

* This appendix contains only the units used throughout this book and should not be regarded as a complete description of the metric system of units.

Table B.1 SI base and supplementary units

Physical quantity	SI unit Name	Symbol
Base units		
length	metre	m
mass	kilogram	kg
time	second	s
temperature	kelvin	K
Supplementary unit		
plane angle	radian	rad

Some derived units have been allocated single word names and special symbols. These are listed in Table B.2, together with their definitions and derivations.

Table B.2 SI derived units with special names

Physical quantity	SI unit Name	Symbol	Definition	Derivation
force	newton	N	kg.m/s^2	
pressure, stress	pascal	Pa	N/m^2	kg/m.s^2
energy, work, quantity of heat	joule	J	N.m	kg.m^2/s^2
power, rate of heat flow	watt	W	J/s	kg.m^2/s^3

There are also many derived units with compound names, i.e. without any special name or symbol, the most common of which are listed in Table B.3, together with their symbols, which reflect their derivation.

B.2 Decimal prefixes

Measurements of much larger or smaller magnitude than the SI units are frequently made. To avoid the inconvenience of using very large or very small numbers, the system allows for decimal multiples or submultiples of the SI units to be used.

Table B.3 Examples of SI derived units with compound names

Physical quantity	*SI unit*	
	Name	*Symbol*
area	square metre	m^2
volume	cubic metre	m^3
volumetric flow rate	cubic metre per second	m^3/s
mass flow rate	kilogram per second	kg/s
speed, velocity (linear)	metre per second	m/s
angular velocity	radian per second	rad/s
acceleration (linear)	metre per second squared	m/s^2
angular acceleration	radian per second squared	rad/s^2
density (mass density)	kilogram per cubic metre	kg/m^3
specific volume	cubic metre per kilogram	m^3/kg
torque, moment of force	newton metre	N.m
momentum	kilogram metre per second	kg.m/s
second moment of area[a]	metre to the fourth power	m^4
mass moment of inertia	kilogram square metre	$kg.m^2$
specific energy[b]	joule per kilogram	J/kg
specific thermal capacity, characteristic gas constant	joule per kilogram kelvin	J/kg.K
local gravitational constant	newton per kilogram	N/kg

a Otherwise known as moment of inertia of an area.
b This may include any form of energy expressed per unit mass, such as latent heat, internal energy, calorific value, etc.

Rather than introducing new names for the units which are larger or smaller than the SI units themselves the system provides

(a) that the multiples and submultiples shall be decimally related to the "parent" unit;
(b) that the decimal relationships shall be indicated by the attachment of prefixes to the parent unit; and
(c) that each prefix shall have a standard value regardless of the unit to which it is attached.

SI prefixes are shown in Table B.4, together with the symbols for use with parent unit symbols.

One should be aware that the fact that a prefixed unit is part of the accepted Australian metric system does not necessarily mean its use is recommended. In general, prefixes for multiples and submultiples involving powers of 1000 (10^3) are preferred. This means that "hecto", "deka", "deci" and "centi" should be avoided.

Table B.4 SI prefixes

Factor	Prefix	Symbol
10^9	giga	G
10^6	mega	M
10^3	kilo	k
10^2	hecto	h
10	deka	da
10^{-1}	deci	d
10^{-2}	centi	c
10^{-3}	milli	m
10^{-6}	micro	μ

In particular, students of engineering must remember that linear measurements in the mechanical engineering industry should be in metres or millimetres—the centimetre is not to be used in technical applications.*

Special attention must be drawn to the SI unit of mass, the kilogram, which, although it contains the prefix "kilo" for historical reasons, is nevertheless an SI unit and not a multiple. However, to avoid the use of two adjacent prefixes, the SI prefixes having their standard values are attached to the stem word "gram" when forming units of mass larger and smaller than the kilogram. Thus the gram itself is a submultiple of the kilogram.

B.3 Other units within Australia's metric system

Units outside the SI, but which have been declared to be within the Australian metric system, are divided into two classes:

1. units which may be used without restriction for the measurement of the particular physical quantity (Table B.5), and
2. units which may be used only for particular, specified purposes (Table B.6).

* It must be noted that while it is not a technical unit, the centimetre is, nevertheless, a legal part of the metric system which is widely used in non-engineering areas of education, for general household purposes including body and clothing sizes, and for establishing the units of area and volume, such as the millilitre which is equal to one cubic centimetre.

Table B.5 Declared units having general application

Description of units			
Name of unit	*Symbol*	*Definition in terms of other metric units*	*Physical quantity*
degree	°	$\frac{\pi}{180}$ rad	plane angle
minute	′	$\frac{\pi}{10.8} \times 10^{-3}$ rad	
second	″	$\frac{\pi}{648} \times 10^{-3}$ rad	
litre	L	10^{-3} m^3	volume
millilitre	mL	10^{-6} m^3	
day	d	86.4×10^3 s	time interval
hour	h	3.6×10^3 s	
minute	min	60 s	
kilometre per hour	km.h^{-1} or km/h	1/3.6 m.s^{-1}	velocity, speed
tonne	t	10^3 kg	mass
tonne per cubic metre	t.m^{-3} or t/m^3	10^3 kg.m^{-3}	density
kilogram per litre	kg.L^{-1} or kg/L	10^3 kg.m^{-3}	
degree Celsius	°C	The Celsius temperature, given by the relationship $t_C = t_K - 273.15$	temperature
degree Celsius	°C	1K	temperature interval

Table B.6 Declared units having restricted application

Name of units	*Symbol*	*Definition in terms of other metric unit*	*Physical quantity*	*Purpose or purposes*
International nautical mile	n mile	1852 m	length	marine navigation, aerial navigation, meteorology
knot	kn	$\frac{1852}{3600}$ m.s^{-1}	speed	marine navigation, aerial navigation, meteorology
millibar	mb	100 Pa	pressure	meteorology
kilowatt-hour	kW.h	3.6×10^6 J	energy	measurement of electric energy

B.4 Units specifically excluded from Australia's metric system

Other metric systems (e.g. the CGS and technical metric systems) contain many units which are not in Australia's metric system. Some of the units specifically excluded from the Australian metric system are given in Table B.7, with the physical quantities they measure. The appropriate units for use in their stead are also shown.

Table B.7 Excluded units

Excluded unit and symbol	*Physical quantity*	*Australian metric unit and symbol*
atmosphere, standard (atm)	pressure, stress	kilopascal (kPa)
bar (b)	pressure, stress	kilopascal (kPa)
calorie (cal)	energy	joule (J)
centimetre of mercury (cm Hg)	pressure	kilopascal (kPa)
kilogram force (kgf)	force	newton (N)
kilogram force per square centimetre (kgf/cm^2)	pressure, stress	kilopascal (kPa)
micron (μ)	length	micrometre (μm)[a]
torr (torr)	pressure	pascal (Pa)

a The term "micron" was an earlier name for the micrometre.

B.5 Coherent units

SI units constitute a coherent system in which the product or quotient of any two unit quantities in the system is the unit of the resultant quantity. It follows that in a coherent system, derived units are formed without the use of numerical constants.

The coherence of SI units is probably the major benefit of the SI as it removes the need for introducing numerical factors into calculations. Thus, if all data are expressed in terms of SI units only, results will be in terms of SI units. This leads to a series of logical steps in carrying out any calculations:

1. express each measurement as a product of a number and an SI unit;
2. perform the calculation, resulting in an answer in the form of a product of an SI unit and a number; and
3. if necessary select an appropriate multiple or submultiple of the SI unit to reduce the number to a convenient order of magnitude.

Example B.1

Find the power necessary to pull a load at a uniform speed of 4.2 kilometres per hour when the load exerts a tension of 6 kN in the pulling rope.

Solution

(a) Convert data to SI units:

$$4.2 \text{ km} = 4200 \text{ m}$$
$$1 \text{ h} = 3600 \text{ s}$$
$$6 \text{ kN} = 6000 \text{ N}$$

(b) Perform calculation:

$$\text{power} = \frac{\text{force} \times \text{distance}}{\text{time}}$$
$$= \frac{6000 \times 4200}{3600} \text{ N.m/s}$$
$$= 7000 \text{ J/s} = 7000 \text{ W}$$

(c) Select appropriate unit for the result:

$$7000 \text{ W} = 7 \text{ kW}$$
$$\therefore \text{ Power} = 7 \text{ kW}$$

A special mention should be made with respect to the unit of mass, the kilogram, which in spite of the inclusion of the prefix "kilo" is a base unit and not a multiple. The kilogram should, therefore, be used in all calculations to take advantage of the coherent property.

This can be illustrated by reference to Newton's second law of motion which, when paraphrased, states that the physical quantity "force" is related to the product of the two physical quantities "mass" and "acceleration". The product of the SI units of mass and acceleration (kg and m/s^2) is the SI unit of force ($kg.m/s^2$), which is given the special name newton (symbol N).

Thus the kilogram, and not the gram, is used for the formation of the newton.

The coherent property of SI units does not *generally* extend to the multiples and submultiples of SI units, nor does it apply to declared units such as the hour, used in the above example, necessitating appropriate conversion to SI units.

However, in some special cases, where repetitive calculations of the same kind are performed, involving very large or very small quantities, it may be convenient to remember a *particular* coherent relationship between suitable prefixed multiple and submultiple units. For example, in Chapter 16 it was suggested that if force is expressed in newtons and dimensions in millimetres, stress can be expressed in megapascals without any conversions. All three units are usually appropriate for the quantities being measured and happen to have a coherent relationship between themselves.

Example B.2

Determine the tensile stress in a metal bar, 20 mm × 15 mm in cross-section, subjected to an axial pull of 7.5 kN.

Solution

$$\text{Stress} = \frac{\text{force}}{\text{area}} = \frac{7500 \text{ N}}{20 \text{ mm} \times 15 \text{ mm}} = 25 \text{ MPa}$$

The advantage of the above calculation is that an appropriate unit for the result is obtained automatically with relatively simple calculations. This advantage is somewhat outweighed by the necessity to memorise which prefixed units enter into a particular coherent relationship.

The choice of a suitable method for solving specific categories of problems is left to the discretion of the student. However, the golden rule is, if in doubt, convert all data to SI units and then perform the calculations, as in Example B.1.

B.6 General rules

The following is a brief summary of the rules recommended for the consistent and uniform use of metric units. If followed, these rules help to achieve clarity of expression and to avoid ambiguity.

Names

1. When written in full, names of units, including prefixed units, are written in lower case (small) letters except at the beginning of a sentence. The sole exception is "degree Celsius".
2. Unit names take a plural "s" when associated with numbers greater than unity, e.g. 1.5 newtons.

Symbols

1. Symbols are internationally recognised mathematical representations of units and are not abbreviations of the unit names. Names and symbols must not be mixed within the same unit expression.
2. As with any other symbols, unit symbols are translated into the names when spoken.
3. Unit symbols are written in lower case letters* except the symbols for units named after people, when the first letter of the symbol is a capital, e.g. pascal, Pa.

---

* The only exception from this rule is the use of capital L for the litre in order to avoid ambiguity—see Chapter 2.

4. Unit symbols do not take a full stop to indicate an abbreviation, nor do they take the plural form.
5. In writing, symbols should be separated from any associated numerical value by a space.

Prefixes

1. When written, the prefix is not separated from the parent unit, forming with it a new name or symbol.
2. All prefixes except those representing a million and more, e.g. mega (M) and giga (G), have lower case symbols.
3. When a prefix is attached to a unit it becomes an integral part of the new unit thus created and is therefore subjected to the same mathematical processes as the parent unit, e.g.

$$\begin{aligned} \text{mm}^2 &= \text{mm} \times \text{mm, i.e. } 0.001 \text{ m} \times 0.001 \text{ m, not } 0.001 \text{ m}^2 \\ &= (10^{-3} \text{ m})^2 \\ &= 10^{-6} \text{ m}^2\text{, not } 10^{-3} \text{ m}^2 \end{aligned}$$

4. No more than one prefix should be included in any unit, e.g. a tonne is a megagram and not a kilokilogram.
5. As a general rule, where a unit is expressed in the form of a product or a quotient the prefixed unit (if any) should preferably be the first occurring unit.*

Powers

1. The raising of a unit to a power is indicated by placing the words "squared", "cubed", "to the fourth power" and so on as appropriate after the unit so raised.**
2. Powers of any unit may be shown by the appropriate indices (superscripts) to the unit symbol.

Products

1. Where a unit is derived by multiplication of two or more different units, the resulting compound name is formed by stating the names of the constituent units in succession. In writing, a space is left between each of the successive names.
2. Products of symbols are indicated by a full stop or a point at mid-height between the symbols.

---

* There are however, exceptions to this rule. In particular, because the kilogram is the SI base unit of mass, when this unit appears in a derived unit it is not considered to be prefixed. It is therefore preferable to write for example, kJ/kg rather than J/g.

** By custom, the word "square" or "cubic" may be placed in front of a unit of length raised to the second or third power, e.g.
square metre, cubic metre, kilogram per cubic metre.

Quotients

1. The quotient of two units is indicated by the word "per" immediately in front of the unit(s) forming the denominator. The word "per" should not occur more than once in any unit name.
2. Quotients are indicated by a solidus (oblique stroke /), or by a horizontal line. The solidus and the horizontal line are directly equivalent to the word "per" used in unit names, so that no more than one solidus or horizontal line may be used in a unit symbol.
3. Alternatively, negative indices may be used to indicate those units which form the denominator of the quotient.

Appropriate units

1. When stating the value of any measurement the appropriate unit must be chosen. No more than one unit name (or symbol) may be included in such a statement.
2. The "appropriate unit" referred to above should generally be so chosen that the numerical value of the statement of measurement lies between 0.1 and 1000, e.g.

   500 kPa or 0.5 MPa, not 500 000 Pa

Pronunciation, spelling, etc.

1. *Gram*—the spellings "gram" and "gramme" are both allowed, but the shorter spelling in preferable.
2. *Joule*—the correct pronunciation rhymes with "pool".
3. *Kilo*—when used as a prefix with any unit, the pronunciation should be "kill-o" with the accent on the first syllable and "o" pronounced as "oh".

   The word "kilo" (pronounced "kee-low"), when used as an abbreviation for kilogram, has no legal standing and its use other than in casual speech should be avoided.
4. *Litre*—the spelling "liter" is not acceptable legally.
5. *Metre*—the American spelling "meter" is not acceptable legally.
6. *Tonne*—the correct pronunciation is "tonn" with "o" as in "Tom".

APPENDIX C

Symbols and formulae

C.1 List of symbols

A area, cross-sectional area
   A_s — area in shear
a acceleration
   a_g — acceleration due to gravity
a, b, c constants, side lengths of plane figures
b base length
C_f calorific value of a fuel
c specific heat capacity of substance
   c_p — specific heat at constant pressure
   c_v — specific heat at constant volume
D diameter
d moment arm, perpendicular distance
E Young's modulus of elasticity
e direct axial strain
   e_s — shear strain
F force
   F_b— buoyant force
   F_c — centrifugal force
   F_f, F_n — frictional and normal forces
   F_g — gravitational attraction
   F_L, F_E, F_F, F_{Th} — load, effort, friction effort and theoretical effort
   F_s, F_t — slack side and tight side belt tensions
   F_s — shear force
   F_w — weight of a body
FS factor of safety
f stress, direct axial stress
   f_A, f_H — axial stress and hoop stress in pressure vessels
   f_b — bending stress
   f_s — shear stress
G universal gravitational constant
G shear modulus of rigidity
g local gravitational constant
H hydraulic head
   H_L — head loss due to viscosity
   H_p — pumping head
   H_t — turbine head

H total enthalpy
h specific enthalpy
h height, elevation above datum level
I second moment of area and mass moment of inertia
KE kinetic energy
k radius of gyration
l length, linear dimension
l specific latent heat of substance
M moment of force, bending moment
M molecular mass
MA mechanical advantage of a machine
m mass
  m_f — mass of solid or liquid fuel
$\dot{m}$ mass flow rate
N speed in revolutions per minute
P power
  P_{in} — power input
  P_{out} — power output
PE potential energy
p pressure
  p_{abs} — absolute pressure
  p_{atm} — atmospheric pressure
  p_{ga} — gauge pressure
Q quantity of heat
R radius of curvature of beam in bending
R characteristic gas constant
RD relative density
r radius
S linear displacement, distance travelled
  S_E — distance moved by effort
  S_L — distance moved by load
s nominal size of weld
T torque
T absolute (Kelvin) temperature
t time
t thickness, weld throat thickness
t practical (Celsius) temperature
U total internal energy
US ultimate strength
  UCS — ultimate compressive strength
  USS — ultimate shear strength
  UTS — ultimate tensile strength
u specific internal energy
V volume
  V_f — volume of gaseous fuel
$\dot{V}$ volume flow rate
VR velocity ratio of a machine

v linear velocity
v specific volume
W mechanical work
x elongation, axial deformation
  x_s — deformation due to shear stress
$\bar{x}, \bar{y}$ position of centroid
y distance from neutral axis to extreme fibre of a beam
α angular acceleration
β one half of wedge angle of V-belt
ϵ coefficient of restitution
η efficiency
  η_t — thermal efficiency
θ angular displacement
θ angle of inclination, angle of contact, angle between two forces
μ coefficient of sliding friction
ρ density
ϕ angle of friction
ω angular velocity

C.2 List of formulae

Chapter reference

Areas of plane figures **2.5**

$$\text{Square } A = a^2$$

$$\text{Rectangle } A = ab$$

$$\text{Triangle } A = \tfrac{1}{2}bh$$

$$\text{Circle } A = \frac{\pi D^2}{4}$$

Volumes of solid figures **2.5**

$$\text{Cube } V = a^3$$

$$\text{Rectangular prism } V = abc$$

$$\text{Cylinder } V = \frac{\pi l D^2}{4}$$

$$\text{Sphere } V = \frac{\pi D^3}{6}$$

Rectangular component of a force **3.5**

$$F_x = F \cos \theta$$

$$F_y = F \sin \theta$$

Addition of rectangular components **3.5**

$$F = \sqrt{F_x^2 + F_y^2}$$

$$\tan \theta = \frac{F_y}{F_x}$$

Chapter reference

Universal gravitation **3.8**

$$F_g = G\frac{m_1 m_2}{d^2}$$
$$G = 66.7 \times 10^{-12}\ \text{N.m}^2/\text{kg}^2$$

Weight of a body on Earth **3.9**

$$F_w = mg$$
$$g = 9.81\ \text{N/kg}$$

Equilibrium of concurrent forces **4.1**

$$\Sigma F_x = 0$$
$$\Sigma F_y = 0$$

Moment of a force **5.1**

$$M = Fd$$

Equilibrium of moments **5.3**

$$\Sigma M = 0$$

Equilibrium of non-concurrent forces **6.1**

$$\Sigma F_x = 0$$
$$\Sigma F_y = 0$$
$$\Sigma M = 0$$

Coefficient of dry sliding friction **7.2**

$$\mu = \frac{F_f}{F_n}$$

Angle of friction **7.3**

$$\tan \phi = \mu$$

Equations of linear motion **8.2**

$$S = t\left(\frac{v_0 + v}{2}\right)$$
$$v = v_0 + at$$
$$S = v_0 t + \frac{at^2}{2}$$
$$2aS = v^2 - v_0^2$$

Gravitational acceleration **8.3**

$$a_g = 9.81\ \text{m/s}^2$$

Newton's Second Law **8.4**

$$F = ma$$

Relationship between angular units **9.1**

$$1\ \text{rev} = 360° = 2\pi\ \text{rad}$$

Chapter reference

Equations of rotational motion — **9.2**

$$\theta = t\left(\frac{\omega_0 + \omega}{2}\right)$$
$$\omega = \omega_0 + \alpha t$$
$$\theta = \omega_0 t + \frac{\alpha t^2}{2}$$
$$2\alpha\theta = \omega^2 - \omega_0^2$$

Relation between rotational and linear terms — **9.3**

$$S = r\theta$$
$$v = r\omega$$
$$a = r\alpha$$

Torque and rotational motion — **9.4**

$$T = I\alpha$$

Centrifugal force — **9.7**

$$F_c = \frac{mv^2}{r} = mr\omega^2$$

Mechanical work — **10.1**

$$W = FS \text{ (linear motion)}$$
$$W = T\theta \text{ (rotation)}$$

Power — **10.2**

$$P = \frac{W}{t}$$
$$P = Fv \text{ (linear motion)}$$
$$P = T\omega \text{ (rotation)}$$

Work and change of velocity — **10.3**

$$W = \frac{m}{2}(v^2 - v_0^2) \text{ (linear motion)}$$
$$W = \frac{I}{2}(\omega^2 - \omega_0^2) \text{ (rotation)}$$

Potential energy — **11.2**

$$PE = mgh$$

Kinetic energy — **11.3**

$$KE = \frac{mv^2}{2} \text{ (linear motion)}$$
$$KE = \frac{I\omega^2}{2} \text{ (rotation)}$$

Chapter reference

Conservation of mechanical energy **11.4**

$$PE_1 + KE_1 = PE_2 + KE_2$$

Work-energy method **11.5**

$$PE_1 + KE_1 \pm W = PE_2 + KE_2$$

Momentum **12.1**

$$\text{Momentum} = mv$$

Impulse **12.2**

$$Ft = mv - mv_0$$

Conservation of momentum during impact **12.3**

$$m_A v_{0A} + m_B v_{0B} = m_A v_A + m_B v_B$$

Coefficient of restitution **12.3**

$$\epsilon(v_{0A} - v_{0B}) = (v_B - v_A)$$

Mechanical advantage **13.1**

$$MA = \frac{F_L}{F_E}$$

Velocity ratio **13.1**

$$VR = \frac{S_E}{S_L}$$

Efficiency of a simple machine **13.2**

$$\eta = \frac{MA}{VR}$$

Friction effort **13.3**

$$F_F = F_E - F_{Th}$$

The law of a machine **13.4**

$$F_E = aF_L + b$$

Limiting efficiency **13.5**

$$\eta = \frac{1}{a.VR}$$

Power transmitted by a rotating component **14.1**

$$P = T\omega = \frac{2\pi NT}{60}$$

Drive efficiency **14.1**

$$\eta = \frac{P_{out}}{P_{in}}$$

Torque transmitted by belt drive **14.4**

$$T = r(F_t - F_s)$$

Chapter reference

Ratio of belt tensions **14.5**

$$\frac{F_t}{F_s} = e^{\mu\theta} \text{ (flat belt)}$$

$$\frac{F_t}{F_s} = e^{\mu\theta/\sin\beta} \text{ (V-belt)}$$

$$e = 2.718$$

Density **15.1**

$$\rho = \frac{m}{V}$$

Ultimate strength **15.2**

$$U.S. = \frac{\text{force causing failure}}{\text{initial area}}$$

Direct axial stress **16.1**

$$f = \frac{F}{A}$$

Factor of safety **16.2**

$$F.S. = \frac{\text{ultimate strength}}{\text{working stress}}$$

Direct axial strain **16.3**

$$e = \frac{x}{l}$$

Hooke's Law & Young's Modulus **16.4**

$$E = \frac{f}{e} = \frac{Fl}{Ax}$$

Shear stress **16.5**

$$f_s = \frac{F_s}{A_s}$$

Shear strain **16.6**

$$e_s = \frac{x_s}{l_s}$$

Modulus of rigidity **16.6**

$$G = \frac{f_s}{e_s}$$

$$G = 0.4\text{E}$$

Efficiency of bolted and riveted joints **17.2**

$$\text{Joint efficiency} = \frac{\text{strength of joint}}{\text{strength of unpunched plate}}$$

Chapter reference

Strength of fillet weld **17.3**

$$F = flt = 0.707fls$$

Stresses in cylindrical shells **18.1**

$$f_H = \frac{pD}{2t} \text{ (Hoop stress)} \quad f_A = \frac{pD}{4t} \text{ (Axial stress)}$$

Stress in a spherical shell **18.2**

$$f = \frac{pD}{4t}$$

Bending stress in a beam **19.6**

$$f_b = \frac{My}{I}$$

Radius of curvature of a beam **19.7**

$$R = \frac{EI}{M}$$

Density **20.3**

$$\rho = \frac{m}{V}$$

Specific volume **20.4**

$$v = \frac{V}{m} \qquad v = \frac{1}{\rho}$$

Absolute pressure **21.1**

$$p = \frac{F}{A}$$

Standard atmospheric pressure **21.2**

$$p_{atm} = 101.3 \text{ kPa} = 760 \text{ mm Hg}$$

Gauge pressure **21.3**

$$p_{ga} = p_{abs} - p_{atm}$$

Pressure due to a column of liquid **22.1**

$$p = \rho gh$$

Buoyant force **22.4**

$$F_b = V\rho g$$

Work done by constant fluid pressure **22.5**

$$W = pAl = p(V_2 - V_1)$$

Volume flow rate **23.2**

$$\dot{V} = Av$$

Chapter reference

Mass flow rate **23.2**

$$\dot{m} = \dot{V}\rho = Av\rho$$

Continuity equation **23.2**

$$A_1v_1\rho_1 = \dot{m} = A_2v_2\rho_2 \text{ (variable density)}$$
$$A_1v_1 = A_2v_2 \text{ (constant density)}$$

Bernoulli's equation for frictionless, incompressible flow

(a) In terms of energy per unit mass **23.3**

$$\frac{p_1}{\rho} + \frac{v_1^2}{2} + gh_1 = \frac{p_2}{\rho} + \frac{v_2^2}{2} + gh_2$$

(b) In terms of hydraulic head **24.1**

$$\frac{p_1}{\rho g} + \frac{v_1^2}{2g} + h_1 = \frac{p_2}{\rho g} + \frac{v_2^2}{2g} + h_2$$

Pumping head **24.2**

$$H_p = \frac{P}{\dot{m}g}$$

Pump shaft input power **24.2**

$$P_{in} = \frac{\dot{m}gH_p}{\eta}$$

Turbine head **24.3**

$$H_t = \frac{P}{\dot{m}g}$$

Turbine shaft output **24.3**

$$P_{out} = \eta\dot{m}gH_t$$

Modified Bernoulli's equation

$$\frac{p_1}{\rho g} + \frac{v_1^2}{2g} + h_1 \pm H = \frac{p_2}{\rho g} + \frac{v_2^2}{2g} + h_2$$

where H_p is pumping head (positive) **24.2**
H_t is turbine head (negative) **24.3**
H_L is head loss (negative) **24.4**

Temperature conversion **25.2**

$$T = t + 273$$

Specific internal energy **25.5**

$$u = \frac{U}{m}$$

Change of internal energy **25.5**

$$\Delta U = U_2 - U_1 = m(u_2 - u_1)$$

Chapter reference

Enthalpy and specific enthalpy **25.5**

$$H = U + pV$$
$$h = u + pv$$
$$h = \frac{H}{m}$$

Change of enthalpy **25.5**

$$\Delta H = H_2 - H_1 = m(h_2 - h_1)$$

Non-flow energy equation (Closed system) **25.7**

$$Q - W = \Delta U$$

Steady-flow energy equation (Open system) **25.7**

$$Q - W = \Delta H$$

Constant volume heating **25.8**

$$\Delta U = Q$$

Constant pressure heating **25.8**

$$\Delta H = Q$$

Sensible heat **26.1**

$$Q = mc\Delta t$$

Latent heat **26.3**

$$Q = ml$$

Heat released from fuel **26.6**

$$Q = m_f C_f \text{ (solid and liquid fuels)}$$
$$Q = V_f C_f \text{ (gaseous fuels)}$$

Combined Boyle-Charles' Law **27.2**

$$\frac{p_1 V_1}{T_1} = \text{const.} = \frac{p_2 V_2}{T_2}$$

The equation of state for gases **27.3**

$$pV = mRT$$

Characteristic gas constant **27.3**

$$R = \frac{8.314}{M} \text{ kJ/kg.K}$$

Constant volume heating of a gas **27.4**

$$Q = \Delta U = mc_v\Delta t$$

Constant pressure heating of a gas **27.4**

$$Q = \Delta H = mc_p\Delta t$$

Chapter reference

Joule's Law for gases **27.4**

$$\left.\begin{aligned}\Delta U &= mc_v\Delta t\\ \Delta H &= mc_p\Delta t\end{aligned}\right\}\text{for any process}$$

Thermal efficiency **28.3**

$$\eta_t = \frac{W}{Q}$$

APPENDIX D

Glossary of selected terms*

(Figures after each definition refer to the section or chapter in the book where the major reference to the concept may be found.)

acceleration A measure of the time rate of change in the velocity of a moving body (linear acceleration) or a rotating body (angular acceleration). **[8.1 and 9.1]**

adiabatic process A process which takes place without any loss or gain of heat. **[25.6]**

angle A measure of the inclination of one straight line to another expressed in radians, degrees or related units. **[2.4]**

angle of friction The angle between the resultant of the frictional and normal forces and the direction normal to the surfaces in contact at the moment of impending sliding motion. **[7.3]**

angle of repose The greatest angle to the horizontal which can be made by an inclined plane before an object resting on it would start sliding down the slope. **[7.4]**

Archimedes' principle The law which states that when a body is wholly or partly immersed in a fluid it experiences an upthrust equal to the weight of fluid displaced by the body. **[22.4]**

area A measure of the extent of a surface expressed in square metres or related units. **[2.5]**

bending moment The amount of bending tendency at a cross-section of a beam measured by the summation of moments about the cross-section of all external forces on one side of the cross-section. **[19.3]**

Bernoulli's equation The mathematical expression of Bernoulli's theorem which states that in a moving fluid the total energy per unit mass is constant, being made up of the pressure energy, the kinetic energy and the potential energy. **[23.3]**

buoyancy The loss in the apparent weight of a body when immersed in a fluid, due to the upthrust exerted by the fluid on the body. **[22.4]**

calorific value The amount of heat liberated from a unit quantity of fuel during complete combustion, usually expressed in kilojoules per kilogram for solid and liquid fuels, and in kilojoules per cubic metre for gaseous fuels. **[26.6]**

calorimetry The science and techniques associated with the measurement of thermal constants such as specific heat capacity, latent heat and calorific values. **[Chapter 26]**

---

* The reader may find the selection of terms included in this glossary to be somewhat arbitrary, which is inevitable in a book of this size. Relative importance of a concept and frequency of its use in engineering science have been the major criteria for selection.

Celsius scale A practical scale of temperature measurement on which the temperature of melting ice is chosen to be zero degrees and the temperature of water boiling at atmospheric pressure to be 100 degrees, with equal unit divisions between, above and below these reference temperatures. **[25.2]**

centrifugal force The force acting in the radial direction away from the centre of curvature of the curved path of a body in circular motion. **[9.7]**

centroid The point which is the geometrical centre of the area distribution of a plane area. **[Appendix A]**

characteristic gas constant A constant pertaining to a particular gas which is used in the equation of state to calculate the gas pressure, volume or temperature under specified conditions. **[27.3]**

circular motion Motion of a body or a particle along a circular path. **[9.3]**

closed system A thermodynamic system which contains the same matter throughout the process under investigation, such as gas expanding in an engine cylinder during its working stroke. **[25.4]**

coefficient of friction The ratio of the force of friction and the normal force between two surfaces in contact when sliding motion is about to occur. **[7.2]**

coefficient of restitution A measure of the ability of two bodies to regain their original shape after impact, defined in terms of relative velocities before and after the impact. **[12.3]**

component of a force The resolved part of a force in any particular direction. Any given force can be replaced by its two components in any two directions. Two mutually perpendicular components are the most frequently used. **[3.5]**

concurrent forces Forces which intersect at a common point called the point of concurrence. **[Chapter 4]**

conservation of energy The law which states that in any isolated system the total amount of energy is constant. It means that energy can be converted from one form into another, but cannot be created or destroyed. **[11.4, 23.3 and 25.7]**

continuity equation The mathematical equation expressing the relationship between mass flow rate of a fluid, its density, velocity and cross-sectional area of the pipe or duct. **[23.2]**

couple A pair of forces having the same magnitude, parallel lines of action and opposite sense. A couple produces a turning effect measured in newton-metres, but its resultant force is zero. **[5.5]**

degree A unit of angular measure equal to 1/90th part of a right angle. **[2.4]**

degree Celsius A practical unit of temperature measurement, equal in magnitude to one kelvin, but used in conjunction with the Celsius scale of measurement with the origin corresponding to the melting temperature of ice. **[25.2]**

density The mass per unit volume of a substance, expressed in kilograms per cubic metre or related units **[20.3 and 15.1]**

displacement A measure of the change in the position of a moving body (linear displacement) or the orientation of a rotating body (angular displacement) with respect to fixed coordinates. **[8.1 and 9.1]**

dynamics The branch of the mechanics of solids which deals with bodies in motion and with forces required to produce a given motion. Dynamics is divided into kinematics and kinetics. **[Chapters 8 to 12]**

efficiency A measure of performance of a machine defined as the ratio of the energy output to the energy input. **[13.2]**

elasticity The ability of a material to return to its original size or shape after having been stretched, compressed or deformed. **[15.3 and 16.4]**

energy A physical quantity associated with a body or a substance which is a measure of its ability to do work or to release heat. Energy may be of several kinds: potential energy, kinetic energy, pressure energy, internal energy, chemical energy. **[11.1, 23.3, 25.5 and 26.6]**

energy equations Mathematical expressions used in thermodynamics showing relationships between various components of energy and the quantities "heat" and "work". The most frequently used special cases are the non-flow energy equation which applies to closed systems and the steady-flow energy equation which applies to open systems. **[25.7]**

enthalpy A thermodynamic property defined as the sum of the internal energy and the product of the pressure and volume of a substance, expressed in joules or related units. **[25.5]**

equation of state The mathematical expression, $pV = mRT$, which relates the pressure, volume and absolute temperature of a given mass of gas to its characteristic constant, R. **[27.3]**

equilibrant The force which when added to a system of forces which is not balanced will produce equilibrium. The equilibrant is a force equal and opposite to the resultant of the given system of forces. **[4.3]**

equilibrium The state of a body which is subjected to a system of forces which balance each other out. If a body is in equilibrium, the resultant force and the sum of the moments acting on the body are zero. **[4.1 and 6.1]**

factor of safety The ratio, allowed for in design, between the ultimate strength of a material and the safe working stress in a component or structure made from it. **[16.2]**

fluid A substance, such as a liquid or a gas, which can flow. It differs from a solid in that it can offer no permanent resistance to change of shape. **[20.1]**

fluid mechanics The branch of mechanics which deals with liquids and gases at rest or in motion. Its major subdivisions include fluid statics, hydraulics and pneumatics. **[Chapters 20 to 24]**

fluid statics The study of fluids at rest concerned with pressure and its effects on submerged bodies and surfaces. **[20.1]**

force Any action, usually described as push or pull, which tends to maintain the position of a body, to alter the position of a body or to produce deformation in the size or shape of a body. **[3.1]**

free-body diagram A diagram showing a body subjected to a system of forces as a single object, often represented as a point, in isolation from other objects, and without unnecessary pictorial details. **[4.2]**

friction The resistance to sliding motion between two solid surfaces in contact with each other. **[7.1]**

gravity The force of mutual attraction between any two masses, proportional to the two masses and inversely proportional to the square of the distance between them. **[3.8]**

head A term often used in hydraulics as an indirect measure of energy levels in a flowing fluid, consisting of pressure head, velocity head and static head, which can be interpreted as energy per unit weight, expressed in metres. **[24.1]**

heat A form of energy transfer which takes place due to a difference in temperature between a system and its surroundings, expressed in joules or related units. **[25.1]**

Hooke's law The law which states that in an elastic material, within its limit of proportionality, strain is proportional to stress. **[16.4]**

hoop stress The tensile stress in a longitudinal joint of a pipe or cylindrical container subjected to internal pressure. **[18.1]**

hydraulics The study of incompressible fluids, such as oil or water, in motion, with particular emphasis on the pumping and flow of liquids through pipes or channels and the practical applications of water-power. **[20.1]**

impact A collision between two bodies which occurs in a very short interval of time and involves relatively large forces which the two bodies exert on each other. **[12.3]**

impulse The product of a force and the time during which it acts. **[12.2]**

inertia Reluctance of a body to change its state of rest or of uniform motion, which is the subject of Newton's First Law of motion. **[8.4]**

internal energy A form of energy stored within a substance due to the kinetic energy of its molecules, expressed in joules or related units. Internal energy is a function of temperature and phase of the substance under consideration. **[25.5]**

joint efficiency A measure of effectiveness of a bolted or riveted joint defined as the ratio of the strength of the joint to the strength of unpunched plate. **[17.2]**

joule The unit of energy, work or heat in SI, defined as the equivalent of work done by a force of one newton over a distance of one metre. **[10.1]**

kelvin The unit of absolute or thermodynamic temperature in SI, presently defined as equal to the fraction 1/273.16 of the temperature of the triple point of water (the triple point of water corresponds approximately to the temperature of melting ice). **[25.2]**

kilogram The unit of mass in SI, defined as the mass of the International Prototype Kilogram, i.e. the mass of a cylinder 39 mm diameter and 39 mm high, made from an alloy containing 90 per cent platinum and 10 per cent iridium, kept at the International Bureau of Weights and Measures in Sèvres, France. **[2.4]**

kilowatt-hour A unit of electric energy equivalent to the work done in one hour by a device working at a constant rate of one kilowatt, i.e. at the rate of one kilojoule per second. 1 kW.h = 3.6 MJ. **[Appendix B]**

kinematics The part of dynamics which is concerned with the study of motion without reference to forces causing the motion. **[Chapters 8 to 12]**

kinetic energy A form of mechanical energy possessed by a solid body or by a quantity of fluid by virtue of its motion. **[11.3 and 23.3]**

kinetics The part of dynamics which is concerned with the relation between force and motion. **[Chapters 8 to 12]**

latent heat The heat transfer associated with a phase change of a substance. Latent heat required for melting is called latent heat of fusion and applies equally to melting and freezing. Latent heat required for boiling is called latent heat of evaporation and applies equally to boiling and condensation. **[26.3]**

length A measure of distance between two points along a single straight or curved line, expressed in metres or related units. **[2.4]**

linear motion Motion of a body or a particle along a linear path. Rectilinear motion is motion along a straight line and curvilinear motion is motion along a curve. **[Chapter 8]**

litre A unit of volume equal to one thousandth part of a cubic metre. **[2.5]**

manometry The techniques of measuring pressure, particularly those which involve the use of manometers. **[21.4]**

mass A measure of the quantity of matter in a body as evidenced by its inertia, expressed in kilograms, tonnes or related units. **[2.4]**

mass flow rate The mass of fluid flowing across a given cross-section of duct or pipe per unit time, expressed in kilograms per second or related units. **[23.1]**

mass moment of inertia A geometrical property of mass distribution in a rigid body defined as the sum of all elementary products of mass elements and the squares of their respective distances from an axis. **[Appendix A]**

mechanical advantage The ratio of the load to the effort in a simple machine. **[13.1]**

mechanics The study of the action of forces and of the conditions of rest or motion they produce. It is divided into mechanics of solids, mechanics of fluids and mechanics of machines. **[1.3]**

mechanics of machines The study of the relative motion between the parts of a machine and of the forces which act on those parts. **[Chapters 13 and 14]**

metre The unit of length in SI, presently defined as equal to exactly 1650763.73 wavelengths of the orange line in the spectrum of the krypton-86 atom in an electrical discharge. **[2.4]**

modulus of elasticity A measure of stiffness of a material, also known as Young's modulus of elasticity (see Young's modulus). **[16.4]**

modulus of rigidity A measure of the ability of a material to resist deformation in shape, defined as the ratio of shear stress to shear strain, usually expressed in megapascals or related units. **[16.6]**

molecular mass The mass of a molecule of a substance expressed on a relative scale of molecular mass units for which the mass of a carbon molecule is taken to be exactly 12 units. One unit of molecular mass is approximately equal to 1.66×10^{-27} kg. **[25.3]**

moment of force The product of the force and the perpendicular distance of its line of action from the reference point, expressed in newton-metres or related units. **[5.1]**

moment of inertia of an area A geometrical property of area distribution, also called second moment of area, defined as the sum of all elementary products of area elements and the squares of their respective distances from the centroidal axis. **[Appendix A]**

momentum The product of the mass of a body and its linear velocity, sometimes described as the quantity of motion. **[12.1]**

neutral plane In a beam subjected to bending, the plane passing through the centroid of a cross-section at which stress due to bending is zero. **[19.6]**

newton The unit of force in SI, defined as the force which imparts an acceleration of one metre per second squared when applied to a body having a mass of one kilogram. **[3.2]**

Newton's laws of motion Three fundamental laws of kinetics concerned with:
1. inertia of a body,
2. relationship between force, mass and acceleration.
3. action and reaction forces. **[8.4]**

open system A thermodynamic system in which fluid is continually flowing into and out of the system, such as the water-steam flow through a boiler. **[25.4]**

pascal The unit of stress and pressure in SI, defined as a force of one newton uniformly distributed over an area of one square metre. **[16.1 and 21.1]**

Pascal's principle The law which states that pressure impressed at any place on a confined fluid is transmitted undiminished to all other portions of it. **[22.3]**

phase A term used in thermodynamics (in preference to the more common term "state") to refer to the solid, liquid or gaseous form of a substance. **[26.3]**

pneumatics The study of compressible fluids, such as gases, with particular emphasis on equipment actuated by means of compressed air. **[20.1]**

polygon of forces A method of graphical addition of forces by drawing a polygon with sides parallel to the directions of the forces, with a head-to-tail sequence of directional arrows, and the lengths of the sides of the polygon representing the magnitude of the forces to some suitable scale. **[3.6]**

potential energy A form of mechanical energy possessed by a solid body or by a quantity of fluid due to its position in the gravitational field, i.e. due to its elevation above some datum level. **[11.2 and 23.3]**

power The time rate of doing work. **[10.2]**

pressure A measure of the intensity of normal distributed forces exerted on a surface, defined as force per unit area and expressed in pascals or related units. A distinction is made between absolute, gauge and atmospheric pressure. **[21.1, 21.2 and 21.3]**

pressure energy The component of energy in a moving fluid which is due to its pressure. (See also Bernoulli's equation.) **[23.3]**

pumping head The increase in the total head of a flowing fluid due to the energy supplied to the fluid by a pump. **[24.2]**

radian The unit of angular measure in SI, defined as the angle between two radii of a circle which cut off on the circumference an arc equal in length to the radius. **[2.4]**

radius of gyration The distance from the axis of rotation of a body where all the mass of the body is considered to be concentrated for the purpose of calculations in rotational dynamics. **[Appendix A]**

reactions Forces which exist between a structure, such as a beam or a truss, and its supporting surfaces. The magnitudes and directions of support reactions depend on the magnitudes and distribution of external loads and on the type of supports. **[4.4]**

relative density A comparative measure of density of a substance, defined as the ratio of its density to that of another substance, usually water. **[20.3]**

resultant force The single force which can replace two or more forces acting on a body without changing the effect produced on the body. The resultant is a vector sum of the given forces. **[3.6]**

rotation The type of motion during which a body turns around a fixed axis in such a way that every particle of the body except the axis travels along a circular path. **[Chapter 9]**

second The unit of time in SI, presently defined as the time interval occupied by 9 192 631 770 cycles of a specified energy change in the caesium atom, as measured by the caesium atomic clock. **[2.4]**

second moment of area A geometrical property of area distribution, often called moment of inertia of the area, defined as the sum of all elementary products of area elements and the squares of their respective distances from the centroidal axis. **[Appendix A]**

sensible heat The heat transfer which is accompanied by a temperature rise. **[26.1]**

specific heat capacity The amount of heat required to produce one degree of temperature rise in a unit mass of substance, expressed in joules per kilogram-kelvin. **[26.1]**

specific volume The volume occupied by unit mass of a substance, expressed in cubic metres per kilogram or related units. **[20.4]**

speed A measure of the distance travelled by a moving body per unit time without reference to the direction of the motion. **[8.1]**

statics The branch of the mechanics of solids which deals with bodies and structures at rest under the action of external forces in equilibrium. **[Chapters 3 to 7]**

strain A relative measure of deformation in a material under load, defined in terms of change in dimensions (in tension or compression) or shape (in shear) of a component compared with its original unloaded condition. **[16.3 and 16.6]**

strength of materials The study of solid materials, structures and machine components with particular reference to internal forces and deformations produced by external loads. **[Chapters 15 to 19]**

stress A measure of the intensity of force distribution within the material of a structure or machine component, defined as force per unit area and expressed in pascals. Depending on conditions of loading stresses may be classified as tensile, compressive, shear or bending stress. **[16.1, 16.5 and 19.6]**

Système International d'Unités The International System of Units, with the abbreviation SI, comprising coherent, decimally related units of measurement, adopted in 1970 as the basis of Australia's metric system. **[2.3]**

temperature A measure of the degree of hotness or coldness of a substance with respect to a fixed scale, expressed in kelvins or in degrees Celsius. **[25.2]**

thermal efficiency In heat engines theory, the ratio of work output to the heat input, which is a measure of the fraction of the thermal input converted into mechanical output. **[28.3]**

thermodynamic process A process which takes place within a thermodynamic system involving heat and/or work transfer resulting in a change of thermodynamic state of the system. **[25.7 and Chapter 28]**

thermodynamics The study of the effects of temperature and energy on systems containing fluids and in particular of the conversion of heat into mechanical work. **[Chapters 25 to 28]**

thermodynamic system A term used in engineering thermodynamics to describe a region in space, e.g. the interior of a boiler or engine cylinder, containing a quantity of matter, such as steam or gas, separated from its surroundings by a real or imaginary boundary. **[25.4]**

three-force principle A useful theorem in statics, which states that in the case of equilibrium under the action of three non-parallel forces the lines of action of the three forces must intersect at a common point. **[4.5]**

time A measure of the sequence of events taking place in the physical world, expressed in seconds or related units, such as minutes or hours. **[2.4]**

tonne A unit of mass equal to one thousand kilograms. **[2.4]**

torque The turning effort exerted by mechanical components such as shafts, during continuous rotation, expressed in newton-metres. **[5.4]**

tractive effort In railway engineering, the force exerted by a locomotive at the draw-bar. In general, the force required for the propulsion of a vehicle along a roadway or a track. **[8.5]**

tractive resistance The sum of all frictional resistances, such as bearing friction and air resistance, expressed as a single force opposing the motion of a vehicle, often stated in units of force per unit mass of the vehicle. **[8.5]**

triangle of forces A particular case of the polygon of forces drawn for three forces in equilibrium at a point. **[3.6]**

turbine head The decrease in the total head of a flowing fluid due to the energy transferred from the fluid in the turbine. **[24.3]**

ultimate strength The highest load that can be applied to a material before fracture occurs, expressed as force per unit area. Depending on the conditions of loading materials have different ultimate strengths in tension, compression and shear. **[15.2]**

unit of measurement An agreed-on part of a physical quantity, defined by reference to some arbitrary material standard or natural phenomenon and used as a standard of comparison in the process of measurement. **[2.2]**

velocity A measure of the time rate of change in the position of a moving body (linear velocity) or in the orientation of a rotating body (angular velocity). **[8.1 and 9.1]**

velocity ratio The ratio of the distance moved by the effort on the input side of a simple machine to the distance moved by the load on the output side. **[13.1]**

viscosity Internal friction due to molecular cohesion in fluids, which represents resistance to flow and is a major cause of energy losses in fluid systems. **[20.2 and 24.4]**

volume A measure of the amount of space occupied by an object or substance, expressed in cubic metres, litres or related units. **[2.5]**

volume flow rate The volume of fluid flowing across a given cross-section of duct or pipe per unit time, expressed in cubic metres per second or related units. **[23.1]**

watt The unit of power in SI, defined as the work done at the rate of one joule per second of time. **[10.2]**

weight The force of gravitational attraction exerted by the Earth on an object. On or near the surface of the Earth the weight of any object is approximately equal to 9.81 newtons for every kilogram of its mass. [**3.9**]

work A form of energy transfer which occurs as a combined result of motion and effort. In linear motion, the product of the force applied to a body and the distance moved by the force. In rotational motion, the product of torque and angular displacement. [**10.1**]

Young's modulus A measure of the ability of a material to resist stretching, defined as the ratio of tensile stress in the material to corresponding strain, usually expressed in megapascals or related units. [**16.4**]

Answers

Chapter 2

2.1 (a) 400 m
(b) 0.053 kg
(c) 0.0753 m
(d) 0.045 s
(e) 0.357 kg
(f) 80 kg
(g) 0.734×10^{-3} m
(h) 0.54×10^{9} s

2.2 (a) 12.3 km
(b) 7.5 t
(c) 79 ms
(d) 4.7 mm
(e) 30 g
(f) 85 t
(g) 3 μm
(h) 4.7 mg

2.3 (a) 5×10^{6} mm^2
(b) 0.75×10^{-3} m^2
(c) 0.663×10^{-3} L
(d) 0.5×10^{-3} m^3
(e) 1350 L
(f) $0.75 \times 10^{9} mm^3$
(g) 0.632 m^3
(h) 47×10^{6} mm^3

2.4 (a) 35.25°
(b) 2.618 rad
(c) 90°
(d) 0.048 rad
(e) 143° 14′
(f) 0.451 rad

2.5 (a) 4620 min
(b) 2700 s
(c) 1 h 23 min 20 s
(d) 402 s

2.6 0.2625 m^2
2.7 354.5 mm $\times$ 354.5 mm
2.8 1.143 m^2 81.71 L
2.9 35 rev/s
2.10 5.085 L/h
2.11 5 h
2.12 Tiles, $25
2.13 47.33 mg
2.14 2.4 kg
2.15 $33
2.16 7157 L/h

Chapter 3

3.1 (a) 10 N
(b) 10 N
(c) 5 N, 5 N
(d) 5 N, 5 N
(e) 10 N, 10 N
(f) 10 N, 10 N
(g) 5 N, 5 N and 10 N
(h) 10 N, 5 N and 5 N

3.2 (a) 7 N →
(b) 1 N →
(c) 5 N ∠ 36.9°
(d) 6.08 N ∠ 25.3°
(e) 2.83 N ∠ 48.5°

3.3 96.6 kN
3.4 260 N and 150 N
3.5 A: 4 kN and 6.93 kN
B: −2.5 kN and 4.33 kN
C: 8.66 kN and 5 kN
D: 7.07 kN and −7.07 kN
3.6 470 N and 171 N
3.7 7.07 kN and 7.07 kN
9.66 kN and 2.59 kN
3.8 400 N and 693 N
3.9 (a) 5.39 kN ∠ 21.8°
(b) 330 N ∠ 33.3°
(c) 20.9 kN ∠ 72°
(d) 0
(e) 11 kN ∠ 17°
3.10 As for 3.9
3.11 3.52×10^{22} N

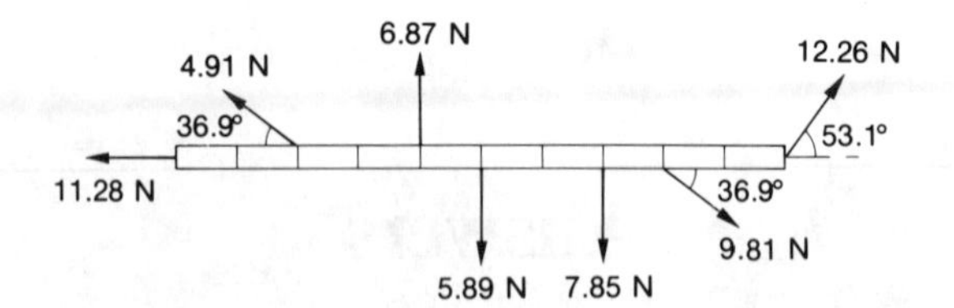

Fig. 1

3.12 (a) 9.81 mN
(b) 9.81 N
(c) 9.81 kN
(d) 39.2 N
(e) 0.49 N
3.13 22.6 kN
3.14 (a) 98.1 N
(b) 98.1 N
(c) 98.1 N
3.15 3.43 kN and 12.8 kN
3.16 12.8 kN at 72.4° to horizontal
3.17 Fig. 1 (Answer)

4.14 123 N and 506 N ∠ 76°
4.15 3.61 kN ∠ 56.3° and 3 kN ↓
4.16 871 N ∠ 73.3° and 250 N ←
4.17 649 N ∠ 30° and 864 N ∠ 49.5°
4.18 4.19 kN and 5.44 kN ∠ 50.4°
4.19 59.5 N ∠ 33.7°
4.20 19.2 kN ∠ 66.8° 0.8 m

Chapter 4

4.1 13.2 N and 10.8 N
4.2 3.68 kN
4.3 46.0 kN 37.5 kN
4.4 2.76 kN 4.51 kN
2.85 kN 4.64 kN
4.5 178 N ∠ 65° No
4.6 2.04 t 0 28.3 kN
4.7 24.0 kN 20.8 kN 20.8 kN 0
16.0 kN 8.0 kN 16.0 kN
8.0 kN 20.8 kN 8.0 kN
4.8 (a) 7.55 kN ∠ 19.8°
(b) 49 kN ∠ 52.1°
(c) 8.82 kN ∠ 74.5°
(d) 0
4.9 192 N 279 N ∠ 20.7°
4.10 (a) 4.19 kN and 8.89 kN, 64.7° ↗
(b) 16.8 kN and 15.4 kN, 10.2° ↗
(c) 16.8 kN and 19.7 kN, 39.5° ↗
(d) 25.2 kN and 24.5 kN, 21.5° ↗
4.11 (a) 9.55 kN ∠ 10.3° and 1.71 kN ↑
(b) 8.31 kN ∠ 22.8° and 3.21 kN ↑
(c) 6.61 kN ∠ 40.9° and 4.33 kN ↑
4.12 406 N ∠ 25° and 520 N ∠ 45°
4.13 384 N and 348 N ∠ 17°

Chapter 5

5.1 13.5 N.m, Perpendicular to lever arm.
5.2 75 N.m
5.3 (a) 3 kN.m ↻ and 3 kN.m ↺
(b) 1.5 kN.m ↻ and 4.5 kN.m ↺
(c) 3 kN.m ↻ and 3 kN.m ↺
(d) 3.5 kN.m ↻ and 2.5 kN.m ↺
(e) 2.6 kN.m ↻ and 2.6 kN.m ↺
5.4 17.3 kN.m ↻
5.5 28.0 kN.m ↻ and 1.32 kN.m ↺
5.6 188 N.m
5.7 33.3 N
5.8 45 N.m
5.9 2 kN
5.10 50 N
5.11 123 N
5.12 5.56 kN
5.13 28 N.m
5.14 $\Sigma M = 0$
5.15 $\Sigma M = 0$
5.16 110 N
5.17 2 kN and 0.5 kN.m ↻
5.18 15 mm
5.19 1.3 kN and 1.95 kN.m ↺
5.20 49.1 N and 31.9 N.m ↻

Chapter 6

6.1 (a) $F_L = 6$ kN ↑ $F_R = 9$ kN ↑
(b) $F = 2$ kN ↑ $M = 3.2$ kN.m ↻
(c) $F_L = 3$ kN ↑ $F_R = 3$ kN ↑
(d) $F_L = 23$ kN ↑ $F_R = 29$ kN ↑
(e) $F = 9$ kN ↑ $M = 16$ kN.m ↻
(f) $F_L = 11$ kN ↑ $F_R = 3$ kN ↑
(g) $F = 3$ kN ↓ $M = 0$
(h) $F = 0$ $M = 12$ kN.m ↺
(i) $F_L = 21.6$ kN ↑ $F_R = 19.4$ kN ↑
(j) $F_L = 2$ kN ↓ $F_R = 2$ kN ↑

6.2 (a) 48 kN ↓ 4.13 m
(b) 550 kN ↓ 4.09 m
(c) 6.75 kN ∠79.5° 3.81 m
(d) 4.53 kN ∠79.9° 3.88 m
(e) 3.08 kN ∠80.5° 3.68 m

6.3 848 N ∠79.8° 4.47 m up along the ladder

6.4 666 N ∠11.9° 73.2 mm

6.5 33.1 kN ∠65°
4.05 m (perpendicular distance)

Chapter 7

7.1 5.49 N
7.2 0.245
7.3 6.87 kN
7.4 0.5 kg
7.5 88.3 N
7.6 40.5 N
7.7 1.88 kN
7.8 144 N.m
7.9 100 N
7.10 Tip
7.11 31°
7.12 0.35 19.3° 106 N
7.13 124 N at 14° to normal
7.14 0.51 17.5 N 8.91 N
7.15 5.49 N
7.16 0.245
7.17 6.87 kN
7.18 0.5 kg
7.19 88.3 N
7.20 40.5 N
7.21 878 N
7.22 8.49 kN
7.23 37.6 N
7.24 34.6 N
7.25 234 N 340 N
7.26 225 kg
7.27 0.313
7.28 22.8 kg 7.2 kg
7.29 1.67 kg 0.6 kg

Chapter 8

8.1 6.96 m/s
8.2 85 km/h 23.6 m/s
8.3 75 km/h
8.4 2.22 m/s^2
8.5 450 m/s
8.6 5625 m
8.7 1.07 m/s^2 13 s
8.8 0.605 m/s^2 25.7 s
8.9 No 2.04 m/s
8.10 2 km 58.8 km/h
8.11 11 s
8.12 22 s 193.6 m
8.13 3.16 s 31 m/s
8.14 62.4 m 3.57 s
8.15 2.50 km 4.54 km 75.8 s
8.16 5 m/s^2
8.17 5 N
8.18 276 N
8.19 125 kN
8.20 420 N
8.21 126 s
8.22 145 kN
8.23 291 s 3.23 km
8.24 24.6 kN
8.25 9.81 kN 11.4 kN
8.41 kN 8.81 kN
8.26 32.3 kN
8.27 23.6 N 19.6 N 15.6 N
8.28 320 N 160 N
8.29 0.866 m
8.30 0.377 m/s^2 9.43 N
8.31 0.318 kg
8.32 11.2 kg
8.33 0.467 m/s^2
8.34 3.05 m/s^2
8.35 17.1 kg, 2.15 s

Chapter 9

9.1 Hours hand: 1.389×10^{-3} rpm, 1.454×10^{-4} rad/s
Minutes hand: 1.667×10^{-2} rpm, 1.745×10^{-3} rad/s
Seconds hand: 1 rpm, 0.1047 rad/s
9.2 20 s 3100 rad
9.3 40 rad/s^2 1308 rev
9.4 152 rad/s 1450 rpm 184 rev
9.5 0.349 rad 244 mm
9.6 0.429 rad/s 0.214 rad/s^2
9.7 25.1 m/s

9.8 7.59 m/s 250 mm
9.9 3.69 rad/s^2 55.4 rad/s 415 rad
9.10 63.6 rev
9.11 5 rad/s
9.12 0.24 N.m
9.13 12.2 s 152.5 rev
9.14 256 rad/s
9.15 187 N
9.16 36 s
9.17 106 N 0.946 m/s^2
9.18 109 N 0.765 m/s^2
9.19 1.55 kN
9.20 1.89 m/s^2 10 m/s
9.21 1.84 kN.m
9.22 3.5 N.m
9.23 3.7 kN
9.24 No
9.25 79.3 km/h
9.26 118 N
9.27 0.1 kg
9.28 56.4 km/h
9.29 14.6 kN 9.9 kN
9.30 1250 rpm
9.31 30.8 N 50.4°
9.32 41.9 rpm

Chapter 10

10.1 240 MJ
10.2 294 kJ
10.3 7.85 kJ
10.4 56.5 kJ
10.5 11 kJ
10.6 65 kW
10.7 31.4 kW
10.8 921 kW
10.9 7.07 kW
10.10 115 N.m
10.11 10.3 kW
10.12 16.4 kW
10.13 200 kJ
10.14 59 N.m
10.15 28.3 kN 236 kW
10.16 22.6 N.m 711 W

Chapter 11

11.1 790 kJ
11.2 226 kJ
11.3 1.35 m
11.4 55.7 MJ
11.5 −28 MJ
11.6 167 kJ 296 kJ 463 kJ
+130 kJ +167 kJ
11.7 90 km/h
11.8 107 kJ
11.9 106 kg.m^2
11.10 +816 J
11.11 7.67 m/s
11.12 62.4 m
11.13 65.2 J 4.32 m/s
11.14 48.9°
11.15 22.5 m/s
11.16 96.5 km/h
11.17 11 m/s
11.18 2.21 m/s
11.19 0.32
11.20 81 km/h
11.21 2.3 m
11.22 859 N.m
11.23 30.4 N.m
11.24 1.76 m/s
11.25 1.1 kg
11.26 22.6 N
11.27 18.4 kg 34.9 kg
11.28 699 N

Chapter 12

12.1 23 300 kg.m/s
12.2 +7780 kg.m/s −7780 kg.m/s
12.3 12.5 s
12.4 1.5 m/s 250 kN
12.5 2.5 m/s 3.33 m/s
12.6 0.0355 N
12.7 589 m/s 31.1 kN
12.8 9.69 km/h
12.9 794 m/s
12.10 7 m/s 2.27 m/s 17.3 kN
12.11 25 m/s
12.12 0.5
12.13 6.44 m/s →
12.14 0.898
12.15 1.04 m/s → 6.32 m/s → 126 J
12.16 1.2 m/s ← 7.6 m/s → 0
12.17 4.4 m/s → 4.4 m/s 197 J
12.18 0.849

Chapter 13

13.1 143 200 67.2 kJ
94.0 kJ 71.5% 97.8 N
13.2 (a) (i) 134, 67% (b) 75%
(ii) 146, 73%

13.3 $F_E = 0.0625F_L + 2$ 67.3%
13.4 5.69 81.3% 9.36 N
13.5 7.5 6 5 N
13.6 59.3%
13.7 5
13.8 577 N
13.9 75%
13.10 1.87 t 40 rev

Chapter 14

14.1 7.59 kW 6.07 kW 80%
14.2 4 rpm
14.3 2.43 t 0.377 m/s
14.4 3 5
14.5 19.2 N.m 57.6 N.m 640 N
14.6 4 kW
14.7 3
14.8 54.6 N.m 6.86 kW
14.9 93.2 N.m 11.7 kW
14.10 199 N.m 497 N.m 1.07 kN

Chapter 15

15.1 32.8 kg
15.2 2.44 kg
15.3 396 N/mm^2
15.4 12 mm
15.5 734 N
15.6 31 N/mm^2
15.7 35.3 kN
15.8 202 kN
15.9 6 mm
15.10 93.8 kN

Chapter 16

16.1 5 MPa
16.2 0.3×10^{-3}
16.3 120 GPa
16.4 73.8 MPa
16.5 33.3 MPa 6.67 mm
16.6 3 MPa 0.75 mm
16.7 401 kN
16.8 29.7 kN 0.185 mm
16.9 20 mm × 20 mm
16.10 0.162 mm 4.8
16.11 90 MPa 4
16.12 8
16.13 6 mm
16.14 24.5 kN
16.15 85.3 MPa
16.16 5.3 kN.m
16.17 8
16.18 30 kN 200 MPa 3.25
16.19 8 mm
16.20 0.24 MPa 0.16 4 mm

Chapter 17

17.1 50 MPa
17.2 80 MPa
17.3 206 MPa No
17.4 63.6 kN 57.8%
17.5 63.7 MPa 31.3 MPa
125 MPa
17.6 14.1 kN 32.1%
17.7 6 mm 2 mm 18 mm
17.8 64.3%
17.9 24 mm 16 mm 8 mm
17.10 46.3%
17.11 2.8 mm 387 N/mm
4.2 mm 580 N/mm
5.7 mm 773 N/mm
7.1 mm 966 N/mm
8.5 mm 1160 N/mm
11.3 mm 1550 N/mm
14.1 mm 1930 N/mm
17.0 mm 2320 N/mm
17.12 97 mm
17.13 1160 kN
17.14 6 mm
17.15 45 mm
17.16 23.6 t
17.17 85 mm 174 mm

Chapter 18

18.1 30 MPa
18.2 2 mm
18.3 36 MPa 18 MPa
18.4 4 mm
18.5 5 mm
18.6 1.89 MPa
18.7 1.51 m^3
18.8 8
18.9 228 kPa
18.10 24 mm

Chapter 19

19.1 Fig. 2 Answer
19.2 Fig. 2 Answer

ANSWERS

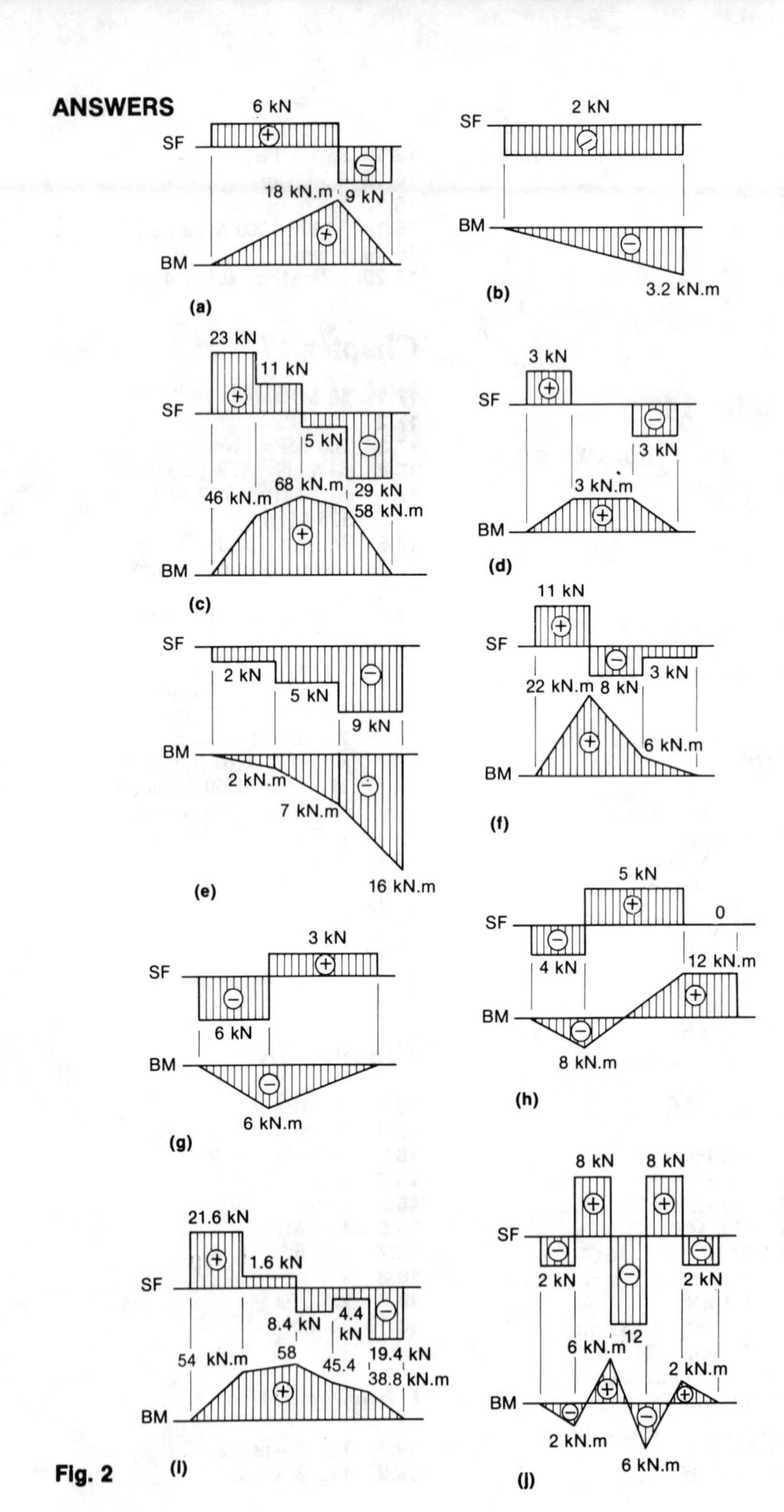

Fig. 2

19.3 72 MPa
19.4 347 m
19.5 20.2 kN
19.6 100 mm
19.7 64 MPa 234 m
19.8 60 MPa 80 MPa 500 m
19.9 $R = \frac{Ey}{f}$ 10 m
19.10 13.3 kN.m 60 m
19.11 6.67 kN.m 30 m
19.12 320 mm × 160 mm

Chapter 20

20.1 0.849 kg/L 0.849 1.18 L/kg
20.2 133 N
20.3 48 t
20.4 412 kg
20.5 7.7 kN
20.6 55.2 kg
20.7 0.136 m^3/kg
20.8 1.95 kg
20.9 0.424 kg/m^3 8 g
20.10 1.69 kg 66.9 m^3/kg

Chapter 21

21.1 500 000 Pa 0.5 MPa 500 000 N/m^2
21.2 26.5 kN
21.3 509 kPa
21.4 102.3 kPa
21.5 713 mm
21.6 578 mm
21.7 801.3 kPa
21.8 76.3 kPa
21.9 70.7 N
21.10 69.4 kPa
21.11 2.94 kPa 104.2 kPa
21.12 2.48 kPa
21.13 97 kPa
21.14 6.76 kPa
21.15 0.196 kPa
21.16 106.8 kPa 5.5 kPa
21.17 133.8 kPa
21.18 RD = 1.78

Chapter 22

22.1 11.8 kPa (ga)
22.2 10.33 m
22.3 0.8
22.4 579 mm
22.5 148 kPa (abs), 29 kN
22.6 130 N
22.7 159 kN
22.8 3.12 t
22.9 1.59 MPa (ga) 781 N
22.10 90.4 mm
22.11 9.81 N 133 N
22.12 9.81 N 0 0
22.13 9.81 N
22.14 165 N
22.15 499 N
22.16 6.64
22.17 1.5 m
22.18 7.49 N
22.19 94.3 N
22.20 5
22.21 0.6 kJ
22.22 0.942 L 637 kPa
22.23 833 kPa 2.94 kJ
22.24 15 mm
22.25 100 80 24.5 kJ
22.26 1.92 kJ
22.27 0.55 kJ
22.28 56.9 J
22.29 13.1 kW
22.30 15.1 kW

Chapter 23

23.1 13 kg/s 15.7 L/s
23.2 700 mm × 700 mm
23.3 81.6 L/s 18.5 m/s 16.3 L/s 33.2 m/s
23.4 10 h
23.5 29.8 kg/min 1.19 m/s 11.1 mm/min
23.6 0.663 m/s 24.5 mm
23.7 150 J 12 J 24.5 J
23.8 −18.3 kPa (ga)
23.9 344 kPa
23.10 6.72 m
23.11 3.11 L/s
23.12 2 min
23.13 2.11 m^3/s
23.14 8.29 m/s 2.6 L/s −39.2 kPa (ga)

Chapter 24

24.1 5.61 m 3.26 m 6.5 m
24.2 27.2 m
24.3 2.18 kW
24.4 501 kW
24.5 73.2%
24.6 2.95 L/s
24.7 47.1 m 6.61 kW 65.5 kPa (ga) 528 kPa (ga)
24.8 139 m 5.1 MW 1.31 MPa −49.1 kPa

Chapter 25

25.1 423 K 273 K 123 K
25.2 227°C 0°C −123°C
25.3 542 K
25.4 Yellow
25.5 $2 \times 1 + 16 = 18$
25.6 181.7 kJ/kg 525.0 kJ/kg
25.7 383.8 kJ/kg 209.3 kJ/kg 3025 kJ/kg
25.8 229.5 kJ/kg
25.9 47.15 MJ 338.5 MJ 44.85 MJ
25.10 0.26 kJ 2.76 kJ
25.11 15 kJ
25.12 100 kJ (input)
25.13 2 kJ (output)
25.14 30 kW
25.15 1.84 kJ (output)
25.16 145 kJ/s
25.17 225°C
25.18 108 kg/h
25.19 150°C 0.646 kJ
25.20 $1.74

Chapter 26

26.1 355 kJ
26.2 40°C
26.3 5 kg
26.4 1944 kJ
26.5 263 s
26.6 604 kg
26.7 45°C
26.8 25.8°C
26.9 0.45 kJ/kg.K
26.10 25 K
26.11 11.3 MJ
26.12 335 MJ
26.13 43.2%
26.14 26 kg
26.15 2.25 MJ
26.16 3820 kJ/s
26.17 667 kW
26.18 18°C
26.19 1700 MJ
26.20 1.33 kW
26.21 37.5 MJ/kg
26.22 $1.15
26.23 55.5 L/h
26.24 42.2 MJ/kg
26.25 49.5 MJ/kg
26.26 45.2 MJ/kg
26.27 41.5 MJ/m^3

Chapter 27

27.1 24 L
27.2 254°C
27.3 202 kPa (abs)
27.4 5.35 L
27.5 0.416 kJ/kg.K 0.099 kJ/kg.K
27.6 0.591 m^3
27.7 Carbon dioxide
27.8 5.35 kg 860 kPa (ga)
27.9 295
27.10 767 kPa (abs)
27.11 851 kJ
27.12 69 kJ
27.13 47°C 276 kJ/kg
27.14 50°C −152 kJ/kg
27.15 0.72 kJ 1.01 kJ 0.29 kJ
27.16 12.1 K
27.17 147 kJ
27.18 42.6 kJ
27.19 291 kJ 303°C
27.20 61.9 kJ 13.5°C
27.21 195 kJ 195 kJ 274 kJ
27.22 455 kJ 324 kJ 455 kJ

Chapter 28

28.1 113 MJ 113 MJ 323 kJ/kg
28.2 28.1 kJ
28.3 229°C 514 kPa (abs) 155 kJ/kg
28.4 40 MJ/kg
28.5 0.9 kg
28.6 1.55 kJ/kg.K
28.7 80.3 kW

28.8 154 J 114 J
28.9 1.05 g 17.9 kJ/kg
126.3 kJ/kg
175°C 601 kPa (abs)
28.10 17 kJ
28.11 18°C
28.12 15 min
28.13 19.7 MJ/m^3
28.14 141 MJ
28.15 75%
28.16 434 kJ/s
28.17 10.3 kg/min
28.18 488 kJ/s 390 kW
28.19 53.8 kJ/s 78%
28.20 68.4 kW 28%
28.21 23%
28.22 35 L/h
28.23 213 t/h
28.24 330 kW 9.3%

Appendix A

A.1 (a) $\bar{x} = 200$ mm
$\bar{y} = 264$ mm
(b) $\bar{x} = 15$ mm
$\bar{y} = 25$ mm
(c) $\bar{x} = 35$ mm
$\bar{y} = 21.7$ mm
(d) $\bar{x} = 11.4$ mm
$\bar{y} = 14.6$ mm
(e) $\bar{x} = 100$ mm
$\bar{y} = 216$ mm
(f) $\bar{x} = 250$ mm
$\bar{y} = 256$ mm
A.2 (a) 4.91×10^6 mm^4
(b) 1.08×10^6 mm^4
(c) 6.4×10^6 mm^4
(d) 4.17×10^6 mm^4
A.3 (a) 0.944×10^9 mm^4
(b) 0.268×10^6 mm^4
(c) 0.493×10^6 mm^4
(d) 85.7×10^3 mm^4
(e) 0.925×10^9 mm^4
(f) 2.88×10^9 mm^4
A.4 7.44×10^{-3} kg.m^2
15.8×10^{-3} kg.m^2
A.5 21.2 mm 30.9 mm
A.6 1.67 kg.m^2 204 mm

Index

INDEX